S0-CAS-447

ANALYTICAL TECHNIQUES IN COMBINATORIAL CHEMISTRY

ANALYTICAL TECHNIQUES IN COMBINATORIAL CHEMISTRY

edited by

Michael E. Swartz

Waters Corporation
Milford, Massachusetts

MARCEL DEKKER, INC.   NEW YORK • BASEL

ISBN: 0-8247-1939-5

This book is printed on acid-free paper.

Headquarters
Marcel Dekker, Inc.
270 Madison Avenue, New York, NY 10016
tel: 212-696-9000; fax: 212-685-4540

Eastern Hemisphere Distribution
Marcel Dekker AG
Hutgasse 4, Postfach 812, CH-4001 Basel, Switzerland
tel: 41-61-261-8482; fax: 41-61-261-8896

World Wide Web
http://www.dekker.com

The publisher offers discounts on this book when ordered in bulk quantities. For more information, write to Special Sales/Professional Marketing at the headquarters address above.

Current printing (last digit):
10 9 8 7 6 5 4 3 2 1

PRINTED IN THE UNITED STATES OF AMERICA

Foreword: Chemistry Becomes an Information Science

Nothing in recent years has had as great an impact on the process of drug discovery as the rise of genomics and combinatorial chemistry. Nothing has been more urgently needed. In the United States today it takes an average of 13 years and over $300 million to develop a drug. Although regulatory hurdles account for a good portion of those costs, major difficulties lie in two other areas.

First, there are not enough drug targets. There are about 6000 known drugs, half of which hit human targets. However, those 3000 human-directed drugs hit only about 500 targets, which means that less than 1% of the human genome (estimated to contain 80,000 to 100,000 genes) has been exploited pharmaceutically. It's even worse for antimicrobial drugs. Consider antifungal agents: almost every marketed antifungal agent hits one of a handful of targets in the same metabolic pathway. Genomics, the science of identifying and sequencing all of the genes in an organism, is changing all this at an astonishing rate. Soon we will have a plethora of targets, for pathogens and people. However, this only makes the second difficulty—that there aren't enough drugs—more acute.

Those 6000 known drugs fall into only around 300 chemical classes (the exact figure depends on how one defines a "class"). Recently, a major pharmaceutical company screened its entire compound inventory—over 400,000 compounds developed over almost 100 years of work—against a new target identified by genomics. They did not find a single hit. This sounds surprising until one examines that inventory closely: almost half the compounds in it could be considered to be derived from a single chemical class.

It is this problem that combinatorial chemistry is designed to solve, and its explosive growth is testimony to both the magnitude of the problem and the early successes of the combinatorial approach. Combinatorial chemistry has many guises. In its purest form it involves the synthesis of all possible compounds from a set of modular building blocks. However, it can also mean high-throughput parallel synthesis of individual pure compounds, simultaneous synthesis of mixtures of compounds free in solution or on solid support, or a number of other variations on these themes. Regardless of the details of the process, the objective is the same: to produce a large number of chemically ''diverse'' compounds as rapidly as possible.

And that it does. Combinatorial methods have rewriten the standards for synthetic productivity. Until about 10 years ago, a good chemist could make and characterize perhaps 50 compounds per year. A combinatorial chemist aims for more like 50,000. Such numbers define a revolution, one that has already transformed the pharmaceutical industry and is likely to eventually make an impact on every other area of industrial chemistry. Analytical chemistry is no exception, which brings us to the subject of this splendid book.

It is, of course, one thing to make 50,000 compounds and quite another to know what one has made. Yet that is essential, because when new molecules are available in such staggering numbers chemistry has become an information science. Consider a chemically ''diverse'' (whatever that means, and nobody really knows yet) library of 50,000 compounds. Now screen that against, say, 50 different drug targets. Then imagine that you don't know what any of the compounds are. The ones that give ''hits'' in your assays won't remain unknown for long: you will purify those and characterize them. Yet in doing so, you will throw information away. If you knew the structure of every compound, then the ones that failed in your assays would be almost as valuable to you as the ones that succeeded because they would define the chemical types that were not likely to work for that particular set of target classes. Combinatorial chemistry promises to provide structure and activity data on a scale never before imagined, but it will do so only if one can characterize what one makes.

The task for the analytical community, then, is to develop methods of separating and characterizing compounds suitable for the assembly-line scale of combinatorial chemistry. This book details the methods currently available and also discusses emerging techniques that could have a major impact. It covers the gamut from a concise introduction to the various methods of combinatorial synthesis (Weller) to a splendidly useful compendium of commercial resources (Brock and Andrews). In between, we are taken through clear and critical descriptions of mass spectrometry (Vouros and Hauser-Fang), I-R (Gremlich), and NMR (Shapiro) methods for high-throughout compound identification, including the difficulties inherent in dealing with mixtures of compounds that most combinatorial synthetic strategies produce. Detailed consideration of the mixture problem is the basis for subsequent chapters on chromatography (Swartz) and capillary electrophoresis (Krull, Gendreau, and Dai). The latter process is of particular interest because it is so open to miniaturization, and there seems little doubt that the laboratory of the future will eventually be on a chip. The remaining chapters treat the information-explosion problem head-on. Nicell discusses techniques for managing the reams of data that an analytical lab will face when it deals with combinatorial libraries, and Kyranos and Chipman outline the various ways of screening libraries in a high-throughput fashion. Although the focus of the book is on analytical techniques, these latter chapters in particular make it an excellent introduction to the entire field of combinatorial chemistry.

It's funny how the wheel comes around for scientific disciplines. At just about the time that chemistry as a whole, and perhaps analytical chemistry in particular, was being written off by some as a ''mature field'' unlikely to produce the kind of cutting-edge excitement of, say, neurobiology or human genetics, along comes combinatorial chemistry. The making, and characterizing, of molecules has once again been thrust into the heart of things, where it has been for almost 200 years. If we are living in the Information Age, then surely it is appropriate that the Central Science should become at last an information science. This book is a manifesto for the next Chemical Revolution.

Gregory A. Petsko
Gyula and Katica Tauber Professor of Biochemistry
Director, Rosenstiel Basic Medical Sciences Research Center
Brandeis University
Waltham, Massachusetts

Preface

Analytical techniques play a critical role in drug discovery by providing valuable information to control the identity, purity, and stability of potential drug candidates. These techniques, commonly in the form of chromatographic and spectroscopic methods, are routinely used in the pharmaceutical laboratory from the initial synthesis of a new chemical entity all the way through the development cycle to the manufacture and sale of a pharmaceutical product.

The use of analytical techniques in the burgeoning field of combinatorial chemistry is anything but routine. Combinatorial chemistry is a term that describes a set of tools for generating extensive chemical diversity rapidly and efficiently. It has been described as having the potential to revolutionize drug discovery. In the drug discovery process, large numbers of compounds are screened for potential biological activity. Combinatorial chemistry does not alter this process, but introduces a new step that greatly increases the molecular diversity available for screening. The unique way in which this diversity is produced, however, places new demands on the analytical methods used to analyze and produce the desired results. In addition, the sheer volume of information generated, and the manual labor that would be required, would be overwhelming if not for the development of information management systems and related techniques critical to the development of this field.

Following the introduction presented in Chapter 1, this book discusses the application and use of specific analytical techniques (mass, infrared, and nuclear magnetic resonance spectrometry, chromatography, and capillary electrophoresis) in the combinatorial chemistry field (Chapters 2–6). It also discusses how to make sense of the vast amounts of data generated (Chapter 7), details how the actual libraries of compounds produced are utilized (Chapter 8), and lists some of the vast commercial resources available to researchers in the field of combinatorial chemistry (Chapter 9).

Analytical Techniques in Combinatorial Chemistry is intended to provide specific details on how analytical techniques are brought to bear on the unique challenges presented in the combinatorial chemistry laboratory. It is aimed primarily at industrial and pharmaceutical chemists faced with the task of developing methods, analyzing the results, and documenting and/or managing the discovery process in a combinatorial setting. Since many major pharmaceutical companies are in the process of staffing combinatorial chemistry departments, this publication could also serve as a training and reference source, or perhaps a graduate-level textbook. While the book is not intended to be an exhaustive literature review, specific citations are examined that highlight the use of analytical techniques and the way in which they are utilized to solve the unique problems encountered. It presents a basic introduction to the field for a novice, while providing detailed information sufficient for an expert in a particular analytical techique.

In closing, I would like to thank several colleagues who have contributed to my efforts in the burgeoning field of drug discovery and combinatorial chemistry. They include Bob Pfeifer, Beverly Kenney, Bob Karol, Ray Crowley, Pat Fowler, Mike Balogh, John Hedon, and Eric Block of Waters Corporation (Milford, Massachusetts) as well as Steve Preece, Mark McDowall, and Andrew Brailsford of Micromass Limited (Manchester, United Kingdom). Special thanks also go to Carol, Kristina, and Robert of MCS (Uxbridge, Massachusetts) for their help and consideration in the preparation of this text.

Michael E. Swartz

Contents

Contributors

Mark Andrews, M.S. Strategic Planning Department, Waters Corporation, Milford, Massachusetts

Mary Brock Market Intelligence Department, Waters Corporation, Milford, Massachusetts

Stewart D. Chipman, Ph.D. Biology Department, ArQule, Inc., Medford, Massachusetts

Hong Jian Dai, M.S. Department of Chemistry, Shuster Laboratories, Quincy, Massachusetts

Christina A. Gendreau, M.S. Chemical Chromatography Division, Waters Corporation, Milford, Massachusetts

Hans-Ulrich Gremlich, Ph.D. CTA/Analytics Department, Novartis Pharma AG, Basel, Switzerland

Annette Hauser-Fang Department of Chemistry, Barnett Institute, Northeastern University, Boston, Massachusetts

Ira S. Krull, Ph.D. Department of Chemistry, Northeastern University, Boston, Massachusetts

James N. Kyranos, Ph.D. Analytical Chemistry Department, ArQule, Inc., Medford, Massachusetts

David Nickell, Ph.D. Discovery Informatics Department, Parke-Davis Pharmaceutical Research, Ann Arbor, Michigan

Michael J. Shapiro, Ph.D. Core Technologies Department, Novartis, Summit, New Jersey

Michael E. Swartz, Ph.D. Market Development Department, Waters Corporation, Milford, Massachusetts

Paul Voúros, Ph.D. Department of Chemistry, Barnett Institute, Northeastern University, Boston, Massachusetts

Harold N. Weller, Ph.D. Combinatorial Drug Discovery Department, Bristol-Myers Squibb Co., Princeton, New Jersey

1

An Introduction to Combinatorial Chemistry

Harold N. Weller
Bristol-Myers Squibb Co.
Princeton, New Jersey

I. INTRODUCTION AND OVERVIEW

The advent of combinatorial chemistry has led to a revolution in the drug discovery process, with impact on many related disciplines. Areas such as analytical chemistry, process development chemistry, and biological assay are reflecting the effects of combinatorial chemistry. However, at the same time, practitioners in those areas often have a vague idea about what combinatorial chemistry really is. Vendors of instruments and consumable supplies recognize the tremendous market potential of this new technology but are unclear about how best to capitalize on that potential. Part of the reason is that the definition of combinatorial chemistry differs from organization to organization and even within organizations depending on the specific drug discovery target. The purpose of this chapter is to provide a brief introduction to the assortment of new technologies collectively referred to as ''combinatorial chemistry.'' It will become clear that no single definition can clearly and concisely describe this new field.

Until the advent of combinatorial chemistry in the early 1990s, the drug discovery process had remained essentially unchanged for many decades. Chemists synthesized new molecular entities (compounds), biological properties were determined, and the chemists then incorporated the biological results into the design of the next-generation compound. This serial iterative process was repeated many times until a molecule with suitable drug-like properties was discovered. Through the years, improvements in biological assay provided higher throughput and more precise data on which to make judgments during the iterative design cycle. Similarly, improvements in computational chemistry provided additional rationale and tools for the chemist to use during iterative design. However, one thing remained constant, i.e., serial synthesis of one compound at a time with biological results providing the insight for iterative design of the next generation. With the development of high-throughput in vitro biological assays, the iterative serial synthesis of individual new compounds became the rate-limiting step in the overall drug discovery process.

In order to overcome the synthesis bottleneck in drug discovery, the concept of preparing many compounds at one time (parallel synthesis) rather than one compound at a time (serial synthesis) was born. In its simplest form, this distinction constitutes the definition of combinatorial chemistry. The origin of the concept has been ascribed (1) to Furka and others as early as 1982. Early applications of parallel synthesis methods were primarily in the area of peptide library synthesis and have been extensively reviewed (2,3). In the early 1990s, however, application to small drug-like molecules was reported (4) and the explosion in combinatorial chemistry activity began.

As more and more organizations have become involved in combinatorial chemistry, an increasingly large number of variations on the theme have been developed. To understand this complexity, one must remember that the overall objective of the field (within the pharmaceutical industry) is to accelerate the drug discovery process and that different ''combinatorial'' technologies may be best suited to different phases of the drug discovery process.

Drug discovery begins with a hypothesis involving a target biomolecule (receptor, enzyme, etc.), a proposed molecular intervention (receptor antagonism, enzyme inhibition, etc.), and a proposed clinical outcome of that intervention. To begin a full drug discovery program requires a reasonable target hypothesis, a validated biological assay, and (ideally) a structural starting point for drug design. Once the assay is in place, the first chemistry phase of a drug discovery process is the search for a structural starting point. That starting point is often called a ''lead compound'' and this phase of drug discovery is often called the ''lead discovery'' phase. A suitable lead compound is a small drug-like molecule with measurable and reproducible activity in the primary assay or assays. Even in programs with good starting templates (e.g., a large biomolecule with the desired activity or a known substrate structure), testing of thousands of compounds is often required before a useful lead compound is found. In the absence of a rational starting point, random screening may require testing of many tens of thousands of compounds in order to uncover a lead in the lead discovery phase.

A combinatorial chemistry program designed to support lead discovery has several well-defined characteristics. First and foremost, it provides very large numbers of compounds for screening—perhaps into the tens of thousands or even hundreds of thousands of new compounds. Since screening is done only in a single primary in vitro assay, only a small amount of each compound is required. Since the objective is merely to find activity and not to compare activity among compounds, a high level of purity of individual compounds is not required (providing that impurities do not themselves interfere with the biological assay). Intentional synthesis and testing of mixtures is common in the lead discovery phase, and indeed early usage of the term ''combinatorial library'' to describe synthetic compounds generally referred to synthetic mixtures containing combinations of many compounds (5). Since the objective is to find a single compound with measurable activity, testing of mixtures is often appropriate. A mixture with moderate average activity may contain many individual compounds with moderate activity or a single very active compound and many inactive compounds. In the lead discovery phase, where the probability of finding even a single active compound is slim, the latter case is often the more likely scenario. The practical limit of the number of compounds that may constitute a mixture depends on the assay sensitivity and the desired activity of a presumed lead compound, as well as the degree of control one has on the ratio of compounds in the mixture.

Once a lead compound has been discovered, the drug discovery project moves into an early ''lead optimization'' phase. In this phase, the intrinsic activity of the lead compound is improved by small structural modifications to the original structure. Design of target compounds for synthesis is guided by results from testing of previous compounds in the series. Since compounds are tested in only a single or a limited number of primary in vitro assays, the amount of each compound required is small. Often it is important to compare the activity of one analog with that of another; thus purity must be sufficient to allow such direct comparisons, and interpretation of data from screening of mixtures becomes cumbersome. It should be clear that combinatorial synthesis methods that are highly suitable for the lead discovery phase may be unsuitable for lead optimization.

Several iterative design cycles of early lead optimization may result in one or more compounds that are highly active in the primary assay. At this point, the drug discovery project moves into late-stage lead optimization; testing will begin in a variety of secondary in vitro or in vivo assays such as measurement of physicochemical properties, pharamcokinetics, metabolism, toxicity, cell permeability, etc. These secondary assays may be less highly automated than the primary assay and often require larger amounts of each compound. Secondary assays may have more strict analysis requirements than automated primary assays; thus high-purity compounds are preferred. As the project moves from early stage lead optimization to late stage lead optimization, the combinatorial synthesis focus will shift from making small amounts of large numbers of compounds of moderate purity to making larger amounts of fewer compounds of high purity. Once again, synthesis methods that are highly suitable for early stage lead optimization may be unsuitable for late stage lead optimization.

The final phase of a drug discovery project is candidate selection. In this phase, compounds are screened in an in vivo disease model in addition to the primary and secondary in vitro assays. Once again, the amount of each compound needed is increased, the purity requirement is higher, and the total number of compounds generally required is smaller.

From the discussion above, it should be clear that no single combinatorial synthesis technology can optimally serve all phases of a drug discovery project. In the sections that follow, various combinatorial synthesis methods will be described, as well as where each particular methodology is most useful in the drug discovery process. It is the proliferation of synthesis methods, each suitable for a certain phase of drug discovery, that leads to confusion as to the role and definition of combinatorial chemistry. In truth, combinatorial chemistry is a collection of different methodologies for the synthesis of multiple compounds, each with its own strengths and weaknesses, and each suitable for a slightly different phase of drug discovery. Importantly, to be successful throughout the entire drug discovery process, an organization must have access to more than a single combinatorial chemistry technique.

II. AUTOMATED SOLUTION PHASE SYNTHESIS: APPLICATIONS AND ANALYTICAL CHALLENGES

Organic synthesis is traditionally done in solution. That is, starting materials and reagents are dissolved in a solvent or homogeneous solvent mixture for the reaction to take place. After each step of a multistep chemical synthesis the intermediate reaction products are isolated, purified, and characterized before the next synthetic step is taken. The challenges in automating solution phase synthesis are to find creative ways to avoid tedious reaction workup and product purification after each step. Balancing these challenges is the availability of the full range of standard analytical methods to assay purity and structure throughout the synthetic sequence.

Solution phase organic reactions are typically "worked up" beginning with a series of extractions whereby the mixture is partitioned between an organic phase and an aqueous phase of known pH. In this way, products and reagents are separated from one another by their solubility and ionic nature. While organic/aqueous extraction procedures have been automated using robotic liquid handlers (6–8), difficulties associated with accurately detecting the solvent boundary have limited their applicability. Methods utilizing ion exchange resins to perform separations based on ionic nature are more easily automated, e.g., using commercially available solid phase extraction robots (9). Other methods, including use of solid supported reagents (10,11) and introduction of scavenger resins to reaction mixtures (12), also facilitate separation of products from reagents during automated solution phase synthesis. Because the various methods used for reagent removal are often compound-specific, solution phase combinatorial synthesis is generally directed to synthesis of individual discreet compounds (13), though mixture synthesis has also been reported (14). Solution phase combinatorial synthesis has been used to prepare libraries of up to 20,000 compounds for lead discovery (15) but may be most suited to preparation of moderate numbers of individual discreet compounds for the lead optimization phase of drug discovery.

As seen from the above discussion, the analytical challenges of automated solution phase synthesis lie primarily with product isolation, purification, and characterization. Specific equipment for high throughput or parallel purification of compounds from combinatorial solution phase synthesis is still largely unavailable; thus creative use of existing equipment is often required. For example, disposable solid phase extraction cartridges and robotic apparatus were invented primarily for concentration of biological and environmental samples prior to analysis but are now widely used for preparative scale purification of samples from combinatorial synthesis (16). High-throughput preparative HPLC is also a powerful tool for purification of solution phase synthesis libraries, but systems designed specifically for that purpose have only recently been reported and are not yet widely available (17). Methods for characterization of solution phase libraries are the same as those used for solid phase libraries and will be discussed later.

III. SOLID PHASE SYNTHESIS: APPLICATIONS AND ANALYTICAL CHALLENGES

Synthesis of organic compounds on solid support was first reported for peptide synthesis in the 1960s (18). In solid-supported (or ''solid phase'') synthesis, the starting organic scaffold is covalently bonded to an insoluble polymer support, generally a functionalized polystyrene derivative. Reagents and building blocks are added in solution and the resulting reaction mixture is heterogeneous (reagents in solution, starting scaffold bound to insoluble solid support). As the reaction takes place, the starting scaffold is modified but remains bonded to the solid support, while reagents are converted to byproducts but remain in solution. When the reaction is complete, simple filtration removes soluble reagents and byproducts, leaving the modified substrate still bound to the insoluble polymer. The elaborate workup required for reagent removal in solution phase synthesis (extraction, purification, etc.) is reduced to a simple filtration when synthesis is done on solid phase. In this way, many sequential chemical synthesis steps can be carried out on a single starting material with only filtration (rather than elaborate extraction and purification) between steps. At the end of the multistep chemical synthesis sequence, the desired product will still be attached to the insoluble resin, and all reactants and byproducts will have been washed away. At this point, a chemical reaction is performed that cleaves the covalent linkage between the product and the solid support, thus bringing the product into solution. Filtration following this ''cleavage'' step thus delivers the product in solution and separates it from the solid support to which is was bound during the synthesis sequence. If each step in the solid phase synthesis sequence has gone to 100% completion and delivered no undesired polymer bound side products, then the product cleaved from the resin will be 100% pure. In practice, such high levels of purity are rarely achieved except in the area of biopolymer synthesis.

The obvious advantages for automation of solid phase synthesis over solution phase synthesis have been exploited for many years in the area of biopolymer synthesis. Automated equipment for parallel solid phase synthesis of peptides is commercially available and often delivers products of very high purity. Peptide synthesis generally proceeds by anchoring a single amino acid to polystyrene resin, then adding sequential amino acids via amide bond coupling reactions. Thus there is a single type of bond forming reaction (amide bond formation), a limited set of monomers (20 naturally occurring amino acids), and chemists have been optimizing the process for over 30 years. It is little wonder that products of high purity are commonly obtained directly from resin cleavage. Compare this situation with the broadly based organic synthesis now being attempted on solid phase i.e., many different reaction types, virtually unlimited monomer availability, and only a few years of collective reaction optimization by the chemistry community. In the latter case, products of high purity are not necessarily obtained directly from resin cleavage unless the chemist has spent considerable time optimizing reaction conditions for each step prior to synthesis of a library of compounds.

The chemist setting out to synthesize a library of several dozen to several thousand organic compounds via solid phase synthesis will generally begin by working out synthesis conditions on a small subset of the library. The first step is to find a suitable chemical linker to tether the starting substrate to an insoluble support. The linker is generally attached to the substrate molecule via a labile functional group such as through an ester linkage (which is later cleaved to a carboxylic acid) (19) or an acetal group (which is cleaved to liberate a hydroxyl group) (20). Design of functionalized polymers and linkers for solid phase synthesis is an emerging technology in itself and has been reviewed (21,22). Once the substrate is linked to the solid phase resin, the various building blocks must be added one at a time. Each chemical step must be optimized to determine the best conditions for optimal yield. One advantage of solid phase synthesis is that a large excess of reagents may be used to drive reactions to completion because all excess reagents will be easily removed by filtration following the reaction. The process of developing suitable reaction conditions for each step of a library synthesis generally takes several months of tedious work by the chemist. In contrast, using modern automation techniques, synthesis of the actual library may only require a few days or at most a few weeks after synthesis conditions have been defined. Developing reaction conditions for solid supported synthesis is particularly tedious because of the lack of suitable methods for analysis of resin-bound reaction products. The chemist no longer has direct access to the full range of conventional methods for assaying purity (e.g., chromatography) or structure (e.g., spectroscopy) unless the product is first cleaved from resin for analysis. Solid phase nuclear magnetic resonance (NMR) (23,24) and Fourier transform infrared spectroscopy (FTIR) methods now provide the most useful information about structures bound to resin, and these methods will be reviewed in subsequent chapters in this volume. Opportunities thus exist for the analytical chemist to have a significant impact on a combinatorial chemistry program in two ways: development of methods for analysis of reaction intermediates on solid support, and development of methods for purification of impure products after cleavage from solid support. Either of these methods will allow the chemist to shorten his or her chemical development time and proceed directly to library synthesis with confidence. Many of the chapters in the remainder of this volume deal directly with these important issues.

Solid phase synthesis of organic molecules is exemplified by the pioneering work of Bunin and Ellman on solid phase synthesis of a small benzodiazepine library (29). As shown in Fig. 1, a substituted benzophenone was tethered to a solid support via a substituent (R_1). The solid supported benzophenone was then elaborated through five synthetic steps to the target benzodiazepines with four points of diversity. Cleavage from the solid support re-

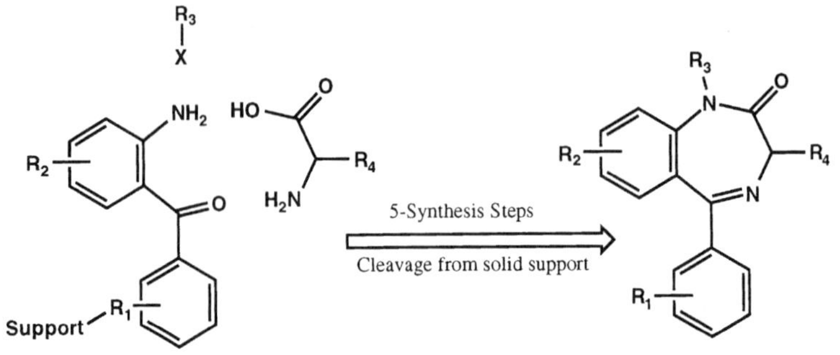

Figure 1 Solid phase benzodiazepine synthesis (29).

leased the products in high yield. This preliminary report was the first widely recognized solid supported synthesis of nonoligomeric compounds and set the stage for rapid development of combinatorial synthesis methods for small organic molecules.

IV. DIFFERENT APPLICATIONS OF COMBINATORIAL SYNTHESIS TO DRUG DISCOVERY

Combinatorial synthesis in its broadest form involves parallel, possibly automated, synthesis of many compounds at once using either solution or solid phase synthesis methods. Beyond that broad definition, though, an impressive variety of specific applications and methods has been developed to suit the particular needs of various projects or corporate sponsors. At one extreme is synthesis of very large mixtures for early phase lead discovery programs, and at the other extreme is parallel synthesis of pure discrete compounds for late-stage lead optimization programs. The following sections will attempt to categorize the main combinatorial synthesis methods into broad groups.

A. Synthesis of Mixtures by Solid Phase Methods

It would seem that the simplest way to prepare a mixture of compounds on solid phase would be to attach a single starting substrate to the polymer and then react that polymer bound substrate with a mixture of building blocks or reagents to produce a mixture of products. The problem with this approach is that the kinetics for reaction of the substrate with each of the diverse building blocks may not be identical. When reaction kinetics are not identical, the resulting product mixture will not be equimolar, and it is possible that some of the presumed products will not be present in the product mixture at all. Furthermore, each individual polymer bead in the mixture will contain a mixture of products rather than a single product, thus complicating analysis and deconvolution (as will be discussed later).

A better way to prepare mixtures by solid phase synthesis is by using the ''split-and-pool'' method. With this method a single substrate is bound to the resin; then the resin is split into a number of equal size pools, where the number of pools represents the number of diverse building blocks to be used in the first step of the synthesis. Each pool is then reacted with its designated building block in a separate reaction vessel under conditions where the reactions are driven to completion (e.g., by using excess reagents). Following reaction, the pools are filtered to remove excess reagents and washed with fresh solvent. All of the polymer from all of the pools is then recombined and mixed thoroughly. The combined resin is then split again into a new set of pools representing the number of diverse building blocks to be used in the second step of the synthesis. At this point, each resin pool should contain an equimolar amount of each of the products created in the first synthesis step. The process is then repeated through as many cycles as needed to complete synthesis of the entire library.

The split-and-pool method is illustrated in Fig. 2 for a two-step synthesis involving three building blocks at each step resulting in three equimolar mixtures of three compounds each. The starting functionalized resin is divided into three equal pools, and each pool is reacted with a specific reagent [R1(a), R1(b), or R1(c)] in a separate container. At the end of this step, each reaction container contains a single specific resin–bound product. Reagents are then removed from the reactions and the solid resins are washed with fresh solvent. All three resin batches are then combined and mixed thoroughly to provide an equimolar mixture of the three possible products. Note that, in principle, one could obtain the same mixture by simply reacting all of the starting resin with a mixture of the three reagents in a single vessel. In practice, though, this process will not provide an equimolar mixture of products due to differential reaction kinetics. It is the desire to obtain equimolar mixtures of products that necessitates the split-and-pool method.

Returning to the hypothetical split and pool synthesis (Fig. 2), the chemist would next split the equimolar mixture of three products into another three identical pools. Reaction of each of these pools separately with three different

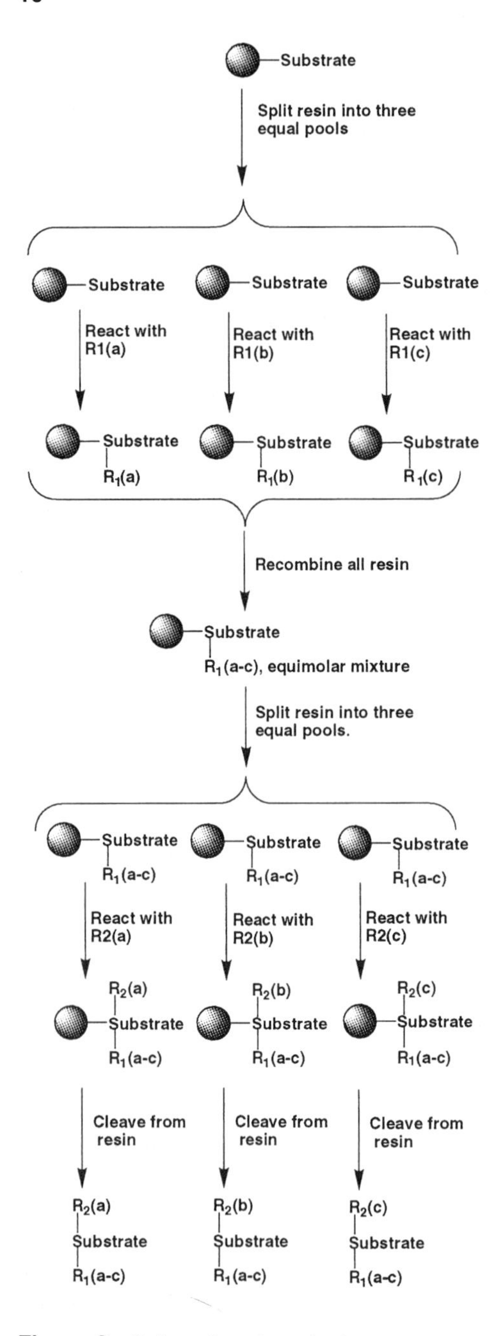

Figure 2 Split-and-pool synthesis.

reagents [R2(a), R2(b), and R2(c)] results in three equimolar mixtures of three products each, for a total of nine new products. The products can be cleaved from the solid support to provide three mixtures of three compounds each.

The split-and-pool method has several interesting ramifications. First, since it does not depend on identical reaction kinetics, it should produce an equimolar mixture of products assuming that all reactions are driven to completion. Second, since any individual polymer particle follows only one synthetic path, all of the product molecules bound to that particle should be the same. That is, one compound is prepared per bead. Third, since one compound is prepared per bead, there must be at least as many polymer beads in the total synthesis pool as there are theoretical products being formed during the synthesis. Imagine, for example, if the split-and-pool synthesis outlined in Fig 2 was begun with only six resin particles! In practice a statistical sampling of the total resin pool is made with each ''split'' and statistical analysis can be done to determine the required number of particles to ensure adequate sampling of a given pool (30,31). Finally, split-and-pool methodology results in significant reduction in the total number of reactions needed to produce the library. Imagine, for example, a library where building blocks are added in three sequential chemical steps and where each building block is represented by 10 different variations. The total library size will be $10 \times 10 \times 10 = 1000$ compounds. To prepare the compounds conventionally would require 1000 individual reactions at each step or 3000 reactions in total. Using split-and-pool methods, only 10 reactions are required at each step or 30 reactions total to prepare the library. Counterbalancing the synthesis efficiency is the effort required to split and pool the resin particles.

At the end of a split-and-pool synthesis, the chemist has the option to recombine all resin particles into a single pool and then cleave products, or to keep the final split mixtures separate and cleave each individually. Using the example in Fig. 2, the chemist could recombine all polymer resin prior to cleavage to obtain an equimolar mixture of all nine compounds in the library. Alternately, the chemist could keep the three pools used in the final step separate and cleave each individually, resulting in three separate mixtures of three compounds each as illustrated in the figure. The choice of which path to take is dictated largely by the target mixture size. If there are too few compounds in the final mixture the synthetic advantages of mixtures are lost, whereas if there are too many compounds in the mixture there is significant risk that a single active compound will be missed during bioassay due to the dilution effect of other mixture members. In the early days of combinatorial synthesis, very large mixture libraries were common, with mixtures containing tens of thousands or even hundreds of thousands of compounds commonly reported.

For example, a library containing 19×10^5 (nearly 2 million) pentapeptides was prepared from 19 of the 20 naturally occurring amino acids in 1991 (32), as was a library of over 34 million hexapeptides consisting of 324 mixtures of 104,976 compounds each (33).

Because of concerns about not detecting low concentrations of active compounds in such large mixtures, the current trend is toward much smaller mixtures containing only hundreds or even dozens of compounds. For example, the Affymax group prepared a library containing mixtures of only 540 compounds and identified potent cyclooxygenase-1 inhibitors after deconvolution (34). The issue of optimum mixture size is not yet settled and discussion will no doubt continue for some time.

When bioassay of a mixture produces apparent activity, one must then ''deconvolute'' the mixture to identify the active component (or components). The most straightforward deconvolution method is independent synthesis of each individual member of the original mixture. For large mixtures, though, such an approach is not practical. Instead, iterative resynthesis of increasingly smaller mixtures is used to ultimately identify the active component. For example, a mixture of 104,976 hexapeptides was deconvoluted by four iterative syntheses of 20 compounds or mixtures each (35). In the first iteration, 20 mixtures of 6859 peptides were synthesized. The most active mixture was then used as the starting point for 20 mixtures of 361 compounds. Next, 20 mixtures of 19 compounds each were synthesized and, finally, the 19 individual compounds of the most active mixture were synthesized to identify the single most active compound in the original mixture.

Iterative deconvolution works best when a single compound of the mixture is much more active than all others. Conversely, deconvolution is difficult when all members of a library have similar activity—as is often the case in the later stages of a drug discovery process. Because of the difficulties associated with mixture deconvolution, split-and-pool synthesis is best suited to the early lead discovery phase of the drug discovery process. Later stages, requiring direct comparison of one compound against another, are best served by other methods described below.

B. Chemically Encoded Libraries

The synthetic advantages of preparing mixtures instead of individual compounds are offset by the need to deconvolute mixture libraries following bioassay. Deconvolution shortcuts, such as synthesis of multiple libraries containing orthogonal pools (36), are sometimes useful but can only be relied on when the activity of the mixture results from a single highly active compound. Keeping in mind that split-and-pool methodology results in only one compound per polymer bead, the difficulties associated with deconvolution and the uncertainties of screening mixtures can be eliminated by the screening of individual compounds from single beads. Methods have been developed for screening products while still attached to the beads. Alternately, following split-and-pool synthesis, the individual beads are separated and the products cleaved individually from the beads. The individual products are then screened in the bioassay. Active compounds discovered in the bioassay are individual compounds rather than components of mixtures; thus, no deconvolution is necessary. How, though, can one identify the structure of the active compounds from the small (subnanomole) amount prepared on a single bead? Direct analysis of the products by sensitive analytical methods has been reported. For example, peptides and oligonucleotides have been analyzed in the 5-pmol range by Edman degradation or DNA sequencing (37). Molecular weights have been obtained for small organic molecules from single-bead synthesis (38), but large libraries may contain more than a single member with the same mass. The solution to the problem of product identification from single bead synthesis lies mainly with encoded libraries.

Chemically encoded libraries are synthesized from bifunctionalized polymers. During each step of the split-and-pool synthesis, two separate chemical reactions take place. The first adds a building block to the polymer-bound substrate and leads to the target product. The second adds a building block to a unique chemical tag that will be used to later identify the product associated with that bead. The method is shown schematically in Fig. 3 using an example of a two-step synthesis with three variations at each step. Following synthesis, products from individual beads are cleaved and screened. When active compounds are found, the original bead is subjected to an orthogonal cleavage step that releases the chemical tag. The chemical tag is subjected to a sensitive and precise analytical method that reveals the structure of the tag and therefore the structure of the product molecule.

Success of an encoded library strategy depends on the availability of a method for screening products from individual beads, and of a suitably precise and sensitive analytical method for decoding the chemical tag. Four primary tagging methods have been reported. As mentioned above, sensitive methods exist for sequencing of peptides and oligonucleotides; thus, peptides and oligonucleotides make suitable chemical tags for encoded library synthesis. For example, Needles and coworkers (39) demonstrated the suitability of oligonucleotide tags for the synthesis of a peptide library using an orthogonal protecting group strategy. The growing peptide chain was protected by the base-labile 9-Fluorenylmethoxycarbonyl (FMOC) protecting group, while the

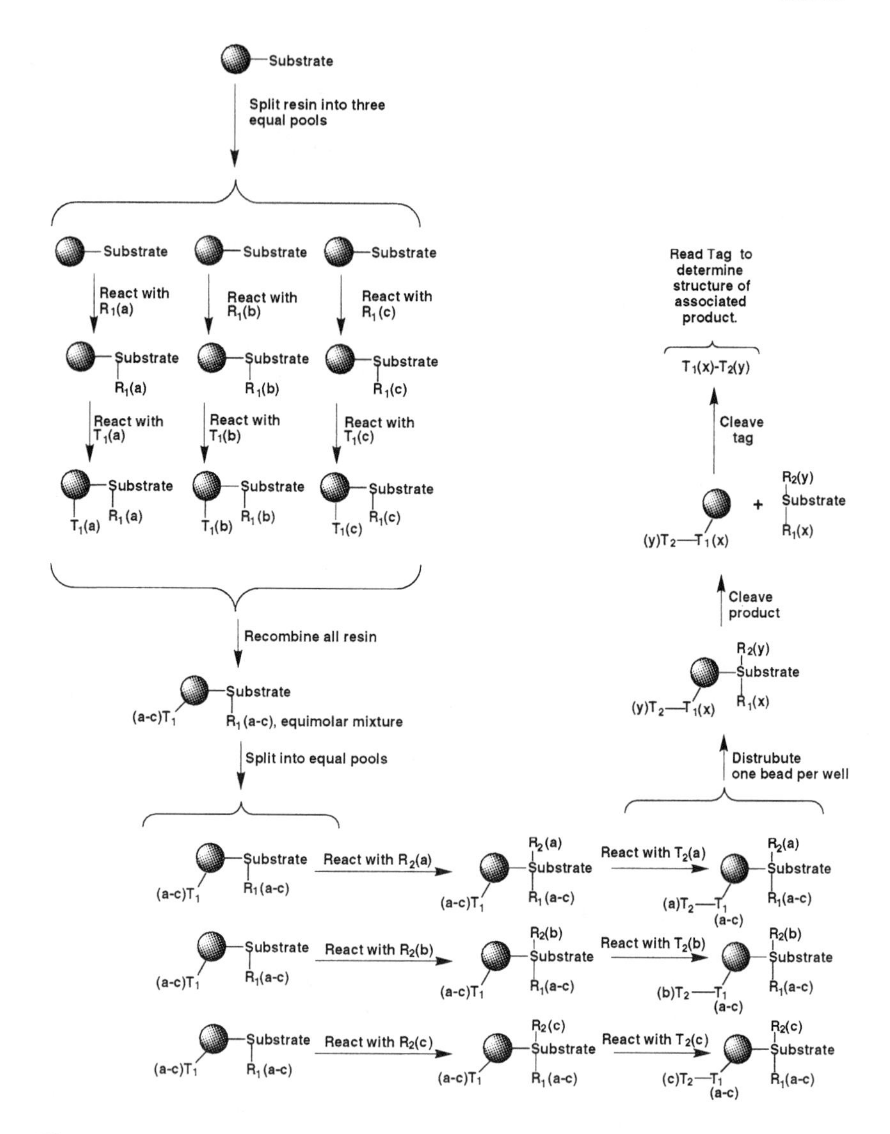

Figure 3 Chemically encoded split-and-pool synthesis.

growing oligonucleotide chain was protected by the acid-labile dimethoxytrityl (DMT) group. A similar strategy should work well for synthesis of small organic molecules instead of peptides, providing the chemical sequence needed for product synthesis is compatible with the oligonucleotide tag. Similar strategies have been reported using peptides as the chemical tag (40). A binary encoding strategy has been reported using substituted haloaromatic tags that can be detected at very low levels by electron capture gas chromatography (41). The tags are attached via a photolabile (42) or oxidatively labile (43) linker, thus allowing most types of chemistry to be used in building the desired product molecules. This method has been reported for synthesis of both peptides (44) and nonpeptide molecules (45). A similar coding method utilizing secondary amines attached via amide linkages has also been reported (46,47). In this case, the tags are hydrolyzed using 6 N HCl, the resulting secondary amines derivatized, separated by HPLC, and detected by fluorescence spectroscopy.

The power of the encoded library strategy is illustrated by an example from Borchardt and Still (48). A 50,000-compound encoded acyl tripeptide amide library was prepared using the split-and-pool method. Beads from the library were mixed with a dye-linked synthetic receptor. After 24 hours, beads containing compounds that bind tightly to the synthetic receptor were stained deep red. Stained beads were mechanically selected, their binary coded tags were cleaved by photolysis, and the codes were read by electron capture gas chromatography to determine the structures of the tight-binding substrates. In this way, 50 compounds with high receptor affinity were selected from the library of over 50,000 compounds.

Chemically encoded libraries are a very powerful method for preparing large numbers of individual compounds whose structures can be determined. Successful application of a chemically encoded library requires access to a suitably sensitive high-throughput analytical method for reading the chemical code. Since code reading itself can become time consuming if many members of a library are found to be active, and the amount of each compound produced is limited to the amount that can be synthesized on a single bead, encoded libraries are well suited to lead discovery and early phase lead optimization but may be less well suited as the drug discovery process approaches candidate selection.

C. Mechanically Encoded Libraries

The main limitations of chemically encoded libraries are the necessity to perform an extra chemical reaction at each step to introduce the coding tag, the limitations introduced by the presence of the tag on the chemistries that may be performed on the target substrate, and the limited amount of material that can be obtained from a single polymer bead. These limitations can all be overcome using mechanically encoded libraries instead of chemically encoded libraries. With mechanically encoded libraries, polymer resin is held in a porous container that allows soluble reagents to pass through but not resin beads. The container is labeled with a mechanical tag (perhaps as simple as a printed label) and the containers of resin are treated in the same way that individual beads are treated in the split-and-pool method. While the split-and-pool method provides one compound per bead, the mechanically labeled split-and-pool method provides one compound per container of beads. If the synthetic path of each container is recorded at each synthetic step, then simply reading the container label will identify the product contained within. In practice, a database is generally created prior to execution of the synthesis outlining the precise synthetic pathway intended for each resin container. After each synthetic step the containers are retrieved, their labels read, and they are pooled for the next synthetic step according to the predetermined sequence from the database. This method is referred to as ''deterministic'' split-and-pool synthesis as opposed to ''statistical'' split-and-pool synthesis. Because the fate of each container is determined ahead of time, statistics play no role and the number of containers required is exactly equal to the number of products being synthesized.

The first example of a mechanically encoded library was reported by Houghten in 1985 (49). Resin was contained in a polypropylene mesh packet resembling a tea bag. The label was mechanically inscribed on the packet and was read visually with manual sorting of the packets. In this way, 247 analogs of a 13-amino-acid peptide were individually prepared. Synthesis of a 500,000-compound mixture library was recently reported using this method (50). The packets used were porous polyethylene tubes with printed labels that could be automatically read by optical character recognition (OCR) software. Each packet contained a mixture of 21 different functionalized resins, and the effort resulted in synthesis of approximately 26,000 individual containers each theoretically containing a mixture of 21 compounds for a total library of 551,070 compounds. The reader will recall, however, that in syntheses of this type each packet will contain an equimolar mixture of 21 products only if the reaction kinetics at each step are identical and that this was the reason for moving away from direct mixture synthesis and toward split-and-pool mixture synthesis in the first place. Two groups have recently reported use of radiofrequency tags in place of visually readable tags (51,52). These machine-readable tags offer the possibility of inexpensive automated reading and sorting. All of the above methods require the use of a nonreactive porous resin container.

Radiofrequency tags encapsulated in grafted functionalized polymers have recently been reported (53), as have laser optical–encoded ceramics with grafted polystyrene supports (54). These devices have the properties of ''really big beads'' that contain the identifying tag and thus combine the features of chemically encoded libraries with the advantages of mechanically encoded libraries.

Mechanically encoded libraries are best suited to synthesis of individual compounds rather than mixtures. The amount of each compound produced can be small or large and is limited only by the size of the packets and related physical limitations. Since reactions can be carried out in conventional glassware, mechanically encoded libraries offer an inexpensive entry into large-library synthesis. The method is applicable throughout most of the drug discovery process with the possible exception of the very earliest discovery phase and the very latest candidate selection phase.

Mechanically encoded libraries offer their greatest advantage over individual compound synthesis when the library consists of a dense symmetrical array. Consider, for example, synthesis of a 100-compound library with 10 structural variations at each of two positions (a 10×10 array). Individual compound synthesis would require that all 100 compounds undergo two chemical synthesis steps (not counting cleavage from the resin) or a total of 200 chemical reactions. Synthesis of the same library using mechanical tags would require only 20 reactions (10 reaction vessels each containing 10 packets for the first step, and the same for the second step). For a 100-compound library consisting of a 2×50 array, though, individual synthesis would still require 200 reactions whereas the mechanically encoded library would require 52 reactions (two reactions each with 50 packets in the first step, and 50 reactions each with two packets at the second step). Thus the advantage over individual synthesis is diminished. The advantage is also reduced for synthesis of ''sparse'' arrays. Consider again the 10×10 array, but suppose that sophisticated library design software has excluded some of the possible combinations so that there are ''holes'' in the array and a total of only 50 targets. Individual synthesis would require 50 reactions at each of 2 steps or 100 reactions total, while the encoded library would still require all 10 reactions at each step (though with less than 10 packets at each step) for a total of 20 reactions. For these reasons, encoded library synthesis will probably take its place alongside individual compound synthesis and split-and-pool mixture synthesis (rather than replace them) in an overall drug discovery program.

D. Automated Parallel Synthesis

The simplest conceptual manifestation of combinatorial chemistry is that of parallel synthesis, which is simply synthesis of several compounds in distinct reaction vessels at once (in parallel) rather than sequentially (in series). Any chemist who has had several reaction flasks stirring at once has practiced a crude form of parallel synthesis. Parallel synthesis results in individual compounds that can be treated in downstream handling steps identically to compounds synthesized conventionally. Thus it is a conceptually easy first entry into combinatorial chemistry. The tedium of handling many reaction flasks at a time has quickly led to a wide variety of apparatus for automating, or partially automating, parallel synthesis. The first equipment available was modeled after the venerable peptide synthesizer and offered complete automation of a multistep reaction sequence. Such equipment is now available commercially from vendors such as Advanced Chemtech (55) and Arognaut Technologies (56). The advantages of such systems (total automation) are offset by relatively complex software and relatively low synthesis capacity. The latter is a result of the fact that the actual synthesis reaction vessels must remain mated to apparatus for reagent addition throughout the process in order to meet the needs of total automation. A number of modular workstations have now been reported, and some are now available commercially from vendors such as Diversomer Technologies (57), Bohdan Automation (58), and Tecan (59). One of the first workstations reported was the Diversomer apparatus from Parke-Davis (60,61). Solid phase resin is contained in glass gas dispersion tubes that are arrayed in solvent containment wells. Reagents are added via robotic liquid handler through a septum that seals the top of the apparatus. A number of other reaction blocks have been reported. For example, an enclosed reaction block that can be heated and cooled and to which reagents are added robotically was described by the Ontogen group (62).

Regardless of the source, reactors for automated parallel synthesis have certain common characteristics. They all offer a group of reaction vessels that can be handled as a unit. All offer a means of liquid addition either via robotic liquid handler or via a closed pneumatic or pumping system. All offer a means of separating soluble reagents from insoluble polymer-supported products, and all offer a means of collecting soluble products after cleavage from the polymer resin. Systems that offer total automation of a multistep chemical sequence are generally referred to as "synthesizers." With synthesizers, the reactions generally take place entirely within the confines of the synthesizer and proceed completely unattended. Throughput per run is limited to the capacity of the synthesizer. Systems that require manual intervention and movement of the reactions from place to place in assembly line fashion are referred to as "workstations." In this case, a group of reaction vessels, perhaps contained in a single "reaction block," is manually moved from station to station as the synthesis progresses. For example, the reaction block may be loaded with polymer resin and moved to a liquid handler for reagent addition. It is then moved to an incubation station where agitation, atmosphere control, and temperature control are maintained while the chemical reaction takes place. This process frees up the liquid handler workstation to accept another reaction block in assembly-line fashion. The reaction block then moves on to various stations for reagent removal, resin washing, product cleavage, and so on. The workstation approach has been well described by Cargill and coworkers (63) who describe synthesis of thousands of individual compounds per month. While the workstation approach is not totally automated and requires manual intervention, it also offers higher overall throughput at lower cost than the synthesizer approach.

A number of novel approaches to parallel synthesis have been reported. For example, synthesis of peptides on grafted polyethylene pins spatially arrayed in 96-well microtiter plate format has been described (64). In this way, a carrier containing 96 grafted pins is inserted into a 96-well microtiter plate to expose the polymer-bound substrate to soluble reagents. The reagents are removed by simply lifting the pin carrier from the reagent solutions in the microtiter plate. Application of this method to synthesis of nonpeptides has also been reported (65).

The Affymax group has reported a novel photolithography method for preparing spatially dispersed peptide arrays (66). Using this method, the starting substrate is bound to a functionalized glass substrate, and the reactive terminus of the starting peptide is protected with a photolabile protecting group. Using photolithography, the peptide is deprotected only on selected sections of the sheet. Reagents are then introduced to add a protected amino acid to the deprotected peptide. The sequence is repeated again and again until all positions on the sheet have been reacted. This system is very powerful, but it has the drawbacks that (1) only substrates that can be suitably protected can be used and (2) the synthesis is essentially a serial rather than a parallel process because only one type of amino acid can be introduced at a time.

Another new approach to very-high-throughput parallel synthesis is the ChemSheet (67). This is a flat polymer sheet containing 2304 8-μl reaction wells. Ink jet dispensing technology allows high-speed delivery of reagents and solvents to the very small reaction wells. While this technology is not yet proven in actual chemical synthesis, it illustrates the directions that parallel synthesis may take in the future.

From the discussion above it should be clear that parallel synthesis provides individual compounds for assay over a wide range of compound classes (68), numbers (of compounds), and scale (mg of each compound), thus making it suitable throughout much of the drug discovery process. Split-and-pool methods and encoded synthesis methods, however, provide greater numbers of compounds with less effort and thus may be more suitable than parallel synthesis in the earliest phases of drug discovery.

V. POSTSYNTHESIS PROCESSING AND SAMPLE HANDLING

A. Format Changes and Liquid Handling

Much of the discussion and focus of combinatorial chemistry is on the synthesis process itself. However, most practitioners in the field will attest that post-synthesis processing is far more time consuming than synthesis and that post-synthesis processing is often the rate-limiting step in an overall combinatorial chemistry project. Early design of synthesis apparatus paid scant attention to the format in which the products are delivered. For example, products delivered into test tubes may have to be individually moved to a concentrator to remove solvent, then individually moved to a liquid handler for dissolution and distribution to microtiter plates for bioassay. Each ''format change'' (change of vessel type or vessel carrier array) may require manual intervention and thus will slow the overall process.

Transfer of samples from one vessel to another is generally done by dissolving the sample in an appropriate solvent and then aliquoting a fixed amount of solution from the first container into the second container. Commercial liquid handlers are available to automate this process, but their effectiveness is limited by the original format design. For example, transferring liquid from vessels in an array to another set of identical vessels in an identical array can be done using a simple multichannel pipetting device. This is most commonly observed when transferring samples from one 96-well microtiter plate to another. Transferring samples from an array into another nonidentical array, however, may require transfer of single samples using a single-channel pipettor. The situation is further complicated when the number of samples in one array is not equal to the number of samples in the other. Consider, for example, transferring samples from purification on a Zymark Benchmate solid phase extraction apparatus into a 96-well microtiter plate. The Benchmate delivers samples into 16 × 100 mm test tubes in a 5 × 10 array. The microplate has 96 wells, in an 8 × 12 array. The liquid handling protocol for this format change will be complex, will not take optimal advantage of multichannel pipettors, and will result in a partially filled microplate. Other common examples include delivery of samples for analysis into autosampler racks of commercial analytical equipment—many of which are round or pie-shaped and not amenable to loading by a multichannel liquid-handling robot. A recent trend toward use of the common 96-well Cartesian format in autosampling and fraction collection equipment will help to minimize format changes in the future. In any event, thoughtful design of a postsynthesis work flow with minimal format changes can greatly enhance throughput in any combinatorial chemistry process. In particular, avoidance of non-Cartesian racks and carriers facilitates optimal use of commercial multichannel liquid-handling devices.

B. Analysis

Analysis of combinatorial libraries is placing new and unprecedented demands on analytical equipment and services. Chemically encoded libraries require high-throughput analysis methods for library deconvolution, whereas parallel synthesis procedures require analysis of individual library members. There has been much discussion about whether it is necessary to analyze every member of a combinatorial library, to analyze representative members only, or to analyze active compounds only. It is probably safe to say, though, that if the tools were available most scientists would prefer to collect as much analytical data as possible. The problem, of course, is that while compounds are being synthesized in parallel the most commonly used analytical methods i.e., gas chromatography (GC), high-performance liquid chromatography (HPLC), and mass spectrometry (MS) are inherently serial methods. Serial processing methods can only keep up with parallel synthesis methods if their speed, capacity, and throughput are fully optimized. Optimization of these processes is the topic of a later chapter in this volume. Alternately, development of suitable parallel analysis methods may someday facilitate analysis of large combinatorial libraries. Thin-layer chromatography and capillary zone electrophoresis may be particularly suited to parallelization for large-library analysis, though reports of automated systems for combinatorial chemistry analysis are sparse. One of the major future challenges for analytical chemistry in support of combinatorial synthesis will be to develop and optimize high-capacity high-speed analysis methods and to develop the required hardware to support those methods.

C. Purification

As the drug discovery process proceeds from early lead discovery through lead optimization and toward candidate selection, the level of purity required of each compound increases. Late-stage compounds may be tested in multiple assays, many of which may involve animal models and may be labor-intensive. Even the best solid phase combinatorial synthesis methods cannot assure that every member of a synthesis library will be delivered in pure form directly from the synthesis procedure. High-throughput automated purification methods are thus required to provide purified compounds for late-stage drug discovery programs. Existing chromatographic methods generally fall into two categories: ''collect-before-detect'' and ''detect-before-collect.'' Collect-before-detect methods are those where product collection is based on a predetermined scheme and is not guided by real-time product detection. Such methods are suitable only when the chromatographic properties of the products can be predicted with confidence. Applications that fall into this category include ion exchange chromatography using solid phase extraction cartridges and apparatus (69). However, similar methods using pure adsorption chromatography provide much less predictable elution and are less amenable to high-throughput purification of a diverse combinatorial library. Detect-before-collect methods rely on feedback from a real-time detection system to guide fraction collection. With real-time detection it is possible to purify diverse libraries with confidence that the products will be collected. A variety of detection methods have been reported, including ultraviolet absorbance (70), evaporative light scattering (71), and even single mass detection (72,73). While chromatographic methods are inherently serial in nature, solid phase extraction can be done in parallel using common vacuum box technology. A four-channel parallel preparative HPLC has recently been reported and will be offered commercially in the near future (74). Once again, the challenge for analytical chemistry in the future will be to provide purification methods and apparatus with increasingly large capacity and throughput.

VI. DATA MANAGEMENT

One of the challenges in any combinatorial chemistry effort is managing the large amount of data being generated. Efficient stand alone systems for managing structural information, analytical information, and so forth exist today, but efficient combinatorial synthesis requires integration of data from a variety of sources to provide complete tracking throughout the synthesis process. Completely integrated solutions do not exist commercially, and it may be impossible to provide one due to the variety of instruments and procedures that can exist throughout the industry.

The data management process often begins with creation of a ''virtual library'' of all possible structures that could be generated by combining available reagents with a core structure. The number of possible structures may run into the millions or billions of compounds, thus requiring efficient structure searching, generation, and display. From the virtual library, the chemist will select (perhaps using automated methods) the targets that will actually be synthesized and become part of the real library. This target list will then lead to a list of reagents needed for the synthesis. Reagents will need to be ordered from outside vendors or acquired internally. Since most automated synthesis apparatus handle reagents in solution, reagents will then have to be weighed and diluted with appropriate solvents to produce reagent solutions of standard concentration. Reagent ordering, tracking, weighing, and dilution alone require both significant data handling and creation of instrument scripts (e.g., for weighing and diluting samples).

Once all materials are in place and the synthesis targets have been defined, instrument control scripts for performing the actual synthesis must be generated. While the actual form the scripts can take will depend on the instrument(s) being controlled, all such scripts will require information about the reagents and synthesis targets. Following synthesis, analytical data will be acquired (either on the synthesis products themselves or on chemical tags used for encoding) and data will have to pass to and from analytical instruments. Finally, sample registration (into a corporate database) and distribution will occur.

The challenge for instrument vendors is to provide sufficiently open architecture to allow customers to import and export data seamlessly to and from instruments. The challenge for combinatorial chemists is to develop protocols for transferring data from instrument to instrument through a seamless data flow. This generally involves accepting a data ''export'' from one instrument, performing calculations based on the data, and reformatting the data for ''import'' into the next instrument in the sequential work flow. The topic of data management for combinatorial chemistry will be addressed in detail in a later chapter in this volume.

VII. SUMMARY

The drug discovery process is a complex and dynamic continuum ranging from early phase lead discovery, through lead optimization, to candidate selection. The field of combinatorial chemistry provides a continuum of methods to support drug discovery. These methods range from split-and-pool synthesis of large libraries, through encoded synthesis of medium size libraries, and to parallel synthesis of smaller libraries of highly purified compounds. A successful drug discovery program will match the combinatorial chemistry methods used with the demands of the overall program at any point in time. It is indisputable that the advent of combinatorial chemistry has irreversibly changed the drug discovery process and the way medicinal chemists think about molecules (75). The challenge for analytical chemists is to understand the entire process and provide a family of specific tools to support each phase of drug discovery and each type of combinatorial chemistry that may be practiced.

REFERENCES

1. Review: F Balkenhoh, C von dem Bussche-Hunnefeld, A Lansky, C Zechel. Combinatorial Synthesis of Small Organic Molecules. Angew Chem Int Ed Engl. 35:2288–2337, 1996.
2. Review: G Jung, AG Beck-Sickinger. Multiple peptide synthesis methods and their applications. Angew Chem Int Ed Engl. 31:367–383, 1992.
3. Review: MA Gallop, RW Barrett, WJ Dower, SPA Fodor, EM Gordon. Applications of combinatorial technologies to drug discovery. 1. Background and peptide combinatorial libraries. J Med Chem 37:1233–1251, 1994.
4. Review: EM Gordon, RW Barrett, WJ Dower, SPA Fodor, MA Gallop. Applications of combinatorial technologies to drug discovery. 2. Combinatorial organic synthesis, library screening strategies, and future directions. J Med Chem 37: 1385–1401, 1994.
5. RA Houghten, C Pinilla, SE Blondelle, JR Appel, CT Dooley, JH Cuervo. Generation and use of synthetic peptide combinatorial libraries for basic research and drug discovery. Nature 354:84–86, 1991.
6. CD Garr, JR Peterson, L Schultz, AR Oliver, TL Underiner, RD Cramer, AM Ferguson, MS Lawless, and DE Patterson. Solution phase synthesis of chemical libraries for lead discovery. J Biomol Screen 1:179–186, 1996.
7. SH DeWitt. Automated parallel purification methods. Solid and Solution Phase Combinatorial Synthesis, New Orleans, 1997.
8. JP Whitten, YF Xie, PE Erickson, TR Webb, EB DeSouza, DE Grigoriadia, JR McCarthy. Rapid microscale synthesis, a new method for lead optimization using robotics and solution phase chemistry: application to the synthesis and optimization of corticotropin-releasing factor receptor antagonists. J Med Chem 39:4354–4357, 1996.
9. RM Lawrence, OM Fryszman, MA Poss, A Biller, and HN Weller. Automated Preparation and Purification of Amides. Proceedings of the International Symposium on Lab Automation and Robotics, Zymark Corp., Hopkinton, MA, 1995 pp. 211–220.
10. NM Yoon, HJ Lee, JH Ahn, J Choi. J Org Chem 59:4687–4688, 1994.
11. JJ Parlow. Simultaneous multistep synthesis using polymeric reagents. Tetrahedron Lett 36:1395–1396, 1995.
12. SW Kaldor, MG Siegel, JE Fritz, BA Dressman, PJ Hahn. Use of solid supported

nucleophiles and electrophiles for the purification of non-peptide small molecule libraries. Tetrahedron Lett 37:7193–7196, 1996.
13. DL Boger, CM Tarby, PL Myers, LH Caporale. Generalized dipeptidomimetic template: solution phase parallel synthesis of combinatorial libraries. J Am Chem Soc 118:2109–2110, 1996.
14. T Carell, EA Wintner, AB Hashemi, J Rebek Jr. A novel procedure for the synthesis of libraries containing small organic molecules. Angew Chem Int Ed Eng 33:2059–2064, 1994.
15. CD Garr, JR Peterson, L Schultz, AR Oliver, TL Underiner, RD Cramer, AM Ferguson, MS Lawless, DE Patterson. Solution phase synthesis of chemical libraries for lead discovery. J Biomol Screen 1:179–186, 1996.
16. RM Lawrence, SA Biller, OM Fryszman, MA Poss. Automated synthesis and purification of amides: exploitation of automated solid phase extraction in organic synthesis. Synthesis 553–558, 1997.
17. HN Weller, MG Young, SJ Michalczyk, GH Reitnauer, RS Cooley, PC Rahn, DJ Loyd, D Fiore, SJ Fischman. High throughput analysis and purification in support of automated parallel synthesis. Mol Divers 3:61–70, 1997.
18. RB Merrifield. Solid phase peptide synthesis. I. The synthesis of a tetrapeptide. J Am Chem Soc 85:2149–2154, 1963.
19. GB Fields, RL Noble. Solid phase peptide synthesis utilizing 9-fluorenylmethoxycarbonylamino acids. Int J Peptide Protein Res 35:161–214, 1990.
20. LA Thompson, JA Ellman. Straightforward and general method for coupling alcohols to solid supports. Tetrahedron Lett 35:9333–9336, 1994.
21. Review: JS Fruchtel and G Jung. Organic chemistry on solid supports. Angew Chem Int Ed Eng 35:17–42, 1996.
22. Review: PHH Hermkens, HCJ Ottenheijm, D Rees. Solid-phase organic reactions: a review of the recent literature. Tetrahedron 52:4527–4554, 1996.
23. SK Sarkar, RS Garigipati, JL Adams, PA Keifer. An NMR method to identify nondestructively chemical compounds bound to a single solid-phase-synthesis bead for combinatorial chemistry applications. J Am Chem Soc 118:2305–2306, 1996.
24. R Garigipati, B Adams, JL Adams, SK Sarkar. Use of spin echo magic angle spinning 1H NMR in reaction monitoring in combinatorial organic synthesis. J Org Chem 61:2911–2914, 1996.
25. B Yan, G Kumaravel. Probing solid-phase reactions by monitoring the IR bands of compounds on a single ''flattened'' resin bead. Tetrahedron 52:843–848, 1996.
26. B Yan, JB Fell, G Kumaravel. Progression of organic reactions on resin supports monitored by single bead FTIR microspectroscopy. J Org Chem 61:7467–7472, 1996.
27. TY Chan, R Chen, MJ Sofia, BC Smith, D Glennon. High throughput on-bead monitoring of solid phase reactions by diffuse reflectance infrared Fourier transform spectroscopy (DRIFTS). Tetrahedron Lett 38:2821–2824, 1997.
28. DE Pivonka, R Russel, T Gero. Tools for combinatorial chemistry: in situ infra-

red analysis of solid-phase organic reactions. Appl Spectros 50:1471–1478, 1996.
29. BA Bunin, JA Ellman. A general and efficient method for the solid phase synthesis of 1,4-benzodiazepine derivatives. J Am Chem Soc 114:10997–10998, 1992.
30. K Burgess, AI Liaw, N Wang. Statistical sampling of resin pools. J Med Chem 37:2985–2987, 1994.
31. P-L Zhao, RB Nachbar, JA Bolognese, KT Chapman. Two new criteria for choosing sample size in combinatorial chemistry. J Med Chem 39:350–352, 1996.
32. KS Lam, SE Salmon, EM Hersh, VJ Hruby, WM Kazmierski, RJ Knapp. A new type of synthetic peptide library for identifying ligand-binding activity. Nature 354:82–84, 1991.
33. RA Houghten, C Pinilla, SE Blondelle, JR Appel, CT Dooley, JH Cuervo. Generation and use of synthetic peptide combinatorial libraries for basic research and drug discovery. Nature 354:84–86, 1991.
34. GC Look, JR Schullek, CP Holmes, JP Chinn, EM Gordon, MA Gallop. The identification of cyclooxygenase-1 inhibitors from 4-thiazolidinone combinatorial libraries. Bioorg Medicin Chem Lett 6:707–712, 1996.
35. RA Houghten, C Pinilla, SE Blondelle, JR Appel, CT Dooley, JH Cuervo. Generation and use of synthetic peptide combinatorial libraries for basic research and drug discovery. Nature 354:84–86, 1991.
36. B Deprez, X Williard, L Bourel, H Coste, F Hyafil, A Tartar. Orthogonal combinatorial chemical libraries. J Am Chem Soc 117:5405–5406, 1995.
37. KS Lam, VJ Hruby, M Lebl, RJ Knapp, WM Kazmierski, EM Hersh, SE Salmon. The chemical synthesis of large random peptide libraries and their use for the discovery of ligands for macromolecular acceptors. Bioorg Medicin Chem Lett 3:419–424, 1993.
38. C Chen, LA Ahlberg Randall, RB Miller, AD Jones, MJ Kurth. Analogous organic synthesis of small-compound libraries: validation of combinatorial chemistry in small-molecule synthesis. J Am Chem Soc 116: 2661–2662, 1994.
39. MC Needles, DG Jones, EH Tate, GL Heinkel, LM Kochersperger, WJ Dower, RJ Barrett, MA Gallop. Generation and screening of an oligonucleotide-encoded synthetic peptide library. Proc Natl Acad Sci USA 90:10700–10704, 1993.
40. JM Kerr, SC Banville, RN Zuckermann. Encoded combinatorial peptide libraries containing non-natural amino acids. J Am Chem Soc 115:2529–2531, 1993.
41. HP Nestler, PA Bartlett, WC Still. A general method for molecular tagging of encoded combinatorial libraries. J Org Chem 59:4723–4724, 1994.
42. MHJ Ohlmeyer, RN Swanson, LW Dillard, JC Reader, G Asouline, R Kobayashi, M Wigler, WC Still. Complex synthetic chemical libraries indexed with molecular tags. Proc Natl Acad Sci USA. 90:10922–10925, 1993.
43. HP Nestler, PA Bartlett, WC Still. A general method for molecular tagging of encoded combinatorial libraries. J Org Chem 59:4723–4724, 1994.
44. P Eckes. Binary encoding of compound libraries. Angew Chem Int Ed Eng 33: 1573–1575, 1994.

45. JJ Baldwin, JJ Burbaum, I Henderson, MHJ Ohlmeyer. Synthesis of a small molecule combinatorial library encoded with molecular tags. J Am Chem Soc 117:5588–5589, 1995.
46. ZJ Ni, D Maclean, CP Holmes, B Ruhland, MM Murphy, JW Jacobs, EM Gordon, MA Gallop. Versatile approach to encoding combinatorial organic syntheses using chemically robust secondary amine tags. J Med Chem 39:1601–1608, 1996.
47. D Maclean, JR Schullek, MM Murphy, ZJ Ni, EM Gordon, MA Gallop. Encoded combinatorial chemistry: synthesis and screening of a library of highly functionalized pyrrolidines. Proc Natl Acad Sci USA 94:2805–2810, 1997.
48. A Borchardt, WC Still. Synthetic receptor binding elucidated with an encoded combinatorial library. J Am Chem Soc 116:373–374, 1994.
49. RA Houghten. General method for the rapid solid-phase synthesis of large numbers of peptides: specificity of antigen–antibody interaction at the level of individual amino acids. Proc Natl Acad Sci USA 82:5131–5135, 1985.
50. EJ Roskamp. The Design and Synthesis of 500,000 Compounds. Molecular Diversity and Combinatorial Chemistry. San Diego, October, 1996.
51. EJ Moran, S Sarshar, JF Cargill, MM Shahbaz, A Lio, AMM Mjalli, RW Armstrong. Radio frequency tag encoded combinatorial library method for the discovery of tripeptide-substituted cinnamic acid inhibitors of the protein tyrosine phosphatase PTP1B. J Am Chem Soc 117:10787–10788, 1995.
52. KC Nicolaou, X-Y Xiao, Z Parandoosh, A Senyei, MP Nova. Radiofrequency encoded combinatorial chemistry. Angew Chem Int Ed Eng 34:2289–2291, 1995.
53. AW Czarnik. No static at all: using radiofrequency memory tubes without (human) interference. Laboratory Automation '97. San Diego, January, 1997.
54. X-Y Xiao, C Zhao, H Potash, MP Nova. Combinatorial chemistry with laser optical encoding. Angew Chem Int Ed Eng 36:780–782, 1997.
55. Advanced Chem Tech, 5609 Fern Valley Road, Louisville, KY 40228.
56. Argonaut Technologies, Inc., 887 Industrial Rd., Suite G, San Carlos, CA 94070.
57. Diversomer Technologies, 2800 Plymouth Road, Ann Arbor, MI 48105.
58. Bohdan Automation, Inc., 1500 McCormick Blvd., Mundelein, IL 60060.
59. Tecan U.S., SLT Lab Instruments, P.O. Box 13953, Research Triangle Park, NC 27709.
60. SH DeWitt, JS Kiely, CJ Stankovis, MC Schroeder, DW Reynolds-Cody, MR Pavia. "Diversomers": an approach to nonpeptide, nonoligomeric chemical diversity. Proc Natl Acad Sci USA. 90:6909–6913, 1993.
61. SH DeWitt and AW Czarnik. Combinatorial organic synthesis using Parke-Davis's DIVERSOMER method. Acc Chem Res 29:114–122, 1996.
62. J Cargill and RR Maiefski. Methods and apparatus for the generation of chemical libraries. U.S. Patent 5,609,826, 1997.
63. JF Cargill and RR Maiefski. Automated combinatorial chemistry on solid phase. Lab Robot Autom 8:139–148, 1996.
64. HM Geyson, H Meloen, SJ Barteling. Use of peptide synthesis to probe viral

antigen for epitopes to a resolution of a single amino acid. Proc Natl Acad Sci USA 81:3998–4002, 1984.

65. BA Bunin, MJ Plunkett, JA Ellman. The combinatorial synthesis and chemical and biological evaluation of a 1,4-benzodiazepine library. Proc Natl Acad Sci USA. 91:4708–4712, 1994.
66. SPA Fodor, JL Read, MC Pirrung, L Stryer, AT Lu, D Solas. Light-directed, spatially addressable parallel chemical synthesis. Science 251:767–773, 1991.
67. AV Lemmo, JT Fisher, HM Geysen, DJ Rose. Characterization of an inkjet chemical microdispenser for combinatorial library synthesis. Anal Chem 69: 543–551, 1997.
68. JA Ellman. Design, synthesis, and evaluation of small-molecule libraries. Acc Chem Res 29:132–143, 1996.
69. RM Lawrence, SA Biller, OM Fryszman, MA Poss. Automated synthesis and purification of amides: exploitation of automated solid phase extraction in organic synthesis. Synthesis 553–558, 1997.
70. HN Weller, MG Young, SJ Michalczyk, GH Reitnauer, RS Cooley, PC Rahn, DJ Loyd, D Fiore, SJ Fischman. High throughput analysis and purification in support of automated parallel synthesis. Mol Divers 3:61–70, 1997.
71. CE Kibbey. Quantitation of combinatorial libraries of small organic molecules by normal-phase HPLC with evaporative light scattering detection. Mol Divers 1:247–258, 1995.
72. DB Kassel. A fully automated mass spectrometry based system for the rapid analysis and purification of combinatorial libraries. Solid and Solution Phase Combinatorial Synthesis, New Orleans, 1997.
73. L Zeng, L Burton, K Yung, B Shushan, DB Kassel. An automated analytical/preparative HPLC/MS system for the rapid characterization and purification of compound libraries. J Chromatogr 794:3–13, 1998.
74. P Coffey. Parallel purification for combinatorial chemistry. Lab Autom News 2: 7–13, 1997.
75. J Ellman, B Stoddard, J Wells. Combinatorial thinking in chemistry and biology. Proc Natl Acad Sci USA. 94:2779–2782, 1997.

2

The Use of Mass Spectrometry

Annette Hauser-Fang and Paul Voúros
Northeastern University
Boston, Massachusetts

Combinatorial chemistry has evolved as a major area of interest for the pharmaceutical industry and has created a need for methods of characterization that are distinctly different from the classical tools of analytical organic chemistry like nuclear magnetic resonance (NMR) and C, H, and N analysis (1,2). There are currently several approaches for combinatorial chemistry that differ from each other in many ways (3–14). One can generate large solvated mixtures of compounds with as many as 10^6 different components in one sample, the so-called true libraries (15), or one can use the microtiter plate approach that is used to synthesize similar numbers of compounds individually on microtiter plates within small wells, so that each well contains only one or very few components. Clearly, for the second approach, synthesis and analysis must be automated to be efficient. Alternatively, many combinatorial libraries have been synthesized on solid supports (16–20) using polymeric beads. Advantages for solid phase chemistry include more efficient and simplified sample cleanup because washing of the beads usually removes all excess reagents and byproducts. This is especially true for the so-called mix-and-split synthesis (21) where each bead carries one component and which has been used with success by synthetic chemists. Single-bead analysis has been a good match for matrix-assisted laser desorption ionization (MALDI) and several publications have shown that this approach can provide excellent data as is shown below.

Within the context of these new developments in drug discovery, mass spectrometry (MS) has begun to assume a leading role (22,23) that, in addition to the traditional analysis of synthetic samples, appears to enter into the examination of molecular recognition and screening for lead drug candidates. This chapter focuses on a review of the use of MS techniques for the characterization and the biological screening of combinatorial libraries, often in combination with on-line separation methods.

I. CHARACTERIZATION OF COMBINATORIAL LIBRARIES BY MASS SPECTROMETRY

With the development of combinatorial chemistry it has become necessary to develop methods of analysis that are applicable to the characterization of components out of mixtures of molecules. It is often not sufficient to rely on biological screening procedures and later identification of active compounds, but it may also be necessary to identify synthetic products and verify the generation of a certain diversity within the combinatorial library. Establishing the integrity of a library can be important in order to avoid false positives and/or false negatives that may result from the presence of impurities (24) or the absence of products anticipated in the synthesis. When opting for certain approaches, the overall effort, the cost of purchasing and maintaining expensive equipment, and the superior contribution of certain methods toward the analysis have to be considered. The following sections give an overview of the applicability of some of the most popular techniques.

A. Low-Resolution CE-MS AND CE-MS/MS Methods Using Triple Quadrupole Mass Spectrometers or Quadrupole Ion Traps

A number of publications have focused on the characterization of combinatorial libraries by direct infusion, liquid chromatography (LC), or capillary electrophoresis (CE) (25–29) coupled to electrospray ionization-mass spectrometry (ESI-MS), electrospray ionization-tandem mass spectrometry (ESI-MS/MS) (30–37), nanoelectrospray-MS (38,39), and atmospheric pressure chemical ionization-mass spectrometry (APCI-MS) (40–42) using triple quadrupole instruments. Results obtained by direct infusion of a library into an ESI-MS system may be fraught with uncertainty even when dealing with small libraries because of the low resolving power of the triple-quadrupole instrument. It is not uncommon to find peaks at virtually every single mass unit depending on the library size. In fact, even relatively small libraries with, for example, 30 components give rise to at least 60 peaks due to isotopes. Counting in a few impurities one can easily confuse the spectrum and generate many false positives. Moreover, the ionization efficiencies of individual library components may vary widely whether in the positive or negative ionization mode. Establishing a ''hit'' from a given ion mass signal may thus be a questionable proposition and even more questionable can be the assessment of molecular ratios of library constituents. Accordingly, determination of ionization efficiencies of library members is necessary before the presence of nonisobaric compounds can be fully established. The combined use of positive and negative ion detection may often provide considerable flexibility in the characterization of diverse libraries. Figure 1 shows an example by Dunayevskiy et al. (27,30) where switching between positive and negative ionization can help distinguish real library components from impurities. Ideally, under low resolution MS conditions prior to separation of the library mixture by CE or LC is advisable along with the need to aquire MS/MS spectra of every peak. But even CE-MS and LC-MS only help separate a mixture for the identification of isobaric components. They do not provide a fingerprint of a compound and, unless the retention time of the compound of interest is known, it is still possible that an isobaric impurity has been identified as a library component. In low-resolution MS, fragmentation spectra are always the most definitive proof of a compound's existence. Figure 2 shows an application where the three dimensional presentation of LC-MS and LC-MS/MS data is used to detect structural trends within the eluted peaks, therefore permitting, for example, the easy identification of library components belonging to certain subgroups (43). On the other hand, it is frequently desirable to only determine the reproducibility of the synthesis of a specific library. In such a case, the ESI-MS profile may be deemed sufficient for general library characterization.

The new development of the quadrupole ion trap technology with its possibility of automated recognition of peaks and subsequent fragmentation using MS^n has simplified the process of fingerprinting of components by fragmentation. It is possible to get completely automated MS^n information from every HPLC peak just on the basis of programming the instrument to fragment the nth most intense ion in the spectrum until the MS^n step. This is helpful for unattended analysis runs (e.g., overnight) and improves the chances of complete identification of components from their fragmentation spectra (44).

Theoretical and experimental results by Blom (45) have focused on the precision requirements for the mass spectrometric analysis of combinatorial library mixtures. In his work, Blom has used discrete mass filters in combination with mass-MS/MS, M + 1/M, and M + 2/M isotope ratio filters to determine library components specifically from peptide libraries. As pointed out below, while nonpeptide libraries have mass distributions that are completely random, peptide library building blocks are limited in their diversity,

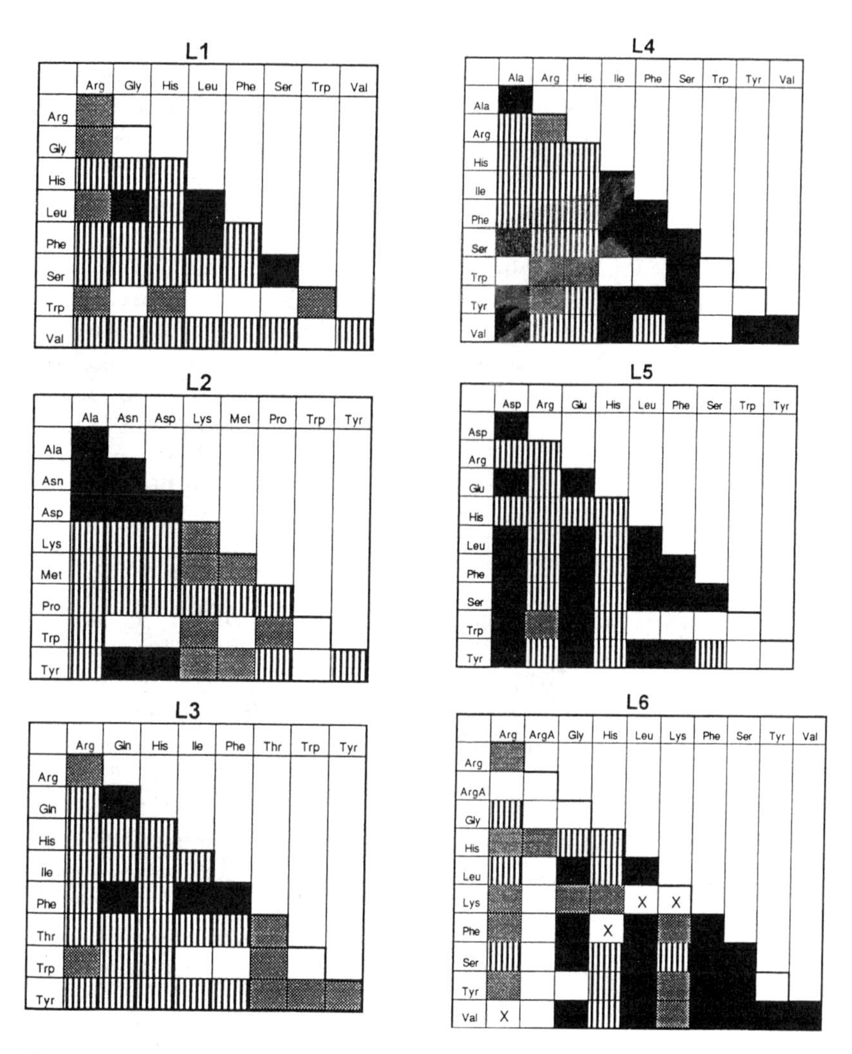

Figure 1 Mass spectrometric analysis of combinatorial libraries L1–L6: gray, detected in positive ion ESI; black, detected in negative ion ESI; vertical stripes, detected in both modes; X, detected in MS/MS experiment. (Reprinted from Ref. 30.)

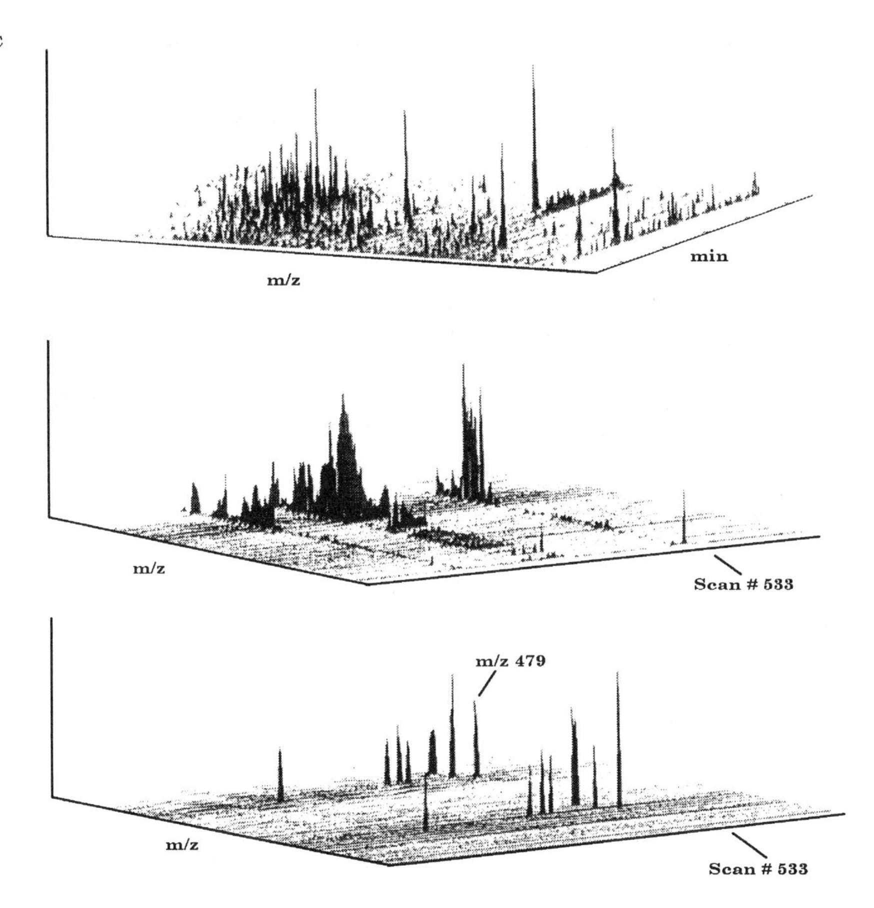

Figure 2 (Top) 3-D surface rendering of a full scan LC-MS dataset obtained on a 625-combinatorial-peptide library. (Middle) 3-D extracted ion surface of m/z 496, indicating three major regions. (Bottom) 3-D software neutral loss (NL-264) surface, indicating components that contain substructure V-F-COOH. (Reprinted from Ref. 43.)

and this leads to clustering of the library components around certain m/z values. The libraries used in this example were complex and in the best case only 6% of their components could be characterized by their integer masses. Use of a mass filter alone clearly produced too poor a fingerprint for unambiguous characterization of all components. Therefore all available mass spectrometric filters were made to work in a complementary fashion in order to recognize impurities and distinguish peptide components from one another. For the experimental part of the work by Blom, a software-modified triple-quadrupole mass spectrometer was employed that allowed automated switching between MS and MS/MS modes whenever the peak intensity of an expected candidate peak was of such a level as to trigger fragmentation.

Automation (37,41,42,46–53) has not only enabled researchers to analyze their library components more efficiently but has also helped in the synthetic process. Many companies (50,54) have adopted the approach of synthesizing one or two components of a library separately in a small well using microtiter plates as reaction vessels and robots for the addition of reagents and mixing. The microtiter plates with the individual library components are then automatically analyzed by low resolution MS (54) or high-resolution Fourier transform ion cyclotron-mass spectrometry (FTICR-MS) (55). Most setups include HPLC equipment with an autosampler to do continuous flow injections and molecular weight analysis of the contents of each well on the microtiter plate. No MS/MS is needed because one mass filter is usually enough for the confirmation of the one expected molecular weight and maybe a few impurity peaks. Samples can be analyzed overnight for maximum efficiency. A typical setup is shown in Fig. 3 (56). However, it should be pointed out that the absence of additional peaks in the mass spectrum does not guarantee 100% sample purity in the titer well. The possibility of false negatives always needs to be addressed by an alternative method of detection because some components (or impurities) may be transparent to ESI ionization.

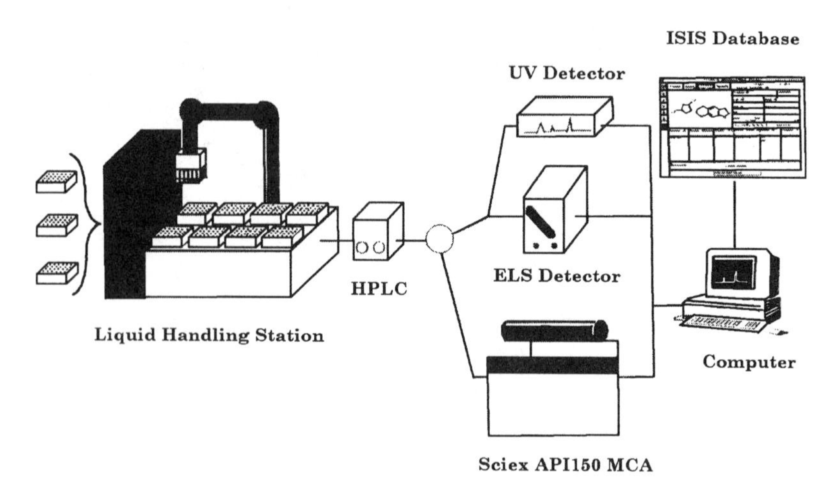

Figure 3 High-throughput QA schematic. (Reprinted from Ref. 56.)

B. Matrix-Assisted Laser Desorption Ionization and Other Desorption Ionization Techniques for the Identification of Combinatorial Libraries

There have been many publications focusing on matrix assisted laser desorption ionization (MALDI) (49,57–59, 60–72), secondary ion mass spectrometry (SIMS) (73–76), and ^{252}Cf Plasma desorption mass spectrometry (77–79) as a tool for the analysis of combinatorial libraries using different kinds of mass spectrometers. MALDI is generally not an ionization technique that can be easily coupled to any separation method but, since many libraries today consist of physically separated compounds on beads or within wells on microtiter plates, separation is often not necessary. It is important though, to understand the limitations and advantages of the MALDI approach because only libraries with relatively high molecular weight components ($m/z > 500$) are suitable for analysis due to matrix interferences at the lower mass ranges in the spectrum. That may be one of the reasons why a significant part of the research using MALDI has focused on the analysis of peptide libraries (58–63, 80–83). MALDI has been used with both time-of-flight (TOF) (65,75) and Fourier transform ion cyclotron resonance mass spectrometry (49, 84) for the analysis of split pool and encoded combinatorial libraries (66–68,70,85,86).

It has been shown by Egner et al. (64,65) that it is possible to analyze compounds directly from prepared polystyrene beads. This approach has been useful to verify and improve the synthetic method for the generation of combinatorial libraries. Assuming that there are approximately 10^6 beads in 1 gram of polystyrene resin, and the substitution factor for the solid phase chemistry is 0.4 mmol/g or more, there is about 400 pmol of compound attached to each single bead. MALDI-TOF analysis has detection limits in the femtomole region, which makes it compatible with single-bead analysis (76).

In preparation for the experiments, the library components were removed from the bead by exposing it to trifluoroacetic acid vapor for 30 min. After the cleavage reaction was complete an internal standard was added together with the matrix solution for the laser desorption. The matrix was formed around the bead within 15–30 min and MALDI-TOF analysis was performed directly from the sample well. Results are shown in Fig. 4 where a variety of peptides was monitored in addition to bradykinin, which acted as the internal calibrant. As for all MALDI experiments, it was critical that the right matrix be chosen for the analysis. After initial examinations of different matrices under a stereo microscope followed by the MALDI-TOF experiment, dihydroxy-benzoic acid (DHP) was found to produce the largest crystals and the best results.

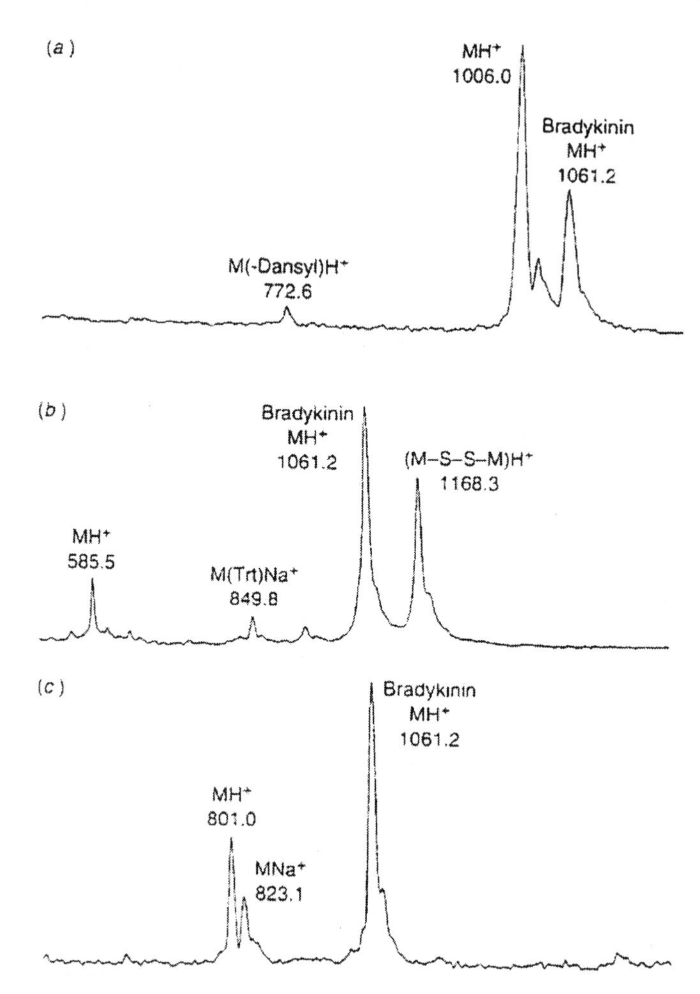

Figure 4 (a) Analysis of dansyl-Ile-Thr(O Bu^t)-Pro-Gln-Trp-Lys(Boc)-Wang-linker-resin, giving peptide dansyl-Ile-Thr-Pro-Gln-Trp-Lys (MH^+, 1006.0). The peak at 772.6 represents a small amount of undansylated material. (b) Analysis of Fmoc-Cys(Trt)-Lys(Boc)-Ile-HMPB-linker-resin, giving Fmoc-Cys-Lys-Ile (MH^+, 585.5), Fmoc-Cys(Trt)-Lys-Ile (MH^+, 849.8), and Fmoc-Cys-Lys-Ile disulfide [(M-S-S-M)H^+, 1168.3]. (c)Analysis of the disulfide of Fmoc-Cys-Asn-Cys-Lys(Boc)-Ile-HMPB-linker-resin giving the disulfide of Fmoc-Cys-Asn-Cys-Lys-Ile (MH^+, 801.0). (Reprinted from Ref. 64.)

Youngquist et al. (62,63) went one step further with their resin bead analysis by not only getting molecular weight information on their peptide based library components but sequencing information as well in the same experiment. Their strategy employed capping reagents at each step of the library synthesis so that beads would carry not only the library component but also a small percentage of peptides of different chain lengths. Sequencing information was obtained from the mass differences between termination products. Figure 5 shows the principle of the generation of the peptide libraries and the use of the capping reagents at every step. Termination products ideally differed by one amino acid (87).

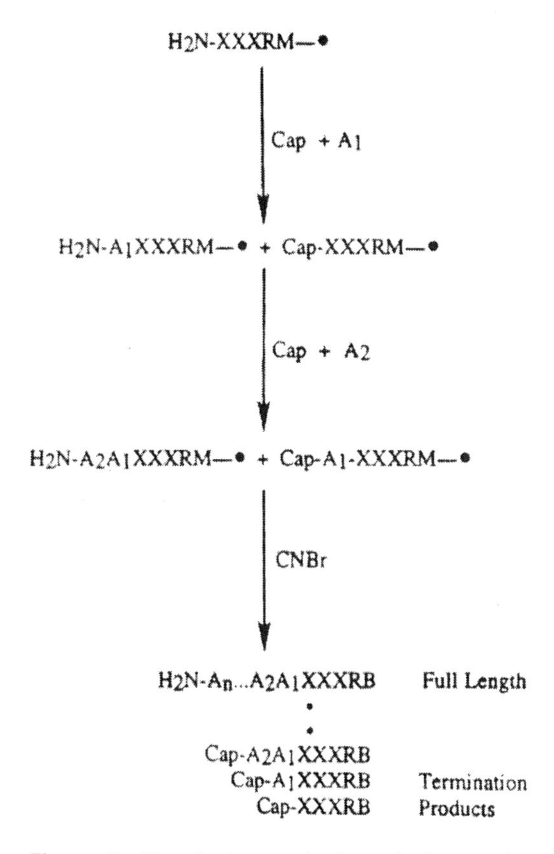

Figure 5 Termination synthesis method to produce sequence-specific families of peptides on each resin bead in the library. After screening, active peptides are isolated and the peptide products are released from the bead (—●) by CNBr digestion. (Reprinted from Ref. 62.)

Other important factors for a successful analysis have included the use of a linker between the bead and the synthesized peptides. This ensured efficient release from the bead and increased the molecular weight of the analytes to m/z values greater than 500, so that there was no interference with the low molecular weight noise that was produced by the desorption of the MALDI matrix. With the linker in place, the method has been sensitive enough to use only 1% of the material from an 88-μm bead for sequencing. Decreasing the bead diameter makes automated library sorting possible. Results were obtained from beads as small as 17 μm in diameter. The sequencing speed that was obtained using this method was 25–30 residues per hour and 77 of 80 beads were sequenced successfully with only one instance where the order of two amino acids could not be determined.

For both of the above examples it was important that the method could identify byproducts. When the reaction did not go to completion, so called deletion peptides were generated that were identified from their mass differ-

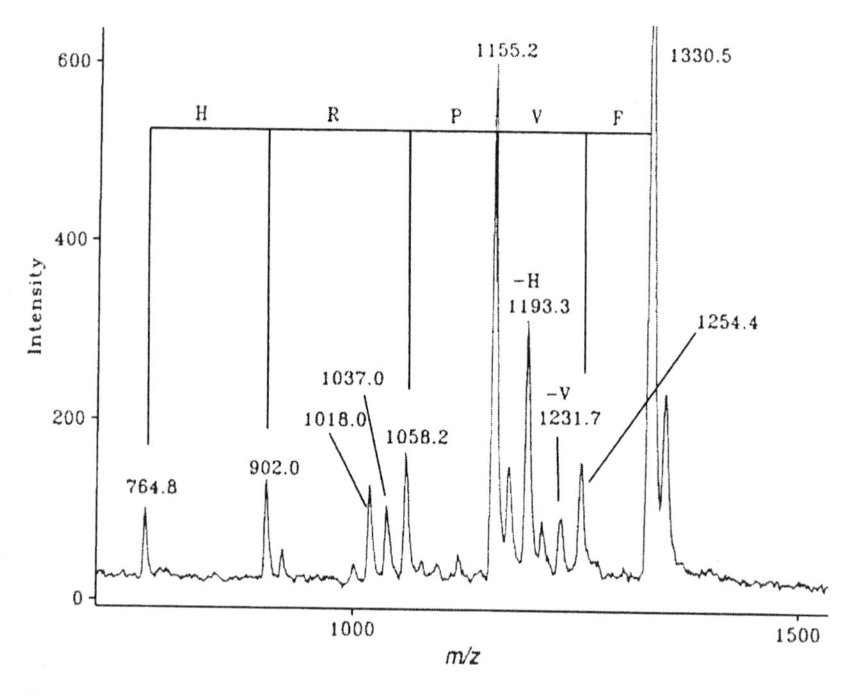

Figure 6 MALDI mass spectrum of an unknown peptide. A 5% portion of the products isolated from a single bead was analyzed. (Reprinted from Ref. 62.)

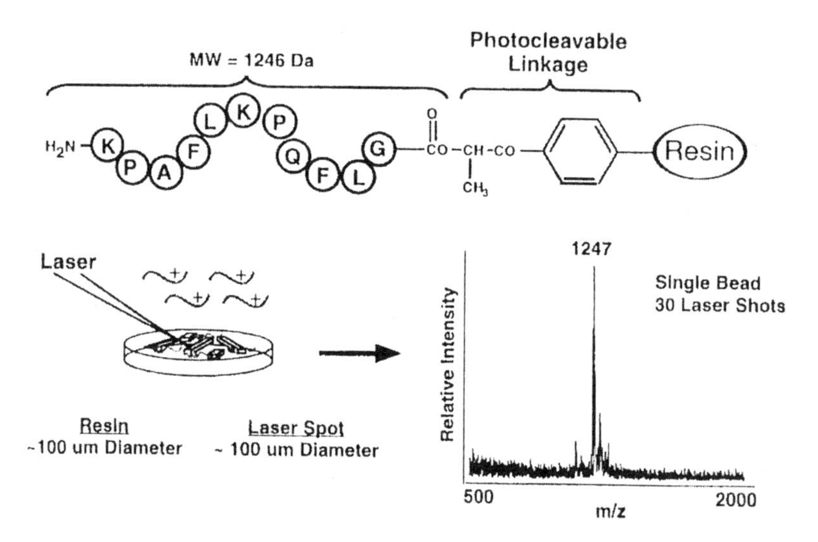

Figure 7 Direct analysis of reaction products from beads without the use of prior cleavage reactions. (Reprinted from Ref. 71.)

ences compared to other known peaks. Those specific mass differences also pointed towards modifications that happened during the synthesis, enabling the synthetic chemist to optimize conditions. A MALDI mass spectrum of an unknown peptide isolated from a single bead is shown in Fig. 6. Only 5% of the generated products were used for the MALDI analysis.

Future routine uses of MALDI mass spectrometry for the detection of combinatorial library components could include techniques that enable the analytical chemist to directly analyze reaction products from beads without using prior cleavage reactions, as is shown in Fig. 7. This means that standard linker molecules would have to be designed in such a way that cleavage from the bead is obtained by the laser irradiation used for the ionization process as has been employed by Oda et al. and others (71,78) for the identification of peptides bound to a resin (72).

C. Fourier Transform Ion Cyclotron Resonance-Mass Spectrometry

Fourier transform ion cyclotron resonance mass spectrometry (FTICR-MS) has been used for years by the petroleum industry to characterize large mixtures of compounds. Both the high mass accuracy and resolution ($m/\Delta m$ of

20,000–40,000) that is routinely achieved on these instruments can help identify components of a mixture simply by comparing their measured mass to their theoretical m/z value (89). An ideal base for the analysis of combinatorial libraries, this technique has been applied by several research groups (49,55,84,90–96). In one recent example by Fang et al. (90,91), combinatorial libraries consisting of mixtures of 36, 78, and 120 components were identified using ESI-FTICR-MS. The three libraries were analyzed in both positive and negative ionization modes. Depending on the acidic or basic character of the library constituents they were ionized preferentially either in the positive or the negative ionization mode, although, a number of components were seen in both ionization modes. Figure 8a shows a negative ionization ESI spectrum of the 36-component library with some of its assigned peaks (X stands for the core molecule). Compounds were identified with mass errors of 5 ppm or less. The spectrum of the 120 component library in Fig. 8b shows three isobaric components that could be resolved easily with this high resolution technique. It was possible to not only distinguish the library constituents from each other but also to identify isotope and impurity peaks. The method is very powerful but has limitations with respect to the number of ions that can be injected into the ICR cell at a given time. Overloading of the ICR cell gave rise to space–charge effects that compromised resolution and mass accuracy. Even though it should be possible to theoretically inject several thousand components at the same time, it was difficult to find conditions that would allow every library component to be ionized efficiently without overloading the ion trap. The solution to this problem may be found in using a segmented approach where certain mass ranges are analyzed while impurity peaks and peaks from other mass ranges are ejected from the trap.

In a publication by Winger and Campana (94), ESI-FTICR was applied to selectively identify several components from an octapeptide mixture. All of the 17 expected peaks were observed by direct infusion using ESI with an average resolving power of $m/\Delta m$ = 25,000. It was possible to distinguish an isobaric impurity from one of the peptide components by performing accurate mass measurements. Comparisons to some low-resolution LC-MS data gave clear indications of the advantages of FTICR.

Other approaches by Nawrocki et al. to the use of FTICR-MS for the identification of combinatorial libraries have focused on comparing the theoretical and observed mass distributions of a synthetic mixture to provide the possibility for a quick check on the progress of a reaction. If the envelope of masses was identical to the theoretical distribution, the assumption was made that the synthesis was successful. It has become increasingly important to verify the existence of expected combinatorial chemistry products by compar-

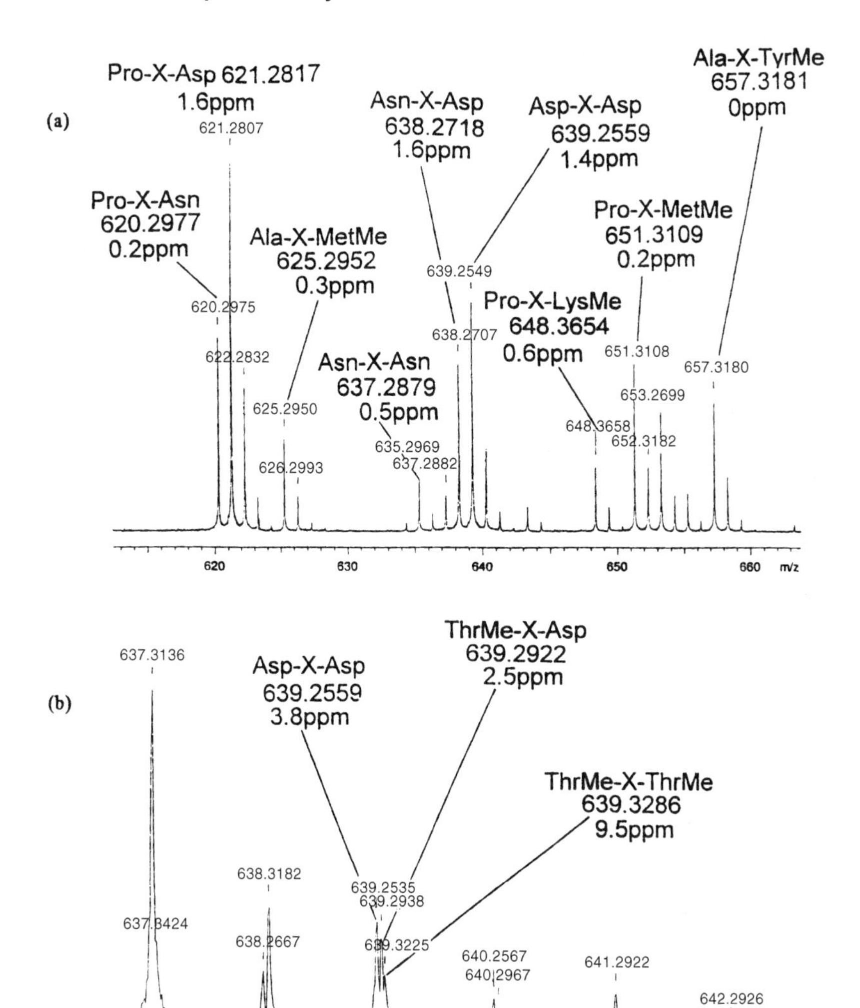

Figure 8 (a) Negative ion ESI spectrum of the 36-component library sample showing the ppm differences between theoretical and found m/z values for the assigned peaks. (b) Negative ion ESI spectrum of the 120-component library sample: resolving nominally isobaric peaks. (Reprinted from Ref. 91.)

ing the results of the analysis with a computer-simulated spectrum. The match between the theoretical and the experimental values is a measure of the completion of the synthesis. Figure 9 shows the obviously very different outcome of a synthesis believed to generate a peptide library that included a phosphorylated tyrosine and the simulated spectrum. The mass envelope of the simulated spectrum is shifted to m/z values that are about 300 Da higher than those of the experimental spectrum. Computer simulation with automated recognition and interpretation of results is becoming an essential tool for the analytical chemistry community and is especially useful when dealing with the very complex direct infusion spectra of combinatorial libraries.

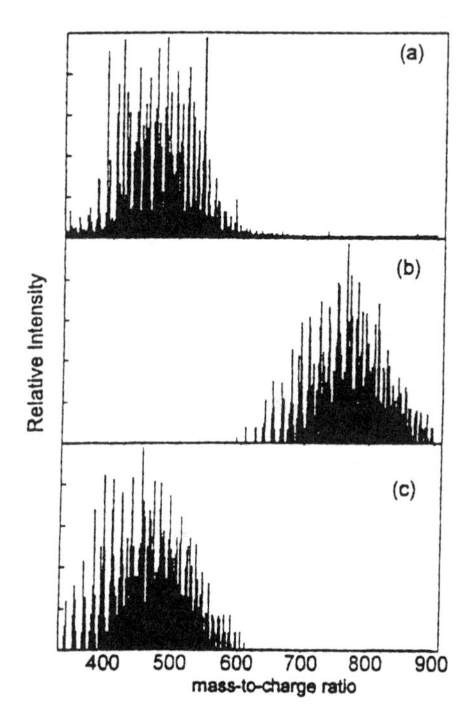

Figure 9 (a) ESI-FTICR broad-band mass spectrum of a library thought to be H-Gly-pTyr-Xxx-Xxx-Xxx-Cys-OH where Xxx can be any one of the naturally occurring amino acids with the exception of Cys and Trp. (b) A simulated mass spectrum of H-Gly-pTyr-Xxx-Xxx-Xxx-Cys-OH. (c) A computer-simulated spectrum of library H-Xxx-Xxx-Xxx-Cys-OH where Xxx can be any one of the naturally occurring amino acids with the exception of Cys and Trp. (Reprinted from Ref. 92.)

D. Predicting Mass Spectra of Combinatorial Libraries

Steinbeck et al. (97) have used a computer generated system to predict the diversity of combinatorial libraries. Using their MASP program, they were able to create libraries with ideal peak overlap in their mass spectra. The program can be employed to optimize and adjust the synthetic process for the creation of true libraries by choosing certain building blocks for the generation of an ideal mass envelope. By minimizing peak overlap, the simultaneous monitoring of reaction products by direct infusion MS or MALDI becomes more realistic, especially for larger libraries. MASP creates theoretical mass distributions calculated from molecular weights of the components and the isotope abundances of the elements contained in the combinatorial library compounds. In addition, the program can adjust its mass envelope according to the input of parameters such as the desorption factor, which is determined experimentally for MALDI. In order to minimize peak overlap and generate libraries that are well suited for mass spectrometric library screening, a ranking system is established whereby those building blocks that produce the least peak overlap together are displayed in a list of suggested reactants (48,98,99).

Other theoretical considerations by Zubarev et al. (100) involving peptide mixtures as model compounds have led to a proposed accuracy requirement of ± 1 ppm for the positive identification of certain peptides that have the same nominal masses. Monoisotopic masses of peptides are not randomly placed throughout a spectrum but rather concentrated around specific points that can be theoretically predicted. An equation by Mann (101) gives the positions of the monoisotopic masses of peptides as $Mp = [\mathrm{M}] + 0.00048M$ (Da) where [M] is the lower integer value of the molecular mass M. Figure 10 shows the theoretical distribution of peptides at a monoisotopic molecular mass of $[\mathrm{M}] = 1000$ Da. The spectrum contains about 50,000 peptides. Of the components, 95% are included within a range of 0.33 ± 0.01 Da, which can be calculated from $Wp = 0.19 + 0.0001M$ (Da), where Wp is the width that includes 95% of all amino acid compositions. Modified amino acids with known modifications can be included in such simulations and are also applicable to the analysis of mixtures of other biopolymers or combinatorial libraries.

II. SCREENING FOR BIOLOGICAL ACTIVITY

Screening combinatorial library mixtures for binding affinities to receptors involves analytical methods that are closely related to mass spectrometric epi-

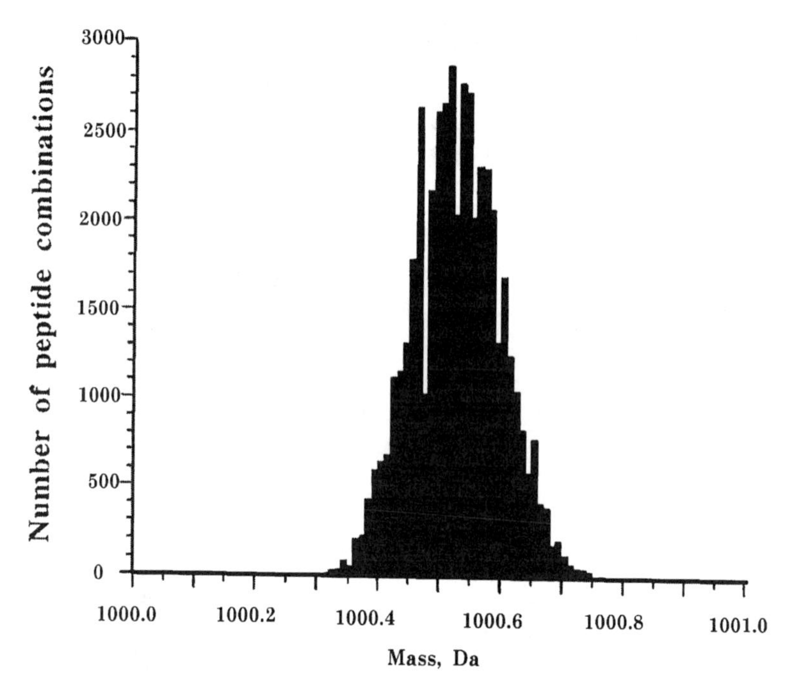

Figure 10 Number of possible amino acid compositions (peptide combinations) as a function of the peptide monoisotopic molecular mass for [M] = 1000 Da ([M] is the nominal molecular mass, i.e., lower integer mass value). The histogram is built with a 10–mDa step. The top density of the distribution is ~230 peptide compositions per mDa or per ppm. (Reprinted from Ref. 100.)

tope mapping (102–104) and other receptor–ligand binding studies. The library mixture is incubated with a protein or other molecule of interest to form complexes with a binding ligand. The complexes are then separated from the nonbinding small molecules by size exclusion chromatography on a gel filtration column, by ultrafiltration, or by some other method. The purified complexes are dissociated and the released ligands are analyzed by CE-MS or LC-MS (79). Depending on what needs to be accomplished in the screening step, it is possible to simply measure the molecular weight of ligands that bind to a receptor or to get further sequencing information from MS/MS. There are a number of research groups that have screened combinatorial library components against vancomycin, benzodiazepine antibodies, and other receptors (12,24,31,68,79,103–131). Others have identified enzyme substrates among combinatorial library components (20,88,132–134). Some examples are given below.

A. Affinity Chromatography–Mass Spectrometry

Results based on HPLC-MS involving benzodiazepine libraries (135) of 19 and 20 components have been published by Nedved et al. (105). A protein G column was employed for immobilizing benzodiazepine antibodies. After injecting the benzodiazepine mixture, the benzodiazepine–antibody complexes that were formed were eluted from the protein G column onto two different reverse phase columns. Selected benzodiazepines were then analyzed using an on-line mass spectrometer. Comparison of fragmentation spectra of the known and unknown benzodiazepines resulted in the identification of an analog of chlordiazepoxide as an active constituent of the unknown 20-component library. This technique of immunoaffinity chromatography/LC/LC/MS (IAC/LC/LC/MS) together with IAC/LC/LC/MS/MS has been very useful for developing automated screening procedures. Kelly et al. have used comparable setups to screen for SH_2–ligand interactions (120,127).

Similar libraries have been used by Wiebolt et al. in their studies of immunoaffinity ultrafiltration coupled on-line to mass spectrometry. Overcoming certain limitations of IAC that may result from the possible alteration of binding characteristics of a receptor after the immobilization process, Wiebolt et al. incubated receptor and ligands in solution prior to the analysis. Ideally, any receptor–ligand binding assay should take place in solution in order to generate an environment that is similar to in vivo conditions. Noncovalent immunoaffinity complexes were formed between antibodies and benzodiazepine libraries and were separated from nonbinding components by centrifugal ultrafiltration using a 50,000 Da molecular weight cutoff filter. The complexes were then dissociated by lowering the pH and the ligands were analyzed by HPLC–ion spray MS. This general approach was extended for screening other small-molecule library/receptor preparations in solution.

Even though closer to in vivo conditions than IAC and a good approach for screening receptor–ligand interactions from solutions, immunoaffinity ultrafiltration methods needed to be modified for situations where the availability of the receptor molecule is extremely limited and/or the price is prohibitive because at this point there is no possibility of receptor recovery after the binding studies are completed. Van Breemen et al. (136) have used a similar experimental setup with their pulsed ultrafiltration MS design using a chamber with a 10,000-Da molecular weight cutoff membrane that enabled them to avoid the problem of receptor loss after each screening analysis. In some of their

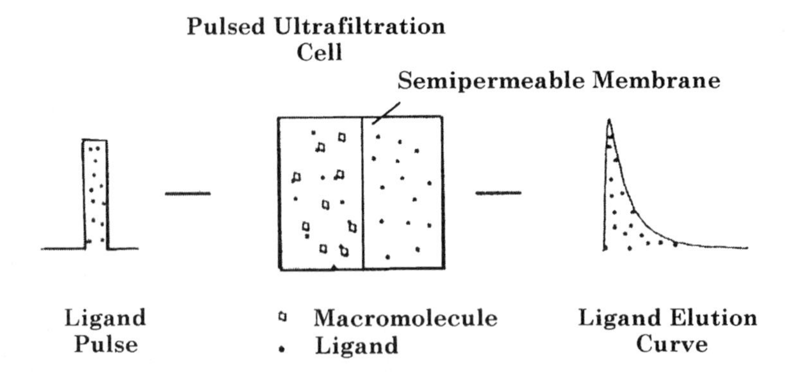

Figure 11 Scheme showing the use of pulsed ultrafiltration–MS for screening a combinatorial library for compounds that bind to a macromolecular receptor. The receptor is trapped in solution by an ultrafiltration membrane, which allows low molecular weight solution-phase compounds in a ''pulse'' of a combinatorial library to pass through. After unbound compounds are washed away, the ''hits'' in the library are eluted from the chamber by destabilizing the ligand–receptor complex using methanol, pH change, etc. (Reprinted from Ref. 136.)

experiments, receptor and ligands were directly injected into the ultrafiltration chamber at low concentrations and binding molecules were extracted from solution by the receptor that was present. Figure 11 shows the general principle of this approach. Their studies have shown that with the model receptor adenosine deaminase it was possible to repeat the binding and release process of their model ligand erythro-9-(2-hydroxy-3-nonyl)adenine (EHNA) three times without adding additional receptor to the chamber. Experiments with human serum albumin (HSA) as a receptor and warfarin as a ligand reproduced this result. The HSA in the ultrafiltration chamber was used for three consecutive binding experiments. The technique was used to identify the highest affinity ligand from a combinatorial library of 20 components.

B. Affinity Capillary Electrophoresis–Mass Spectrometry

The principle of affinity capillary electrophoresis (ACE) is based on the differences in electrophoretic mobilities of the ligand(s) and the receptor–ligand complex. Assuming that the receptor is neutral and not migrating at the chosen buffer pH, the electrophoretic mobility of the ligand would be $\varepsilon = z/m^{2/3}$, where z is the charge and m the molecular mass of the ligand. After complexation with the receptor the electrophoretic mobility of the complex changes to $\varepsilon = z/(M + m)^{2/3}$, where M is the molecular weight of the receptor. Since the electrophoretic mobility of a ligand will always depend on its participation within the receptor–ligand complex, an exact estimate of a possible migration time cannot be made except that non–binding ligand molecules are not retained and migrate as fast as without receptor. Ligands with strong binding affinities are retained according to their interactions with the receptor.

Specific examples for ACE-MS by Chu et al. (108–110,137–140) have shown that peptide libraries of 100 components can be screened for receptor–ligand binding within a CE capillary coupled on-line to MS. Vancomycin, which was used as the receptor in this case, was ''immobilized'' in a fused silica capillary that was coated with a neutral hydrophilic polymer coating to prevent electroosmotic flow during the analysis. The buffer pH was chosen in such a way that the vancomycin remained neutral and did not migrate within the capillary and into the mass spectrometer. The 100-peptide Fmoc DDXX library was injected, and unretained peptides emerged at migration times of 3–4 min, as can be seen in Fig. 12. ACE-UV confirmed the presence of two peaks with much longer retention times of 6–7 min, a good indication of strong binding affinities by the corresponding peptides, but only on-line MS revealed that the peak at 7 min contained two coeluting peptides. Compared to the known ligand of vancomycin that had a migration time of 5 min in this experiment, the two identified binding ligands in the peak at 7 min had higher binding affinities to vancomycin (26).

Additional information on the use of CE can be found elsewhere in this volume (chap. 6).

C. Automated Reaction Screening Using Flow Injection ESI-MS

An approach that employs both high sample throughput and screening without separation is pointed out by Wu et al. (118) in their publication involving the rapid screening of combinatorial libraries of inhibitors of enzymatic reactions. By coupling an autosampler to ESI-MS they were able to develop a screening method that was capable of identifying one potential enzyme inhibitor every 2 min. The experimental setup was as follows: reaction solutions containing reactants, the corresponding enzyme for the reactant, a suspected inhibitor from the library of inhibitor components, and an internal standard were loaded individually into an autosampler tray. The autosampler was connected on-line to a single-quadrupole MS using ESI and was programmed to do continuous flow injections every 2 min. In order to determine enzyme inhibition, product ions were analyzed and their intensity was compared in to the internal stan-

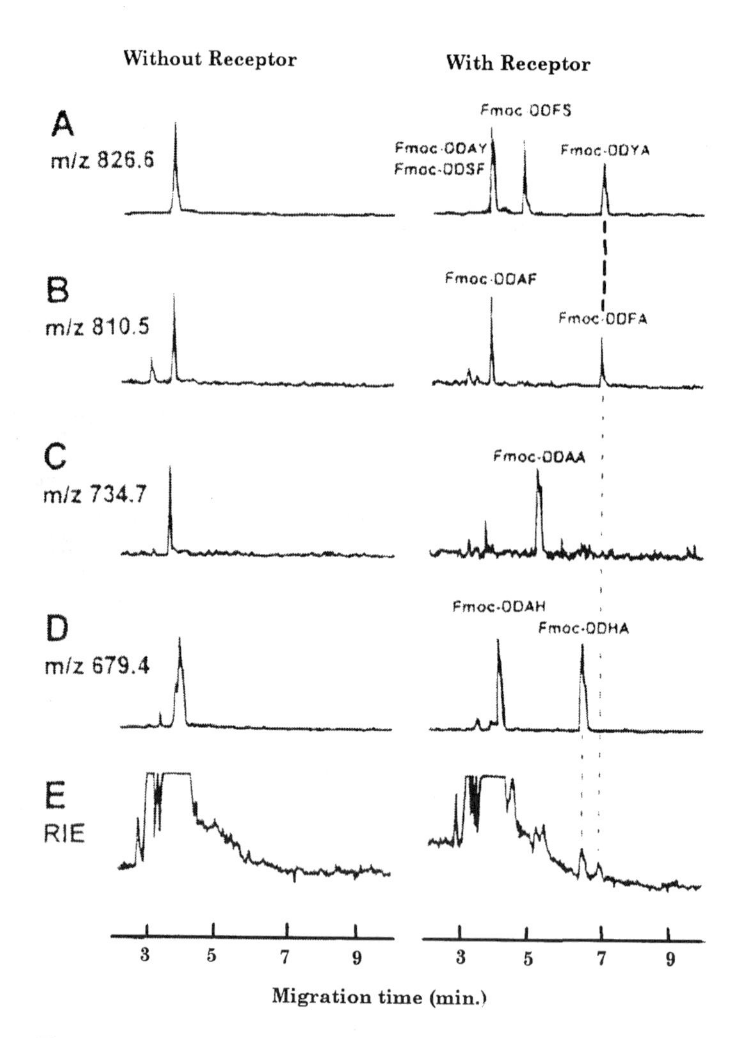

Figure 12 Affinity capillary electrophoresis–mass spectrometry (ACE-MS) of a synthetic all-D, Fmoc-DDXX library of 100 tetrapeptides using vancomycin as the receptor. (A–D) Selected ion electropherograms for the masses indicated. (E) Reconstructed ion electropherogram for runs without (left) and with (right) vancomycin in the electrophoresis buffer. ACE conditions: capillary, 360 μm o.d. × 50 μm i.d. × 38 cm long, coated with a neutral hydrophilic polymer; buffer, 20 mM Tris acetate (pH 8.1) containing no receptor (left) or 70 μM vancomycin (right); electric field, 500 V/cm, 5 μamp; sample, hydrodynamically loaded, 10 cm/8 s/~10 nl at the negative end of the capillary. MS conditions: instrument, Finnigan TSQ-700 with Finnigan API interface operated in positive electrospray ionization (ESI) mode; ESI, + 4.2 kV; gas sheath: 840 cm/min; liquid sheath (2 μl/min), 10 mM Tris acetate (pH 8.1) in H_2O/MeOH (25:75, v/v); MS, scan range, 525–925 amu scan in 2 s. (Reprinted from Ref. 108.)

dard. The least amount of product was generated in three vials that contained the best inhibitory components.

D. Surface Enhanced Affinity Capture and Probe Affinity Mass Spectrometry

One of the advantages of matrix-assisted laser desorption ionization (MALDI) is its sensitivity. As described above, MALDI can be extremely useful for the identification of solid phase combinatorial library components and even sequencing of peptide libraries using a single bead. Therefore, it is also a useful technique for screening combinatorial libraries where sample quantities do not permit the application of immunoaffinity HPLC-MS or ultrafiltration techniques with high flow rates, limited sensitivity, and significant consumption of receptor. Specific examples of applying MALDI to the identification of high-affinity ligands have been published by Hutchens and Yip (125) with their use of surface enhanced affinity capture (SEAC) and others. (128,129). In this approach, a receptor molecule is immobilized on agarose beads, and the analyte is captured directly out of solution. For this first published example, lactoferrin was detected from preterm infant urine by the addition of prepared beads to the urine solution. After removing and washing, the beads were placed on the MALDI target and the ligands detected. Other examples have shown that it is possible to directly detect an anti-monoclonal antibody to cytochrome c from cytochrome c immunoaffinity column (IAC) material. These experiments are useful for concentrating the analyte and providing affinity screening but some interference with the capture medium e.g., the IAC material or the agarose beads, has been observed.

In an alternative approach, Brockman et al. (121) have worked around these problems by immobilizing their analytes directly onto the surface of the MALDI probe (141). Using disposable MALDI probe tips coated with gold, they generated self-assembled monolayers that were then used to immobilize antibodies directly onto the tips. Figure 13 shows how the dextran based probe affinity mass spectrometry (PAMS) surface can screen for binding ligands. The binding molecules in this case were anti-γINF antibodies that were attached to the dextran. It was possible to overcome the problem of nonspecific electrostatic interactions between γINF and the immobilized anti-γINF antibodies by increasing the number of binding sites through oxidation of the dextran molecules and subsequent immobilization of the anti-γINF antibodies, thus avoiding generation of carboxyl groups. This approach may not be applicable to all kinds of analytes but, if both the immobilization chemistry and the

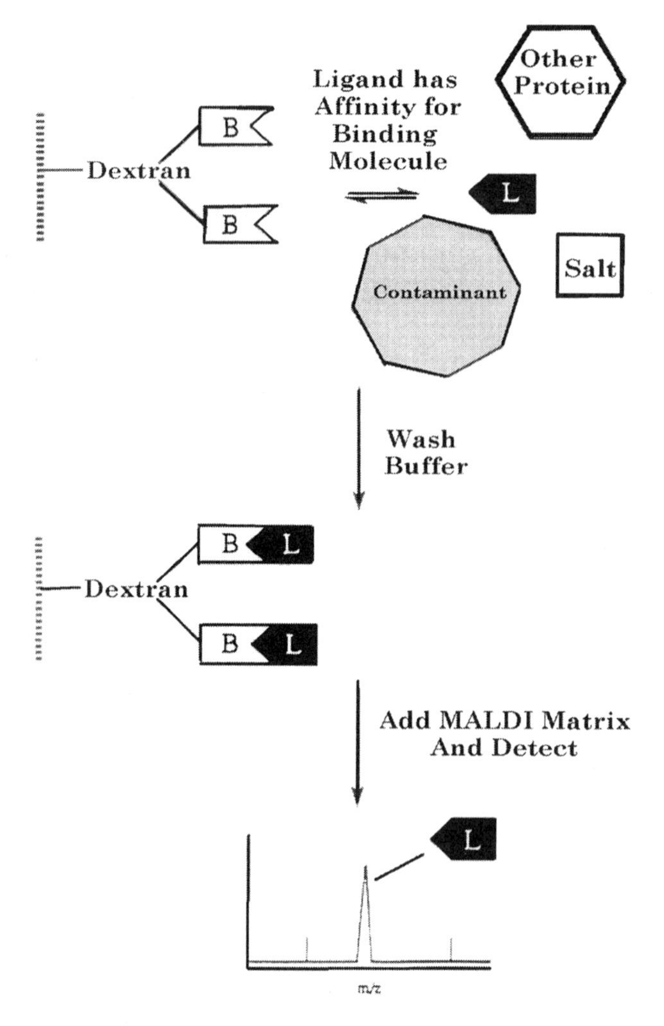

Figure 13 Mechanistic illustration of the probe-based affinity separation. (Reprinted from Ref. 121.)

analyte recognition work, it can be very useful for trace analysis and analyte enrichment on the MALDI target.

Youngquist et al. (63) have used color indicators for the identification of beads carrying active components. Bead-bound combinatorial libraries were subjected to a selection procedure that left dark blue stains on those peptide-containing beads that were binding to the selected antibody. Eight stained and five nonstained beads were chosen from the affinity screening experiment; then the bound peptides were digested and analyzed by MALDI as described in Sec. II.B. The known recognition sequence of the antibody was found in six of the peptide samples isolated from the eight stained beads. Fewer matches were obtained for the two samples derived from the other two stained beads and no matches were found for samples originating from unstained beads.

E. Bioaffinity Characterization–Mass Spectrometry

A method for screening receptor–ligand interactions has been introduced by Smith et al. and Ganem et al. (107,142–146) with their approach of bioaffinity characterization–mass spectrometry (BAC-MS). The method relies on the gas phase interactions present between the receptor and the binding ligands when using ESI. Both ESI and MALDI are soft ionization techniques that produce stable gas phase ions of large molecules without fragmentation. It is also possible to see noncovalent complexes in the mass spectrometer using these techniques (147), but recent research has focused on the specificity of the monitored noncovalent or electrostatic interactions that occur in the gas phase and many questions have remained open as to whether solution and gas phase complexes should be compared. Nevertheless, in an interesting experiment, a receptor and components from a combinatorial library were incubated in solution and after equilibration the liquid was electrosprayed directly into an FTICR high-resolution mass spectrometer. The FTICR instrument allowed the accumulation of selected ions or ion complexes in its ion trap and thereby provided a separation step. The noncovalently complexed molecules were then dissociated by applying additional energy. The remaining ligands were retained in the ion trap and were subjected to further MS/MS studies for characterization. Figure 14 shows the principle of BAC-MS. Bradykinin and bovine ubiquitin were used as model components.

Additional information on library screening can be found elsewhere in this volume (Chap. 8).

III. FUTURE DIRECTIONS FOR THE APPLICATION OF MASS SPECTROMETRY IN COMBINATORIAL CHEMISTRY

The most important reason for the success of combinatorial chemistry is the much greater efficiency when synthesizing, analyzing, and screening compo-

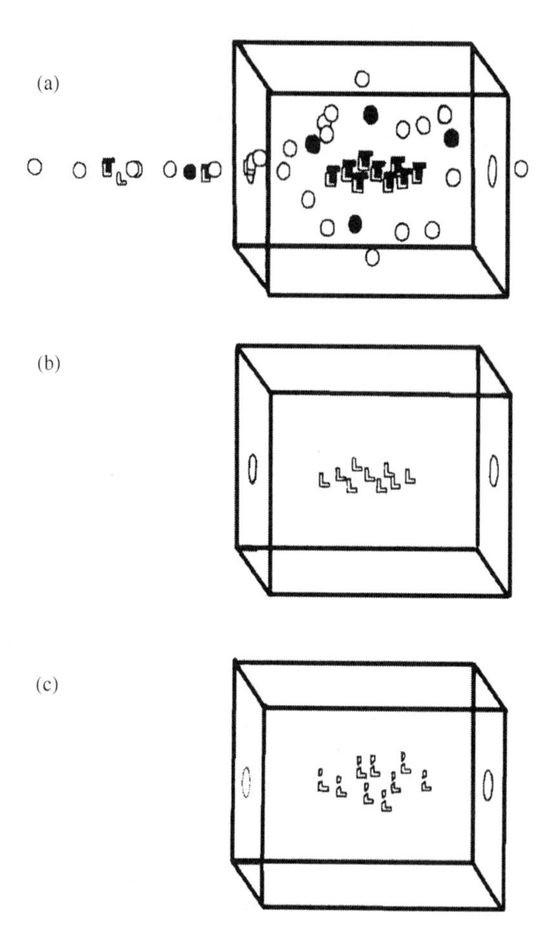

Figure 14 A conceptual representation of the BAC-MS technique. (a) The mixture solution is ionized by electrospray ionization and the complex of interest is selectively accumulated in the FTICR trap. (b) The noncovalent complex is then ''heated'' to liberate the affinity ligand species. (c) The ligand ions are retained for further CAD studies to present structural information. ''T'' represents the target biomolecules, and ''L'' represents ligands that form high-binding-affinity-specific noncovalent complexes with the target molecules. Circles (solid and empty) represent ions from other molecules present in the mixture. (Reprinted from Ref. 142.)

nents. Combinatorial chemistry with its possibility of automation has been found to increase sample throughput per person by a factor of 8 or more since the early 1990s (7) and robotic high-throughput systems are mainly responsible for this dramatic increase (148–150). For even higher sample throughput it is desirable to use a setup for automated synthesis directly together with MS. The analysis can be extended to include other characterization and screening methods like ultraviolet (UV) and nuclear magnetic resonance (NMR) spectroscopy (56,151). Zuckerman et al. (79) used an equimolar peptide mixtures (EPM) synthesizer to generate libraries of up to 36 individual peptides or peptide mixtures. Each well was screened in an enzyme-linked immunosorbent assay (ELISA) and the identified binding ligands were then characterized in a glycerol matrix by secondary ion MS. Boutin et al. (78) have constructed a Zymark-based robot to synthesize their combinatorial libraries with subsequent analysis by MS, CE, and NMR.

For higher resolving power and smaller sample quantities it may be desirable to use a capillary electrochromatography (CEC) system as has been done by Zweigenbaum et al. (44) for their separation of a mixture of steroids. The CEC system was coupled on-line to an ion trap mass spectrometer. CEC is especially useful for the separation of isomers.

Other examples of future developments in combinatorial chemistry related to MS include the use of animals within pharmacokinetic studies (152,153). The coadministration of several components to one animal could dramatically speed up the discovery process for metabolites (154,155). Also, automated combinatorial degradation profiling has been shown to add another dimension to the automated processes of mixture synthesis and analysis (98,156,157).

REFERENCES

1. AM Thayer. Special report: Combinatorial chemistry becoming core technology at drug discovery companies. C&EN (February 12):57–64, 1996.
2. JH Krieger. Special report: Combinatorial chemistry spawns new software systems to manage flood of information. C&EN (February 12):67–73, 1996.
3. SH DeWitt. Molecular diversity strategies. Pharmaceut News 1(4):11–14, 1994.
4. E Gordon, M Gallop, D Patel. Strategy and tactics in combinatorial organic synthesis. Applications to drug discovery. Acc Chem Res 29:144–154, 1996.
5. L Thompson, J Ellman. Synthesis and applications of small molecule libraries. Chem Rev 96:555, 1996.

6. NK Terrett, M Gardner, DW Gordon, RJ Kobylecki, J Steele. Combinatorial synthesis: the design of compound libaries and their application to drug discovery. Tetrahedron 51(30):8135–8173, 1995.
7. Intelligent drug design. Nature 384(Suppl 6604):1–26, 1996.
8. A Czarnik. Special issue on combinatorial chemistry (March). Acc Chem Res 29, 1996.
9. AL Burlingame, RK Boyd, SJ Gaskell. Mass spectrometry. Anal Chem 68: 599R–651R, 1996.
10. F Balkenhohl, Cvd Bussche-Hunnefeld, A Lansky, C Sechel. Combinatorial synthesis of small organic molecules. Angew Chem Int Ed Eng 35:2288–2337, 1996.
11. CY Cho, EJ Moran, SR Cherry, JC Stephaus, SPA Fodor, CL Adams, A Sundaram, JW Jacobs, PG Schultz. An unnatural biopolymer. Science 261: 1303–1305, 1993.
12. KD Janda. Tagged versus untagged libraries: methods for the generation and screening of combinatorial chemical libraries. Proc Natl Acad Sci USA 91: 10779–10785, 1994.
13. G Lowe. Combinatorial chemistry. Chem Soc Rev p 309, 1995.
14. M Lebl, V Krchnak, NF Sepetov, V Nikolaev, M Stankova, P Kocis, M Patek, Z Flegelova, R Ferguson, KS Lam. Synthetic combinatorial libraries: a new tool for drug design. In: M Atassi, E Appella, eds. Methods in Protein Structure Analysis. New York: Plenum Press, 1995, pp 335–342.
15. T Carell, EA Wintner, A Bashir-hashemi, J Rebek Jr. A novel procedure for the synthesis of libraries containing small organic molecules. Angew Chem Int Ed Eng 33:2059–2061, 1994.
16. PHH Hermkens, HCJ Ottenheijm, D Rees. Solid-phase organic reactions: a review of the recent literature. Tetrahedron 52(13):4527–4554, 1996.
17. J Nielson, S Brennar, K Janda. Synthetic methods for the implementation of encoded combinatorial chemistry. J Am Chem Soc 115:9812–9813, 1993.
18. SH DeWitt, JS Kiely, CJ Stancovic, MC Schroeder, DM Reynolds-Cody, MR Pavia. Diversomers-an approach to nonpeptide, nonoligomeric chemical diversity. Proc Natl Acad Sci USA 90:6909–6913, 1993.
19. SH DeWitt, MC Schroeder, CJ Stankovic, JE Strode, AW CZarnik. DIVERSOMER Technology: solid phase synthesis, automation, and integration for the generation of chemical diversity. Drug Dev Res 33:116–124, 1994.
20. M Medal, I Svendsen, K Breddam, F-I Auzanneau. Portion-mixing peptide libraries of quenched fluorogenic substrates for complete subsite mapping of endoprotease specificity. Proc Natl Acad Sci USA 91:3314–3318, 1994.
21. B Deprez, X Willard, L Bourel, H Coste, F Hyafil, A Tarter. Orthogonal combinatorial chemical libraries. J Am Chem Soc 117:5405–5406, 1995.
22. JA Loo. Mass Spectrometry in the combinatorial chemistry revolution. Eur Mass Spectrom. In press.
23. JA Loo. Bioanalytical mass spectromety: many flavors to choose. Bioconj Chem 6:644–665, 1995.

24. DL Flynn, JZ Crich, RV Devraj, SL Hockerman, JJ Parlow, MS South, S Woodard. Chemical library purification strategies based on principles of complementary molecular reactivity and molecular recognition. J Am Chem Soc 119(21):4874–4881, 1997.
25. ED Lee, W Muck, JD Henion, TR Covey. On-line capillary zone electrophoresis–ion spray tandem mass spectrometry for the determination of dynorphins. J Chromatography 458:313–321, 1988.
26. RD Smith, JH Wahl, DR Goodlett, SA Hofstadler. Capillary electrophoresis/mass spectrometry. Anal Chem 65:A574, 1993.
27. Y Dunayevskiy, P Vouros, E Wintner, G Schipps, T Carrell, J Rebek Jr. Application of capillary electrophoresis–electrospray ionization mass spectrometry in the determination of molecular diversity. Proc Natl Acad Sci USA 93:6152–6157, 1996.
28. P Thibault, C Paris, S Pleasance. Analysis of Peptided and proteins by capillary electrophoresis/mass spectrometry using acidic buffers and coated capillaries. Rapid Commun Mass Spectrom 5:484–490, 1991.
29. GA Valadovic, NL Kellher, FW McLafferty. Attomole protein characterizationby capillary electrophoresis–mass spectrometry. Science 273:1199–1201, 1996.
30. Y Dunayevskiy, P Vouros, T Carell, E Wintner, JJ Rebek Jr. Characterization of the complexity of small molecule libraries by electrospray ionization mass spectrometry. Anal Chem 67(17):2906–2915, 1995.
31. T Carrell, EA Wintner, AJ Sutherland, J Rebek, Y Dunayevskiy, P Vouros. New promise in combinatorial chemistry: symthesis, characterization, and screening of small-molecule libraries in solution. Chem Biol 3:171–183, 1995.
32. SC Pomerantz, JA McCloskey, TM Tarasow, BE Eaton. Deconvolution of combinatorial oligonucleotide libraries by electrospray ionization tandem mass spectrometry. J Am Chem Soc 119(17):3861–3867, 1997.
33. JW Metzger, K-H Wiesmüller, V Gnau, J Grünges, G Jung. Mass spectometry of synthetic pepitide libraries. Angew Chem Int Ed Eng 32:894–896, 1993.
34. JW Metzger, K-H Wiesmüller, V Gnau, J Brünjes, G Jung. Ion spray mass spectrometry and multiple sequence analysis of synthetic peptide libraries. Angew Chem Int. Ed Eng 32(6):894–896, 1993.
35. JW Metzger, C Kempter, K-H Wiesm̃uller, G Jung. Electrospray mass spectrometry and tandem mass spectrometry of synthetic multicomponent peptide mixtures: determination of composition and purity. Anal Biochem 219:261–277, 1994.
36. JW Metzger, S Stenanovic, J Brunjes, K-H Wiesmueller, G Jung. Electrospray mass spectrometry and multiple sequence analysis of synthetic peptide libraries. Methods 6:425–431, 1994.
37. WL Fitch, A Lu, K Tsutsui, N Shah. Single bead mass spectrometry for characterization of combinatorial libraries. Proceedings of the 44th ASMS Conference on Mass Spectrometry and Allied Topics, Portlan, OR, 1996, p. 1043.

38. F Bitsch, S Herter, K Mueller, F Gombert. Nonoelectrospray ionization mass spectometry as a tool for mass measuring compounds synthesized on beads. Proceedings of the 45th ASMS Conference on Mass Spectrometry and Allied Topics, Palm Springs, CA, 1997, p. 1263.
39. M Wilm, M Mann. Analytical properties of the nanoelectrospray ion source. Anal Chem 68:1–8, 1996.
40. RG Smith, S Woodard, K Leschinsky. Mixture analysis using LC/APCI/MS. Proceedings of the 44th ASMS Conference on Mass Spectometry and Allied Topics, Portland, OR, 1996, p. 1036.
41. RG McKay, BS Larsen, RS Livingston, JG Doughty. Combinatorial library automation for rapid APCI analysis. Proceedings of the 44th ASMS Conference on Mass Spectometry and Allied Topics, Portland, OR, 1996, p. 1037.
42. LYT Li, JN Kyranos. Automated multi-dimensional HPLC/UV/MS forquantitative method development. Proceedings of the 44th ASMS Conference on Mass Spectrometry and Allied Topics, Portland, OR, 1996, p. 1041.
43. J Whitney, M Hall, R Rourick, K Volk, E Kerns, M Lee. Using scientific visualization software to analyze structural trends in LC/MS profile datasets. Proceedings of the 44th ASMS Conference on Mass Spectrometry and Allied Topics, Portland, OR, 1996, p. 1046.
44. J Zweigenbaum, J Henion. High-performance capillary electrochromatography with ion trap MS^n detection. Proceedings of the 44th ASMS Conference on Mass Spectrometry and Allied Topics, Palm Springs, CA, 1997, p. 1260.
45. KF Blom. Strategies and data precision requirments for the mass spectrometric determination of structures form combinatorial mixtures. Anal Chem 69(21): 4354–4362, 1997.
46. H Muenster, R Pesch, AK Ziberna. Automated MS/MS for combinatorial chemistry library analysis with a hybrid tandem mass spectrometer. Proceedings of the 45th ASMS Conference on Mass Spectrometry and Allied Topics, Palm Springs, CA, 1997, p. 1250.
47. MF Bean, ME Hemling, KK Sonenson, SA Carr. Automated analysis and data processing for combinatorial chemistry. Proceedings of the 45th ASMS Conference on Mass Spectrometry and Allied Topics, Palm Springs, CA, 1997, p. 1254.
48. R Wilgus, M Geysen, D Wagner, F Schoenen, C Wagner, W Bodnar. Automated mass spectral analysis for combinatorial chemistry: The Capture Program. Proceedings of the 45th ASMS Conference on Mass Spectrometry and Allied Topics, Palm Springs, CA, 1997, p. 1255.
49. DC Tutko, KD Henry, BE Winger. Sequential MS and MS^n analysis of combinatorial libraries by using automated MALDI-FT/MS. Proceedings of the 45th ASMS Conference on Mass Spectrometry and Allied Topics, Palm Springs, CA, 1997, P. 1259.
50. V Nicolaev, O Issakova, S Wade, N Sepetov. Automatic LC/MS Analysis of

combinatorial libraries—a rational approach. Proceedings of the 45th ASMS Conference on Mass Spectrometry and Allied Topics, Palm Springs, CA, 1997, p. 1262.

51. J Batt, PA Bott, MA McDowall, SW Preece, JA Rontree. Automated LC-MS for the characterization of combinatorial libraries. Proceedings of the 44th ASMS Conference on Mass Spectrometry and Allied Topics, Portland, OR, 1996, p. 1033.
52. D Wagner, W Bodnar, H Geysen. Automated mass spectral data analysis for combinatorial chemistry. Proceedings of the 44th ASMS Conference on Mass Spectrometry and Allied Topics, Portland, OR, 1996, p. 1034.
53. ME Hail, BM Warrack, AS Arroyo, GC DiDonato, MS Lee. Automated high throughput LC/MS for the rapid profiling of pharmaceutical compounds. Proceedings of the 44th ASMS Conference on Mass Spectrometry and Allied Topics, Portland, OR, 1996, p. 1038.
54. KL Morand, TM Burt, RLM Dobson, LJ Wilson. Development of a mass spectrometry laboratory for support of pharmaceutical combinatorial chemistry programs. Proceedings of the 45th ASMS Conference on Mass Spectrometry and Allied Topics, Palm Springs, CA, 1997, p. 1261.
55. CW Ross, MB Young, DR Patrick, HG Ramjit. the simultaneous characterization of multi-component libraries from single reaction wells by use of FT/ICR/MS. Proceedings of the 45th ASMS Conference on Mass Spectrometry and Allied Topics, Palm Springs, CA, 1997, p. 1257.
56. L Taylor, C Garr, M Kobel, L Burton, L Cameron, W McFee. Quality control of the Optiverse library by mass spectrometry. Proceedings of the 45th ASMS Conference on Mass Spectrometry and Allied Topics, Palm Springs, CA, 1997, p. 1248.
57. R Zambias, DA Boulton, PR Griffin. Microchemical structural determination of a peptoid covalently bound to a polymeric bead by matrix assisted laser desorption ionization time-of-flight mass spectrometry. Tetrahedron Lett 35(25):4283–4286, 1994.
58. MA Gallop, RW Barrett, WJ Dower, SPA Fodor, EM Gordon. Application of combinatorial technologies to drug discovery. 1. Background and peptide combinatorial libraries. J Med Chem 37:1233–1251, 1994.
59. EM Gordon, RW Barrett, WJ Dower, SPA Fodor, MA Gallop. Applications of combinatorial technologies to drug discovery. 2. Combinatorial organic synthesis, library screening strategies, and future directions. J Med Chem 37:1385–1401, 1994.
60. M Bodsnszky, A Bodsnszky. Principle of Peptide Synthesis. Heidelberg: Springer-Verlag, 1984.
61. Á Furka, F Sebestyén, M Asgedom, G Dibó. A general method for the rapid synthesis of multicomponent peptide mixtures. Int J Peptide Protein Res 37: 487–493, 1991.
62. R Youngquist, GR Fuentes, M Lacey, T Keough. Matrix-assisted laser desorption ionization for rapid determination of the sequences of biologically active

peptides isolated from support-bound combinatorial peptide libraries. Rapid Commun Mass Spectrom 8:77–81, 1994.

63. RS Youngquist, GR Fuentes, MP Lacey, T Keough. Generation and screening of combinatorial peptide libraries designed for rapid sequencing by mass spectrometry. J Am Chem Soc 117:3900–3906, 1995.
64. BJ Egner, M Cardno, M Bradley. Linkers for combinatorial chemistry and reaction analysis using solid phase *in situ* mass spectrometry. J Chem Soc Chem Commun pp 2163–2164, 1995.
65. BJ Egner, GJ Langley, M Bradley. Solid phase chemistry: direct monitoring by matrix-assisted laser desorption/ionization time of flight mass spectrometry. A tool for combinatorial chemistry. J Org Chem 60:2652–2653, 1995.
66. JM Kerr, SC Banville, RN Zuckermann. Encoded combinatorial peptide libraries containing non-natural amino acids. J Am Chem Soc 115:2529–2531, 1993.
67. KC Lewis, WL Fitch, D Maclean. Characterization of split/pool combinatorial libraries. Proceedings of the 45th ASMS Conference on Mass Spectrometry and Allied Topics, Palm Springs, CA, 1997, p. 1265.
68. A Borchardt, WC Still. Synthetic receptor binding elucidated with an encoded combinatorial library. J Am Chem Soc 116:373–374, 1994.
69. S Brenner, RA Lerner. Encoded combinatorial chemistry. Proc Natl Acad Sci USA 89:5381–5383, 1992.
70. HP Nestler, PA Barlett, WC Still. A general method for molecular tagging of encoded combinatorial chemical libraries. J Org Chem 59:4723–4724, 1994.
71. MC Fitzgerald, K Harris, CG Shevlin, G Siuzdak. Direct characterization of solid phase resin-bound molecules by mass spectometry. Bioorg Med Chem lett 6(8):979–982, 1996.
72. BB Brown, DS Wagner, HM Geysen. A single-bead decode strategy using electrospray ionization mass spectrometry and a new photolabile linker: 3-amino-3-92-nitrophenyl)propionic acid. Mol Diver 1:4–12, 1995.
73. CL Brummel, INW Lee, Y Zhous, SJ Benkovic, N Winograd. A mass spectometric solution to the address problem of combinatorial chemistry. Science 264: 399–402, 1994.
74. SJ Benkovic, N Winograd, CL Brummel, INW Lee. Method for identifying members of combinatorial libraries. Patent Appl PCT Int Appl WO9525737 1995.
75. C Drouot, C Enjalbal, P Fulcrand, J Martinez, J-L Aubagnac, R Combarieu, Y dePuydt. Step-by-step control by time-of-flight secondary ion mass spectrometry of a peptide synthesis carried out on polymer beads. Rapid Commun Mass Spectrom 10:1509–1511, 1996.
76. CL Brummel, JC Vickerman, SA Carr, ME Hemling, GD Roberts, W Johnson, J Weinstock, D Gaitanopoulos, SJ Benkovic, N Winograd. Evaluation of mass spectrometric methods applicable to the direct analysis of non-peptide bead-bound combinatorial libraries. Anal Chem 68:237–242, 1996.
77. PA vanVeelan, UR Tjaden, j vanderGreef. Direct molecular weight determina-

tion of resin-bound oligopeptides using ^{252}Cf plasma-desorption mass spectrometry. Rapid Commun Mass Spectrom 5:565–568, 1991.

78. J Boutin, P Henning, P Lambert, S Berlin, L Petit, J-P Mahieu, B Serkiz, J-P Volland, JL Fauchere. Combinatorial peptide libraries: robotic synthesis and analysis by NMR, mass spectometry, tandem mass spectrometry and high-performance capillary electophoresis techniques. Anal Biochem 234:126–141, 1996.
79. RN Zuckermann, JM Kerr, MA Siani, SC Banville, DV Santi. Identification of highest-affinity ligands by affinity selections from equimolar peptide mixtures generated by robotic synthesis. Proc Natl Acad Sci USA 89:4505–4509, 1992.
80. RA Houghten. A general method for the rapid solid phase synthesis of large numbers of peptides. Proc Natl Acad Sci USA 82:5131–5135, 1985.
81. RA Houghten, ST DeGraw, MK Bray, SR Hoffmann, ND Frizzell. Simultaneous multiple peptide synthesis. Biotechniques 4:522–528, 1986.
82. RA Houghten, C Pinilla, SE Blondelle, JR Apple, CT Dooley, JH Cuervo. Generation and use of synthetic peptide combinatorial libraries for basic research and drug discovery. Nature 354:84–86, 1991.
83. M Davies, M Bradley. C-Terminally modified peptides and peptide libraries: another end to peptide synthesis. Angew Chem Int Ed Engl 36(10):1097–1099, 1997.
84. T Solouki, L Pas-a-Tološ, GS Jackson, S Guan, AG Marshall. High resolution multistage MS, MS^2 and MS^3 matrix-assisted laser desorption/ionization FT-ICR mass spectra of peptides from a single laser shot. Anal Chem 68(21):3718–3725, 1996.
85. H Geysen, CD Wagner, WM Bodnar, DJ Markworth, GJ Parke, FJ Schoenen, DS Wagner, DS Kinder. Isotope or mass encoding of combinatorial libraries. Chem Biol 3(8):679–688, 1996.
86. Z Ni, D Maclean, CP Holmes, MM Murphy, B Ruhland, JW Jacobs, EM Gordon, MA Gallop. A versatile approach to encoding combinatorial organic synthesis using chemically robust secondary amine tags. J Med Chem 39:1601–1608, 1996.
87. BT Chait, R Wong, RC Beavis, SBH Kent. Protein ladder sequencing. Science 262:89–92, 1993.
88. Y Oda, MR Carrasco, MC Fitzgerald. Screening resin-bound peptides to study the substrate specificity of a proteolytic enzyme. Proceedings of the 45th ASMS Conference on Mass Spectrometry and Allied Topics, Palm Springs, CA, 1997, p. 1251.
89. S Guan, A Marshall, S Scheppele. Resolution and chemical formula identification of aromatic hydrocarbons and aromatic compunds containing sulfur, nitrogen, or oxygen in petroleum distillates and refinery streams. Anal Chem 68: 46–71, 1996.
90. AS Fang, P Vouros, C Stacey, GH Kruppa, FH Laukien, T Carell, EA Wintner, J Rebek. Rapid screening of combinatorial libraries using electospray ionization

Fourier transform ion cyclotron resonance mass spectrometry. Proceedings of the 44th ASMS Conference on Mass Spectrometry and Allied Topics, Portland, OR, 1996, p. 1047.

91. AS Fang, P Vouros, CC Stacey, GH Kruppa, FH Laukien, EA Wintner, T Carell, J Rebek. Characterization of combinatorial libraries using electrospray ionization Fourier transform ion cyclotron resonance mass spectrometry. Comb Chem High Throughput Screen 1(1):22–33, 1998.
92. JP Nawrocki, M Wigger, CH Watson, TW Hayes, MW Senko, SA Benner, JR Eyler. Analysis of combinatorial libraries using electrospray ionization Fourier transform ion cyclotron resonance mass spectrometry. Rapid Commun Mass Spectrom 10:1860–1864, 1996.
93. JP Nawrocki, M Wigger, CH Watson, SA Benner, JR Eyler. Characterization of combinatorial libraries using ESI-FTICR mass spectrometry. Proceedings of the 45th ASMS Conference on Mass Spectrometry and Allied Topics, Palm Springs, CA, 1997, p. 1256.
94. BE Winger, JE Campana. Characterization of combinatorial peptide libraries by electrospray ionization Fourier transform mass spectrometry. Rapid Commun Mass Spectrom 10:1811–1813, 1996.
95. Q Wu, O Issakova, M Stankova, L Zhao. Structural determination using multistage accurate mass spectrometry: an application to combinatorial chemistry. Proceedings of the 45th ASMS Conference on Mass Spectrometry and Allied Topics, Palm Springs, CA, 1997, p. 1258.
96. P Speir, CW Ross III, H Ramjiit, G Kruppa, F Laukien. The potential of FTMS in addressing structural diversity of combinatorial libraries. Proceedings of the 44th ASMS Conference on Mass Spectrometry and Allied Topics, Portland, OR, 1996, p. 1035.
97. C Steinbeck, K Berlin, C Richert. MASP: a program predicting mass spectra of combinatorial libraries. J Chem Inf Comput Sci 37:449–457, 1997.
98. K Volk, R Rourick, J Whitney, E Kerns, M Lee. Combinatorial degradation profiling. Proceedings of the 45th ASMS Conference on Mass Spectrometry and Allied Topics, Palm Springs, CA, 1997, p. 1252.
99. K Volk, R Rourik, J Whitney, T Spears, E Kerns, M Lee. Predictive combinatorial stability profiling. Proceedings of the 44th ASMS Conference on Mass Spectrometry and Allied Topics, Portland, OR, 1996, p. 1040.
100. RA Zubarev, P Håkansson, B Sundqvist. Accuracy requirements for peptide characterization by monoisotopic molecular mass measurements. Anal Chem 68:4060–4063, 1996.
101. M Mann. Useful tables of possible and probable peptide masses. Proceedings of the 43th ASMS Conference on Mass Spectrometry and Allied Topics, Atlanta, GA, 1995, p 639.
102. D Suckau, J Köhl, G Karwath, K Schneider, M Casaretto, D Bitter-Sauermann, M Przybylski. Molecular epitope identification by limited proteolysis of an immobilized antigen–anitbody complex and mass spectrometric peptide mapping. Proc Natl Acad Sci USA 87:9848:9852, 1990.

103. YV Lubarskaya, YM Duanyevskiy, P Vouros, BL Karger. Microscale epitope mapping by affinity capillary electophoresis–mass spectrometry. Anal Chem 69:3008–3014, 1997.
104. M Macht, W Fiedler, K Kürzinger, M Przybylski. Mass spectrometric mapping of protein epitope structures of mycardial infarct markers myoglobin and troponin T. Biochemistry 35:15633–15639, 1996.
105. M Nedved, S habibi-Goudarzi, B Ganem, J Henion. Characterization of benzodiaxepine ''combinatorial'' chemical libraries by on-line immunoaffinity extraction, coupled column HPLC–ion spray mass sprectrometry–tandem mass spectrometry. Anal Chem 68:4228–4236, 1996.
106. T Carell, EA Wintner, J Rebek Jr. A solution phase screening procedure for the isolation of active compounds from a library of molecules. Angew Chem Int Ed Eng 33:2061–2064, 1994.
107. X Cheng, R Chen, JE Bruce, BL Schwartz, GA Anderson, SC Gale, RD Smith,J Gao, GB Sigal, M Mammen. Using electrospray ionization to study competitive binding of inhibitors to carbonic anhydrase. J Am Chem Soc 117:8859–8860, 1995.
108. Y Chu, DP Kirby, BL Karger. Free Solution identification of candidate peptides from combinatorial libraries by affinity capillary electrophoresis/mass spectrometry. J Am Chem Soc 117:5419–5420, 1995.
109. Y-H Chu, LZ Avila, J Gao, GM Whitesides. Affinity capillary electrophoresis. Acc Chem Res 28:461–468, 1995.
110. Y-H Chu, YM Dunayevskiy, DP Kirby, P Vouros, BL Karger. Affinity capillary electrophoresis–mass spectometry for screening combinatorial libraries. J Am Chem Soc 118:7827–7835, 1996.
111. S Kaur, L McGuire, D Tang, G Dollinger, V Huebner. Affinity selection and mass spectrometry–based strategies to identify lead compounds in combinatorial libraries. J Prot Chem 16(5):505–511, 1997.
112. KS Lam, SE Salmon, EM Hersh, VJ Hruby, WM Kazmierski, RJ Knapp. A new type of synthetic peptide library for the identification of ligand binding activity. Nature 354:82–84, 1991.
113. JA Loo, DE De-John, RRO Loo, PC Andrews. Application of mass spectrometry for characterizing and identifying ligands from combinatorial libraries. Ann Rep Medic Chem 31:319–325, 1996.
114. DI Papac, J Hoyes, KB Tomer. Direct analysis of affinity-bound analytes by MALDI/TOF MS. Anal Chem 6:2609–2613, 1994.
115. C Pinilla, JR Appel, P Blanc, RA Houghten. Rapid identification of high affinity pepide ligands using positional scanning synthetic peptide combinatorial libraries. Biotechniques 13:901–905, 1992.
116. SE Salmon, KS Lam, M Lebl, A Kandola, PS Khattri, S Wade, M Patek, P Kocis, V Krchnak, D Thorpe. Discovery of biologically acitve peptides in random libraries. Proc Natl Acad Sci USA 90:11708–11712, 1993.
117. R Wieboldt, J Zweigenbaum, J Henion. Immunoaffinity ultrafiltration with ion spray HPLC/MS for screening small-molecule libraries. Anal Chem 69:1683–1691, 1997.

118. J Wu, S Takayama, C-H Wong, G Suizdak. Quantitive electrospray mass spectrometry for the rapid assay of enzyme inhibitors. Chem Biol 4:653–657, 1997.
119. RN Zuckermann, EJ Martin, DC Spellmeyer, GB Stauber, KR Shoemaker, JM Kerr, GM Figliozzi, DA Goff, MA Siani, RJ Simon. Discovery of nanomolar ligands for seven transmembrane G-protein coupled receptors from a diverse peptoid library. J Med Chem 37:2678–2685, 1994.
120. LC Bock, LC Griffin, JA Latham, EH Vermass, JJ Toole. Selection single stranded DNA molecules that bind and inhibit human thrombin. Nature 355: 564–566, 1992.
121. AH Brockman, R Orlando. New immobilization chemistry for probe affinity mass spectrometry. Rapid Commun Mass Spectrom 10:1688–1692, 1996.
122. YM Dunayevskiy, J-J Lai, C Quinn, F Talley, P Vouros. Mass spectrometric identification of ligands selected from combinatorial libraries using gel filtration. Rapid Commun Mass Spectrom 11:1178–1184, 1997.
123. J Eichler, RA Houghten. Identification of substrate–analog trypsin inhibitors through the screening of synthetic peptide combintorial libraries. Biochemistry 32:11035–11041, 1993.
124. NJ Haskins, DJ Hunter, AJ Organ, SS Rahman, C Thom. Combinatorial chemistry: direct analysis of bead surface associated materials. Rapid Commun Mass Spectrom 9:1437–1440, 1995.
125. TW Hutchens, T Yip. New desorption strategies for the mass spectrometric analysis of macromolecules. Rapid Commun Mass Spectrom 7:576–580, 1993.
126. M Laskowski Jr, I Kato. Protein inhibitors of proteinases. Annu Rev Biochem 49:593–626, 1990.
127. MA Kelly, H Liang, I Sytwu, I Vlattas, NL Lyons, BR Bowen, LP Wennogle. Characterization of SH2–ligand interactions via library affinity selection with mass spectrometric detection. Biochemistry 35:11747–11755, 1996.
128. JR Krone, RW Nelson, D Dogruel, P Williams. Interfacing BIA with MALDI mass spectrometry. Proceedings of the 44th ASMS Conference on Mass Spectrometry and Allied Topics, Portland, OR, 1996, p. 753.
129. RW Nelson, JR Krone, AL Bieber, P Williams. Mass spectrometric immunoassay. Anal Chem 67:1153, 1996.
130. GD Dollinger, VD Huebner. Affinity selection of ligands by mass spectrometry. Patent Appl PCT Int Appl WO 9622530 1996.
131. NJ Hales, E Clayton. Affinity chromatographic method for screening combinatorial libraries. Patent Appl PCT Int App. WO9632642 1995.
132. DB Kassel, MD Green, RS Wehbie, R Swanstrom, J Berman. HIV-1 Protease Specificity Derived from a Complex Mixture of Synthetic Substrates. Anal Biochem 228:259–266, 1995.
133. T Kupke, C Kempter, G Jung, F Gotz. Oxidative decarboxylation of peptides catalyzed by flavoprotein Epi D. J Biol Chem 270:11282–11289, 1995.
134. JH Till, RS Annan, SA Carr, WT Miller. Use of synthetic peptide libraries and phosphopeptide-selective mass spectrometry to probe protein kinase substrate specificity. J Biol Chem 269:7423–7428, 1994.

135. BA Bunin, JA Ellman. A general and expedient method for the solid phase synthesis of 1,4-benzodiazepine derivatives. J Am Chem Soc 114:10997–10998, 1992.
136. RB VanBreemen, CR Haung, D Nikolic, CP Woodbury, YZ Zhao, DL Venton. Pulsed ultrafiltration mass spectrometry: a new method for screening combinatorial libraries. Anal Chem 69:2159–2164, 1997.
137. Y-H Chu, GM Whitesides. Affinity capillary electrophoresis can simultaneously measure binding constants of multiple peptides to vancomycin. J Org Chem 57:3524–3525, 1992.
138. Y-H Chu, LZ Avila, HA Biebuyck, GM Whitesides. Use of affinity capillary electrophoresis to meausure binding constants of ligands to proteins. J Med Chem 35:2915–2917, 1992.
139. FA Gomez, LZ Avila, Y-H Chu, GM Whitesides. Determination of binding constants of ligands to proteins by affinity capillary electrophoresis: compensation for electroosmotic flow. Anal Chem 66:1785–1791, 1994.
140. Y-H Che, WJ Lees, A Stassinopoulos, CT Walsh. Using affinity capillary electrophoresis to determine binding stochiometreis of protein–ligand interactions. Biochemistry 33:19616–19621, 1994.
141. C Bain, J Evall, G Whitesides. Formation of monolayers by the coadsorption of thiols on gold: variation in the head group, and solvent. J Am Chem Soc 111:7155–7164, 1989.
142. JE Bruce, GA Anderson, R Chen, X Chen, DC Gale, SA Hofstadler, BL Schwartz, RD Smith. Bio-affinity characterization mass spectrometry. Rapid Commun Mass Spectrom 9:64, 1995.
143. JE Bruce, L Pasa-Tolic, P Lei, RD Smith, SA Carr, D Dunnington, W Prichett, D Yamashita, J Yen, E Applebaum. Bioaffinity characterization mass spectrometry (BACMS) of the human scr SH2 domain protein. Proceedings of the 44th ASMS Conference on Mass Spectrometry and Allied Topics, Portland, OR, 1996, p. 1403.
144. B Ganem, JK Henion. Detecting noncovalent complexes of biological macromolucules: new applicants of ion-spray mass spectrometry. Chemtracts: Org Chem 6:1–22, 1993.
145. RD Smith, KJ Light-Wahl. The observation of non-covalent interactions in solution by electrospray ionization mass spectrometry: promise pitfalls and prognosis. Biol Mass Spectrom 22:493–501, 1993.
146. J Gao, X Cheng, R Chen, GB Sigal, JE Bruce, BL Schwartz, SA Hofstadler, GA Anderson, RD Smith, GM Whitesides. Screening derivatized peptide libraries for tight binding inhibitors to carbonic anhydrase II by electrospray ionization–mass spectrometry. J Med Chem 39:1949–1955, 1996.
147. D Lafitte, V Benezech, J Bompart, F Laurent, P Bonnet, JP Chapat, G Grassy, B Calas. Characterization of low affinity complexes between calmodulin and pyrazine derivatives by electrospray ionization mass spectrometry. J Mass Spectrom 32:87–93, 1997.
148. FS Pullen, GL Perkins, KL Burton, RS Ware, MS Teague, JP Kiplinger. Putting

mass spectrometry in the hands of the end user. J Am Soc Mass Spectrom 6: 394–399, 1995.

149. R Spreen, et al. Open access MS: a walk-up MS service. Anal Chem News Features pp 414A–419A, 1996.
150. LCE Taylor, et al. Open access atmospheric pressure chemical ionization mass spectrometry for routine sample analysis. J Am Soc Mass Spectrom 6:387–393, 1995.
151. RM Holt, MJ Newman, FS Pullen, DS Richards, AG Swanson. High-performance liquid chromatography/NMR spectroscopy/mass spectrometry: further advances in hyphenated technology. J Mass spectrom 32:64–70, 1997.
152. J Bergman, K Halm, K Adkison, J Shaffer. Simultaneous Pharmacokinetic screening of a mixture of compounds in the dog using API LC/MS/MS analysis for increased throughput. J Med Chem 40:827–829, 1997.
153. TV Olah, DA McLoughlin, JD Gilbert. The simultaneous determination of mixtures of drug candidates by LC APCI MS as in *in vivo* drug screening procedure. Rapid Commun Mass Spectrom 11:17–23, 1997.
154. DA McLoughlin, TV Olah, JD Gilbert. A direct technique for the simultaneous determination of 10 drug candidates in plasma by liquid chromatography atmospheric pressure chemical ionization mass spectrometry interfaced to a Prospekt solid-phase extraction system. Pharm Biomed Anal 15(12):1893–1901, 1997.
155. E Taylor, et al. Intestinal absorption screening of mixtures from combinatorial libraries in the Caco-2 model. Pharm Res 14(5): 572–577, 1997.
156. RA Rourick, JL Whitney, SW Fink, EH Kerns, MS Lee. Applications of automation to combinatorial degradation profiling. Proceedings of the 45th ASMS Conference on Mass Spectrometry and Allied Topics, Palm Springs, CA, 1997, p. 1264.
157. WM Bodnar, DS Wagner, M Geysen. Pharmacokinetic characterization of combinatorial libraries. Proceedings of the 44th ASMS Conference on Mass Spectrometry and Allied Topics, Portland, OR, 1996, p. 1044.

3
Infrared and Raman Spectroscopy

Hans-Ulrich Gremlich
Novartis Pharma AG
Basel, Switzerland

I. INTRODUCTION

Since 1905, when William W. Coblentz obtained the first infrared spectrum (1), vibrational spectroscopy has become an important analytical tool in research and in technical fields. In the late 1960s, infrared spectrometry was generally believed to be an instrumental technique of declining popularity that was gradually being superseded by nuclear magnetic resonance (NMR) and mass spectrometry (MS) for structural determinations and by gas and liquid chromatography for quantitative analysis.

However, the appearance of the first research grade Fourier transform infrared (FTIR) spectrometers in the early 1970s initiated a renaissance of infrared spectrometry. After analytical instruments (in the late 1970s) and routine instruments (in the mid-1980s) became available, dedicated instruments have now become available at reasonable prices. With its fundamental multiplex or Fellgett's advantage and throughput or Jacquinot's advantage (2), FTIR offers a versatile approach to measurement problems that is often superior to other techniques. Furthermore, FTIR is capable of extracting from samples information that is difficult to obtain or even inaccessible for NMR and MS.

Raman and IR spectra give images of molecular vibrations that complement each other, i.e., the combined evaluation of both spectra yields more information about molecular structure than their separate evaluation. Raman spectroscopy (2) offers a high degree of versatility, and sampling is easy because no special preparation is required. As it is a scattering technique, samples are simply placed in the laser beam and the backscattered radiation is analyzed. While in mid-IR spectroscopy Fourier transform instruments are used almost exclusively, both conventional dispersive and Fourier transform techniques have their applications in Raman spectroscopy (2).

By the early 1990s, combinatorial chemistry had revived solid phase organic synthesis. For this technology, IR spectroscopy has always been an important analytical tool. It allows spectral analysis directly on the solid phase, thus obviating the tedious cleavage of compounds from the solid support. In contrast, thin-layer chromatography (TLC), high-performance liquid chromatography (HPLC), gas chromatography (GC), and MS techniques are clearly not feasible unless the compound of interest is first cleaved from the polymeric support. Cleaving the compound from the resin for analysis renders these methods inherently destructive, to say nothing of the uncertainty involved in the cleavage chemistry.

As described in the following sections, IR and Raman spectroscopy, using modern Fourier transform techniques such as IR microspectroscopy, offer excellent analytical tools for the burgeoning field of combinatorial chemistry.

II. INFRARED TRANSMISSION SPECTROSCOPY

Transmission spectroscopy (2) is the simplest sampling technique in IR spectroscopy and is used for routine spectral measurements on diverse samples. Resin samples such as polystyrene or TentaGel (3) beads are usually prepared as a potassium bromide disc (pellet). A small amount, usually 1–3 mg, of finely ground solid sample is mixed with approximately 400 mg powdered potassium bromide and then pressed in an evacuated die under high pressure. The resulting discs are transparent and yield good spectra.

With diagnostic absorption bands in starting materials or products, IR spectroscopy in general is efficient in monitoring each reaction step directly on the solid phase, i.e., without cleaving the substance from the solid support. The direct functional group analysis provided by IR spectroscopy permits qualitative analysis of each reaction step throughout the synthesis. An example is shown in Fig. 1, the success of the addition of adipinic acid monoamide to a polystyrene resin is immediately seen by overlaying the corresponding normalized spectra.

Furthermore, IR spectra from KBr pellets of polymer samples can also be used to study reaction mechanisms in combinatorial synthesis. For instance,

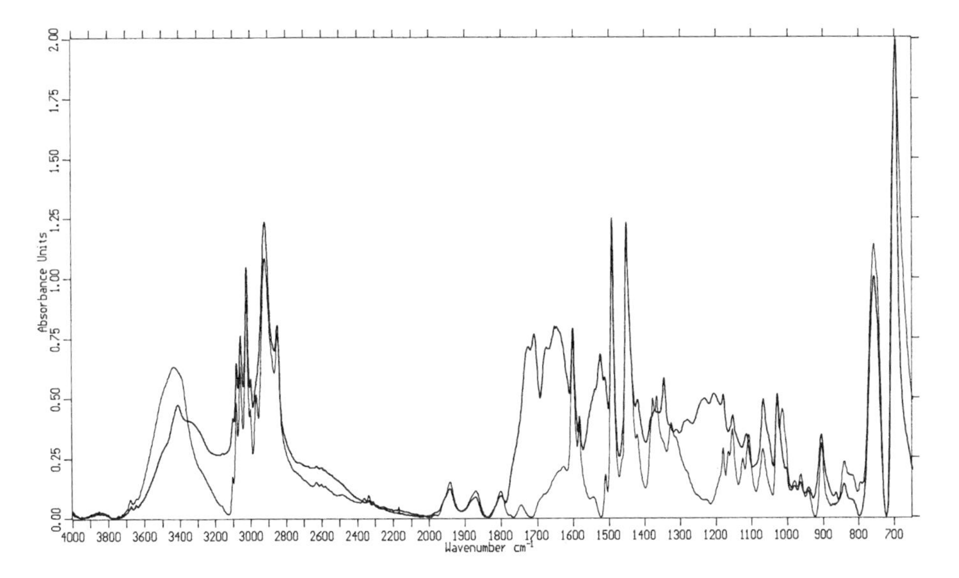

Figure 1 KBr transmission spectrum of polystyrene resin (lower curve) and of polystyrene resin loaded with adipinic acid monoamide (upper curve). The carbonyl bands of the acid (at 1720 cm^{-1}) and of the amide (at 1650 cm^{-1}) are clearly seen in the overlay plot. The spectra were obtained, in the 4000–400 cm^{-1} range with 32 scans at 2 cm^{-1} resolution, by means of a Bruker IFS 66 spectrophotometer.

the mechanism of the solid phase synthesis of 1,4-dihydropyridines was elucidated by observing the IR bands of the ester carbonyls (4).

III. INFRARED INTERNAL REFLECTION SPECTROSCOPY

Internal reflection spectroscopy (2), also known as attenuated total reflectance (ATR), is a versatile, nondestructive technique for obtaining the IR spectrum of the surface of a material or the spectrum of materials either too thick or too strongly absorbing to be analyzed by standard transmission spectroscopy. The technique goes back to Newton who, in studies of the total reflection light at the interface between two media of different refractive indices, discovered that an evanescent wave in the less dense medium extends beyond the reflecting interface. Infrared spectra can conveniently be obtained by measuring the interaction of the evanescent wave with the external less dense medium.

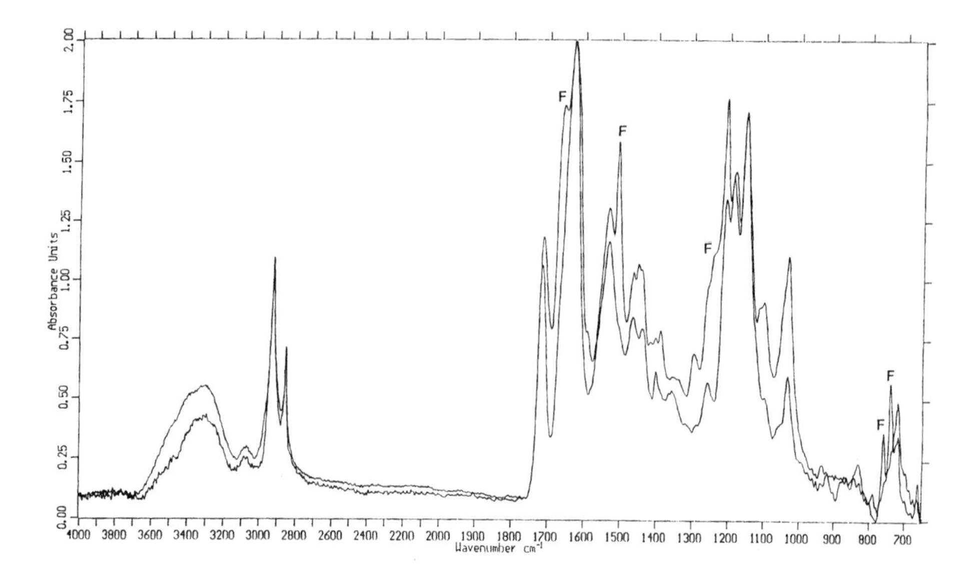

Figure 2 Internal reflection spectra of Mega Crowns with linker, Fmoc-protected (upper curve), and Fmoc-deprotected (lower curve). The bands due to the Fmoc protecting group are marked ''F.'' The spectra were obtained, in the 4000–650 cm^{-1} range with 32 scans at 4 cm^{-1} resolution, by means of a Bruker IFS 28 spectrophotometer equipped with a SplitPea accessory.

The sample is placed in contact with the internal reflection element, the light is totally reflected, and the sample interacts with the evanescent wave resulting in the absorption of radiation by the sample. The internal reflection element is made from a material with a high refractive index, e.g., zinc selenide (ZnSe) or silicon (Si).

As to combinatorial chemistry, internal reflection spectroscopy is especially suited for the analysis of the surface of solid polymer substrates, known as ''pins'' or ''crowns'' (5). With FTIR accessories like the SplitPea (6–8) or the Golden Gate (9), which combine internal reflection and microspectroscopy, direct observation of the chemical synthesis on pins is possible (10). This is shown in Fig. 2 where the spectra of a Mega Crown with linker, Fmoc[1]-protected (upper curve), and of the same crown after the removal of the pro-

[1] 9-Fluorenylmethoxycarbonyl.

tecting group (lower curve) are overlaid. In the lower curve it is clearly seen that the bands due to the Fmoc protecting group, marked "F," have completely disappeared.

This spectroscopic method is very fast because no further sample preparation is required, and it is absolutely nondestructive. Thus each step of the synthesis can be spectroscopically checked with the use of the same pin.

Furthermore, the internal reflection spectra of these reactions are much less distorted by the absorption bands of the solid support than are the transmission spectra recorded from KBr pellets made of polystyrene or TentaGel beads. In Fig. 3 the spectra of polystyrene resin (upper curve) and of a polystyrene Mega Crown (lower curve), both loaded with adipinic acid monoamide, are shown. The disturbing polystyrene bands are marked "P."

With all of these advantages, micro internal reflectance spectroscopy with the SplitPea accessory is an excellent tool for checking successive steps of a combinatorial synthesis on pins. By overlaying the corresponding spectra, it is also extremely helpful for the elaboration of a synthesis when reaction

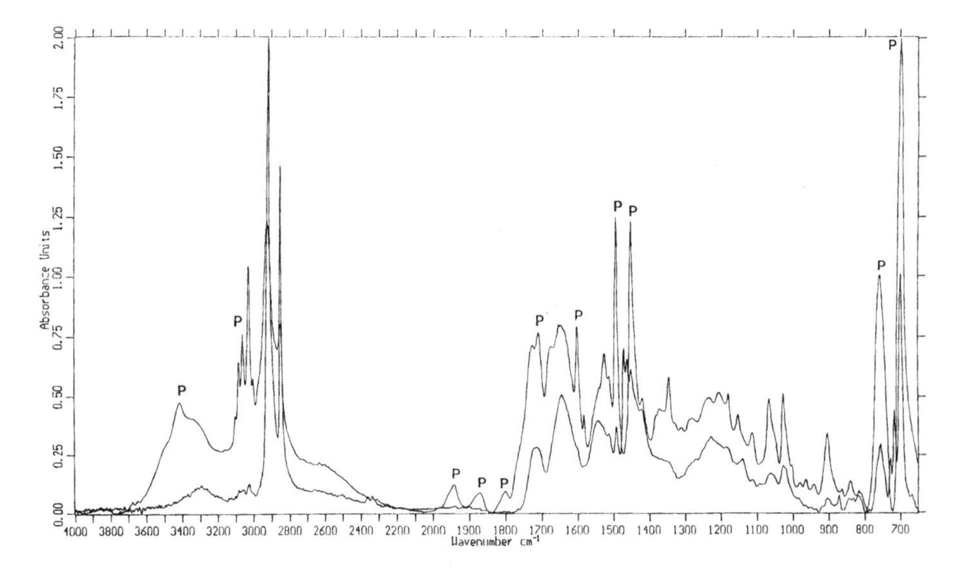

Figure 3 KBr transmission spectrum of polystyrene resin (upper curve) and internal reflection spectrum of a polystyrene Mega Crown (lower curve), both loaded with adipinic acid monamide. The polystyrene bands are marked "P." The spectra were obtained, in the 4000–650 cm^{-1} range with 32 scans at 4 cm^{-1} resolution, by means of a Bruker IFS 28 spectrophotometer equipped with a SplitPea accessory.

conditions are to be optimized regarding reagents, reaction temperature, or reaction time. Furthermore, if biological testing of compounds is to be directly done with the pins, i.e., without cleaving the substance from the support, it is very important to have an analytical method like internal reflectance spectroscopy that is directly applicable to the pins.

Since it can be applied to both pins and beads, micro internal reflectance spectroscopy with accessories like the SplitPea or Golden Gate provides a fast and reliable analytical tool at a reasonable price for the rapidly growing field of combinatorial chemistry.

IV. INFRARED MICROSPECTROSCOPY

By coupling an optical microscope to an FTIR spectrometer, infrared microspectroscopy (2) provides unsurpassed sensitivity, permitting analysis of samples in the picogram range. Infrared microscopes use reflecting optics such as Cassegrain objectives instead of IR lenses, which are poor in performance. As the visible optical train is collinear with the infrared light path, it is possible to position the sample and to isolate and aperture the area for analysis visually. Thus IR microscopy combines in an ideal way the capabilities to visually inspect a sample, e.g., using a video camera (11), and at the same time to acquire information about the sample at the molecular level. Being a nondestructive method, IR microspectroscopy requires virtually no sample preparation.

In the combinatorial chemistry laboratory, IR microspectroscopy lends itself to monitoring reaction products and reaction kinetics directly on a single resin bead (12) without having to cleave the molecule from the bead.

The IR microspectra of single beads can be obtained either in transmission mode (12,13) or with use of an ATR objective (11,14). The ATR objective utilizes a small ATR crystal that is placed on the sample with a contact surface of less than 150 μm in diameter. With a penetration depth of just a few micrometers useful spectra can be obtained from the surface of the resin.

Micro-IR spectra were used for real-time monitoring of solid phase reactions (12,15,16) and, utilizing deuterium isotope containing protecting groups, for quantitative infrared analysis of solid phase, resin-bound chemical reactions (13).

V. PHOTOACOUSTIC SPECTROSCOPY

Complementing other established techniques such as ATR, photoacoustic spectroscopy (PAS) (2) is a quick, nondestructive method for analyzing various materials in the gas, liquid, or solid state with no sample preparation. For photoacoustic measurements, the sample is sealed in a small-volume cell with windows for optical transmission. The cell contains a nonabsorbing gas such as helium. The sample is illuminated with modulated radiation from the interferometer of a FTIR spectrometer. At wavelengths where the sample absorbs a fraction of the incident radiation, a modulated temperature fluctuation at the same frequency as that of the incident radiation is generated in the sample. This modulation is not necessarily with the same phase as the incident radiation. The surrounding inert gas produces periodic pressure waves in the sealed cell. These fluctuations in pressure can be detected by a microphone because the modulation frequency of the incident beam usually lies in the acoustic range, e.g., between 100 and 1600 Hz. FTIR spectrometers are ideal for photoacoustic measurements in the mid-IR, where mirror velocities on the order of 0.05–0.2 cm s^{-1} provide modulation frequencies in the acoustic range (2).

Due to the fact that thermal diffusion length (the distance a thermal wave travels in the sample before its magnitude is reduced by 1/e, where e is the base for natural logarithms) and hence intensity of the photoacoustic signal are increased by lowering the modulation frequency, photoacoustic spectroscopy lends itself to depth profiling (2). This is especially so when step-scan interferometers, accommodating any desired modulation frequency, constant over the entire wavelength range, are used, e.g., in the depth profiling of beads.

Photoacoustic spectra of resin samples have also been reported (16,17). The photoacoustic spectrum of TentaGel S beads coupled via a trityl linker with Fmoc-protected tryptophan is shown in Fig. 4. Compared with the photoacoustic spectrum of native aminoethyl TentaGel S beads, the appearance of diverse carbonyl bands is clearly seen: the ester band of the linker at 1750 cm^{-1}, the carbamate band of Fmoc protecting group at 1723 cm^{-1}, and the amide I band at 1660 cm^{-1}.

Because no sample preparation is required, photoacoustic spectroscopy can be used to examine a sequence of reactions on the same resin sample without product loss (18). Concerning sensitive spectroscopic analysis of solid phase organic chemistry, photoacoustic spectroscopy offers a convenient, nondestructive alternative method to other IR techniques.

VI. LIQUID CHROMATOGRAPHY–INFRARED SPECTROSCOPY

As in other organic synthesis laboratories, high-performance liquid chromatography (HPLC) is also intensively used for separations in combinatorial chemistry laboratories. Using standard detectors such as ultraviolet (UV),

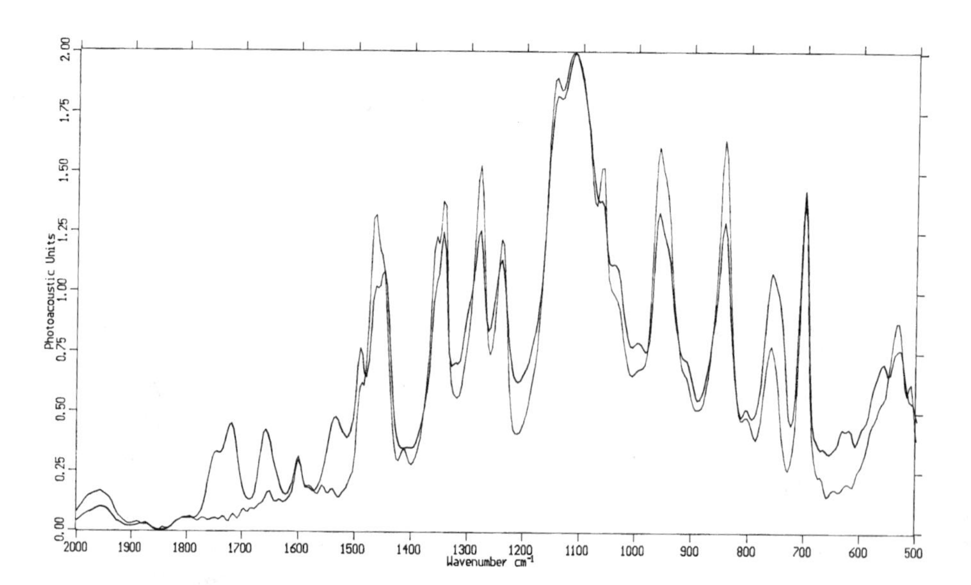

Figure 4 Photoacoustic spectrum of native aminoethyl TentaGel S beads (lower curve) and of TentaGel S beads coupled via a trityl linker with Fmoc-protected tryptophan (upper curve). The carbonyl bands of the ester (at 1750 cm^{-1}), of the carbamate (at 1723 cm^{-1}), and of the amide (at 1660 cm^{-1}) are clearly seen in the overlay plot. The spectra were obtained with 50 scans at 8 cm^{-1} resolution by means of a Bruker IFS 55 spectrophotometer equipped with a MTEC photoacoustic detector.

chromatography is useful in providing quantitative data, when the identities of the mixture components are known. Often the provisional identity of a chromatographically eluting compound can be established through the combination of its retention time and the corresponding UV spectrum.

Due to the unique fingerprinting capability of IR spectroscopy, the combination of chromatographic separation with spectroscopy significantly improves the analysis of complex mixtures. The crucial point in LC-IR methods, however, is the interferences caused by the liquid phase. This is overcome by the revolutionary LC-Transform system (19), which utilizes a high-performance ultrasonic nebulizer or a pneumatic linear capillary nozzle to remove the mobile phase from samples as they elute from the chromatograph. The capillary is surrounded by hot, flowing sheath gas that provides sufficient thermal energy to evaporate the mobile phase and to contain the nebulized spray in a tight, focused cone as the spray emerges from the nozzle. The sample is deposited on a germanium sample collection disk as spots with diameters between 0.5 and 1 mm. Good-quality IR spectra can be obtained from these spots, which contain microgram to submicrogram mg amounts, by the use of a specially designed optics module, which basically is a beam condenser.

HPLC-FTIR has been successfully applied to the rather difficult identification of anabolic steroids (20); difficult, because steroids encompass a large number of natural and synthetic compounds with minor variations in molecular configuration. Despite the structural similarities of fluoxymesterone, testosterone, methyl testosterone, and epitestosterone, the distinct spectral signatures of these four compounds, which are all 17-hydroxysteriods with identical backbone structure, were recognizable.

This methodology easily can be extended to the analysis of combinatorial chemistry samples. In combinatorial chemistry, the combination of chromatography and MS has proved a powerful method for substance identification based primarily on molecular weight. Analogously, HPLC-FTIR provides powerful identification possibilities, but based on molecular structure. As identifying a substance by MS requires knowledge about its possible molecular structure, the two techniques form a highly complementary system of identification.

If—as it is usually the case in combinatorial chemistry—analysts have an idea about what an eluted compound is, MS can confirm the component's identity precisely. On the other hand, if confirmation is not possible for some reason, IR spectroscopy generally will provide more valuable insight. A high degree of confidence can be placed on substance identification that simultaneously matches IR spectral characteristics and chromatography elution time. This is especially so when molecules that have various conformational isomers are to be distinguished.

Since HPLC-FTIR is a high-throughput automatable hyphenated technique, it will be a powerful analytical tool in the burgeoning field of combinatorial chemistry.

VII. RAMAN SPECTROSCOPY

Complementary to infrared, Raman spectroscopy (2) provides unique information about molecular structure. Whereas polar groups and antisymmetrical vibrations of molecular fragments are better detected by IR spectroscopy, Raman spectroscopy is more suitable for the identification of unpolar groups and symmetrical vibrations of molecular fragments. However, dispersive Raman spectroscopy did not find general acceptance in the analytical laboratory during the last 10–20 years because Raman spectra excited with visible lasers are often masked by fluorescence of the sample or of impurities.

With the advent of Fourier transform–Raman (FT-Raman) spectroscopy, this situation has changed drastically. FT-Raman systems virtually eliminate fluorescence by using near-IR laser sources with 1064 or 785 nm excitation (21). Although near-IR lasers produce much weaker Raman signals than visible lasers, FT-Raman technology provides the signal-to-noise advantage necessary to overcome this low signal level, with the consequence that most laboratory samples give useful FT-Raman spectra.

A general advantage of Raman spectroscopy is the extended spectral range, and the Stokes shift (2) in FT-Raman spectra is usually recorded from 3500 to 50 cm^{-1}. Like IR internal reflection spectroscopy (see Sec. III) and IR microspectroscopy (see Sec. IV), Raman spectroscopy is a nondestructive technique requiring little or no sample preparation.

For combinatorial chemistry applications, high-quality FT-Raman spectra can be obtained directly from resin beads, i.e., no cleavage of the molecules from the polymeric support is necessary. This is shown in Fig. 5, where the spectra of TentaGel S beads coupled via a trityl linker with Fmoc-protected tryptophan and the native aminoethyl TentaGel S beads are overlaid. As expected, significant differences in the spectra occur in the spectral region between 1620 and 1500 cm^{-1} where aromatic rings show pronounced Raman activity.

With all of these properties, FT-Raman spectroscopy—similar to usual vibrational spectroscopy—may be a valuable supplementary technique to IR spectroscopy in the field of combinatorial chemistry.

VIII. CONCLUSIONS

Infrared and Raman spectroscopy allow direct spectral analysis of the solid phase, thus avoiding the tedious cleavage of compounds from the solid support. With diagnostic bands in starting materials or products, IR and Raman spectroscopy in general are efficient in monitoring each reaction step directly on the solid phase.

While IR transmission spectroscopy is a general analytical method for resin samples, internal reflection spectroscopy is especially suited for solid polymer substrates known as "pins" or "crowns." Single-bead analysis is best done by IR microspectroscopy, whereas photoacoustic spectroscopy allows totally nondestructive analysis of resin samples.

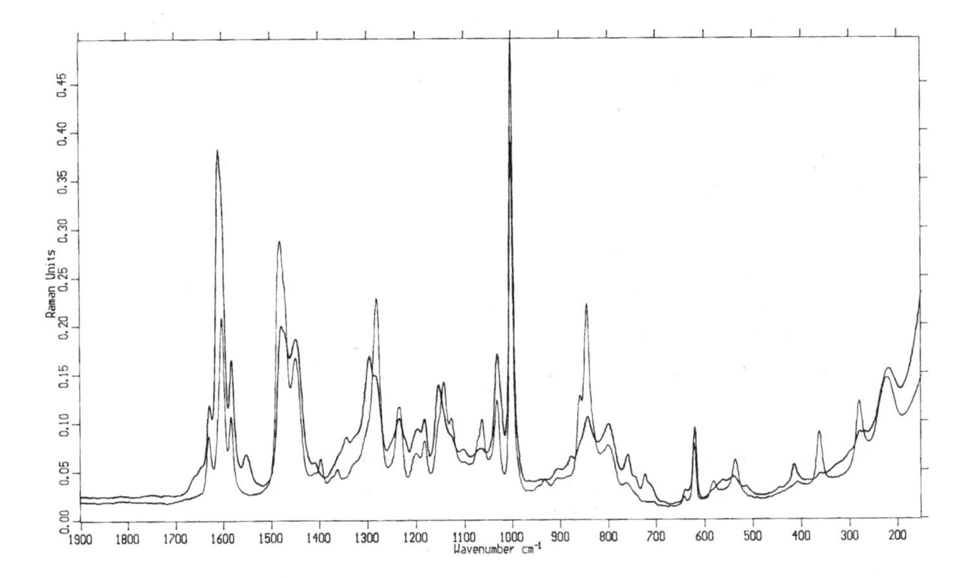

Figure 5 FT-Raman spectrum of native aminoethyl TentaGel S beads (lower curve) and of TentaGel S beads coupled via a trityl linker with Fmoc-protected tryptophan (upper curve). The spectra were obtained with 300 scans at 4 cm^{-1} resolution by means of a Bruker IFS 55/S spectrophotometer equipped with a Bruker FRA 106 FT-Raman accessory. For excitation, a Nd:YAG laser working at 1064 nm was used with a power of 390 mW.

Providing identification based on molecular structure, HPLC-FTIR is therefore complementary to LC-MS.

Additionally, Raman spectroscopy as a complement to IR spectroscopy can be applied to resin samples and—using a Raman microscope—to single beads.

REFERENCES

1. WW Coblentz. Investigations of infrared spectra. Washington: Carnegie Institution, 1905. Republished Norwalk: The Coblentz Society, 1962.
2. HU Gremlich. Infrared and Raman spectroscopy. In: Ullmann's Encyclopedia of Industrial Chemistry. Weinheim: VCH, 1994, B5, pp. 429–469.

3. Available from RAPP Polymere GmbH, Ernst-Simon-Strasse 9, D-72072 Tuebingen, Germany.
4. MF Gordeev, DV Patel, EM Gordon. Approaches to combinatorial synthesis of heterocycles: a solid-phase synthesis of 1,4-dihydropyridines. J Org Chem 61: 924–928, 1996.
5. NJ Maeji, AM Bray, RM Valerio, W Wang. Larger scale multipin peptide synthesis. Peptide Res 8:33–38, 1995.
6. Available from Harrick Scientific Corporation, 88 Broadway, Ossining, NY 10562, U.S.A.
7. NJ Harrick, M Milosevic. US Patent 5,210,418, 1993; M Milosevic, NJ Harrick, CR Wisch. US Patent 5,308,983, 1994.
8. NJ Harrick, M Milosevic, SL Berets. Advances in optical spectroscopy: the ultrasmall sample analyzer. Appl Spectrosc 45:944–948, 1991.
9. Available from Graseby Specac Inc., 301 Commerce Drive, Fairfield, CT 06432, U.S.A.
10. HU Gremlich, SL Berets. Use of FT-IR internal reflection spectroscopy in combinatorial chemistry. Appl Spectrosc 50:532–536, 1996.
11. Bruker Report 143:14–17, 1996.
12. B Yan, G Kumaravel, H Anjaria, A Wu, RC Petter, CF Jewell Jr, JR Wareing. Infrared spectrum of a single resin bead for real-time monitoring of solid-phase reactions. J Org Chem 60:5736–5738, 1995.
13. K Russell, DC Cole, FM McLaren, DE Pivonka. Analytical techniques for combinatorial chemistry: quantitative infrared spectroscopic measurements of deuterium-labeled protecting groups. J Am Chem Soc 118:7941–7945, 1996.
14. PA Martoglio, DW Schiering, MJ Smith, DT Smith. Nicolet Application Note 9693, 1996.
15. B Yan, JB Fell, G Kumaravel. Progression of organic reactions on resin supports monitored by single bead FTIR microspectroscopy. J Org Chem 61:7467–7472, 1996.
16. B Yan, H Gstach. An indazole synthesis on solid support monitored by single bead FTIR microspectroscopy. Tetrahedron Lett 37:8325–8328, 1996.
17. W Rapp, J Metzger. Synthesis and identification of peptide and nonpeptide libraries by hyphenated techniques. Proceedings of the 24th European Peptide Symposium, Edinburgh, pp 743–744, 1996.
18. F Gosselin, M DiRenzo, TH Ellis, WD Lubell. Photoacoustic FTIR spectroscopy, a nondestructive method for sensitive analysis of solid-phase organic chemistry. J Org Chem 61:7980–7981, 1996.
19. Available from Lab Connections Inc., 201 Forest Street, Marlborough, MA 01752, U.S.A.
20. JL Dwyer, AE Chapman, X Liu. Analysis of steroids by combined chromatography-infrared spectroscopy. LC-GC 13:240–250, 1995.
21. C Lehner, J Sawatzki, NT Kwai. FT-Raman spectroscopy in the industrial environment. Proceedings of the 15th International Conference on Raman Spectroscopy, Pittsburgh, 1996, pp. 1094–1095.

4
NMR Methods

Michael J. Shapiro
Novartis
Summit, New Jersey

I. DIRECT ANALYSIS ON SOLID PHASE

The realm of combinatorial chemistry is covered by a wide umbrella of techniques that include both solution phase and solid phase synthesis. The revival of solid phase synthesis has led to great interest in the development of analytical methods to monitor the reaction directly on the polymer support during the course of combinatorial syntheses. These techniques have a significant advantage compared to cleave-and-analyze characterization, particularly for optimizing reaction conditions.

Nuclear magnetic resonance (NMR) spectroscopy has been shown to be useful in this regard. The two procedures available to obtain NMR data on resin are gel phase NMR and a variant of this methodology using magic angle spinning (MAS) (1). While Fourier transform infrared (FTIR) (2,3) and mass spectrometry (MS) (4–6) are faster and more sensitive techniques than NMR, only NMR can give information concerning the detailed structure of a molecule.

II. GEL PHASE NMR

Gel phase ^{13}C NMR is an established technique that was first performed in 1971 (7). Gel phase NMR involves taking an NMR spectrum using a standard high resolution NMR spectrometer. It is nondestructive and the sample can be readily recovered. Spectra are generally obtained in a standard 5-mm NMR tube by swelling a resin with an appropriate solvent. Due to signal broadening, gel phase NMR is generally limited to heteronuclei such as ^{13}C, ^{19}F and ^{31}P where there is a great deal of chemical shift dispersion.

One of the first applications of gel phase ^{13}C NMR showed that the build up of a peptide chain and deprotection of Boc groups could be readily followed (8). An example of the gel phase ^{13}C NMR of a peptide is shown in Fig. 1. Similarly ^{13}C gel phase NMR was used to evaluate and develop new side chain deprotection cocktails (9). Liebfritz and Geralt, describing the use of polyethylene and polystyrene supports, showed the general application of ^{13}C gel phase NMR methodology in the solid phase synthesis of peptides (10,11).

Application of gel phase NMR to organic solid phase synthesis was first reported by Manatt in 1980 (12). Jones and Leznoff extended this work to show that ^{13}C NMR data could be obtained from polystyrene-supported pheromones and a wide variety of organic substrates (13). The application of ^{13}C gel phase NMR to the study of functional group interconversion of polymer-bound steroids has also been reported (14). High-resolution ^{1}H gel phase NMR has been reported for an octapeptide (15). The NMR data were obtained by use of a deconvolution method to enhance the spectral resolution.

^{13}C gel phase NMR is generally impractical for the monitoring of reactions in real time. The typical ^{13}C gel phase NMR experiment takes thousands of transients (several hours) to get sufficient signal-to-noise. This is especially true for obtaining resonances in the carbonyl region of the spectrum. In order to obviate the use of gel phase ^{13}C NMR as a more practical method, a strategy exploiting the use of ^{13}C labeling to monitor the reaction and to further enhance the experiment sensitivity was developed (16,17). Optimizing the location of a sample by using an insert in the NMR tube, a sample could be prepared with only 20 mg of resin. ^{13}C gel phase NMR data could be obtained in this manner in only 15–30 min.

^{19}F makes an excellent probe to monitor solid phase reactions; most solid supports do not contain fluorine, ^{19}F NMR is almost as sensitive as proton NMR, the range of ^{19}F chemical shifts is large, and the resonance can be monitored even when the fluorine atom is quite remote from the site of reaction. The buildup of the peptide chain could be monitored by incorporating ^{19}F into a peptide-protecting group (18).

In the case of the SNAr nucleophilic displacement reaction shown in Fig. 2, ^{19}F was found to be an useful probe nucleus (19). Since fluorine is the leaving group, the disappearance of the ^{19}F signal offers a window through which to observe reaction progress. Shown in Fig. 2 is a typical spectrum

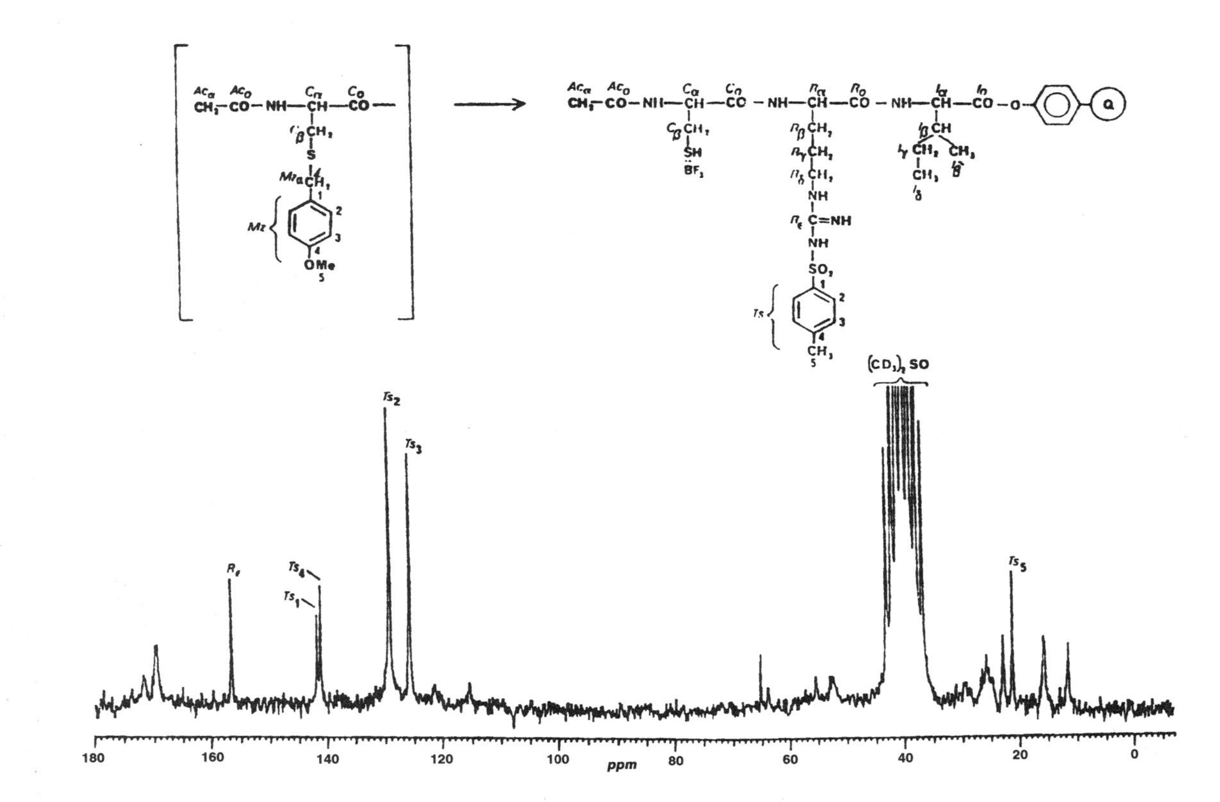

Figure 1 Typical peptide ^{13}C Gel phase NMR spectrum in DMSO-d_6. Product shown is obtained after deprotection reaction by BF_3.

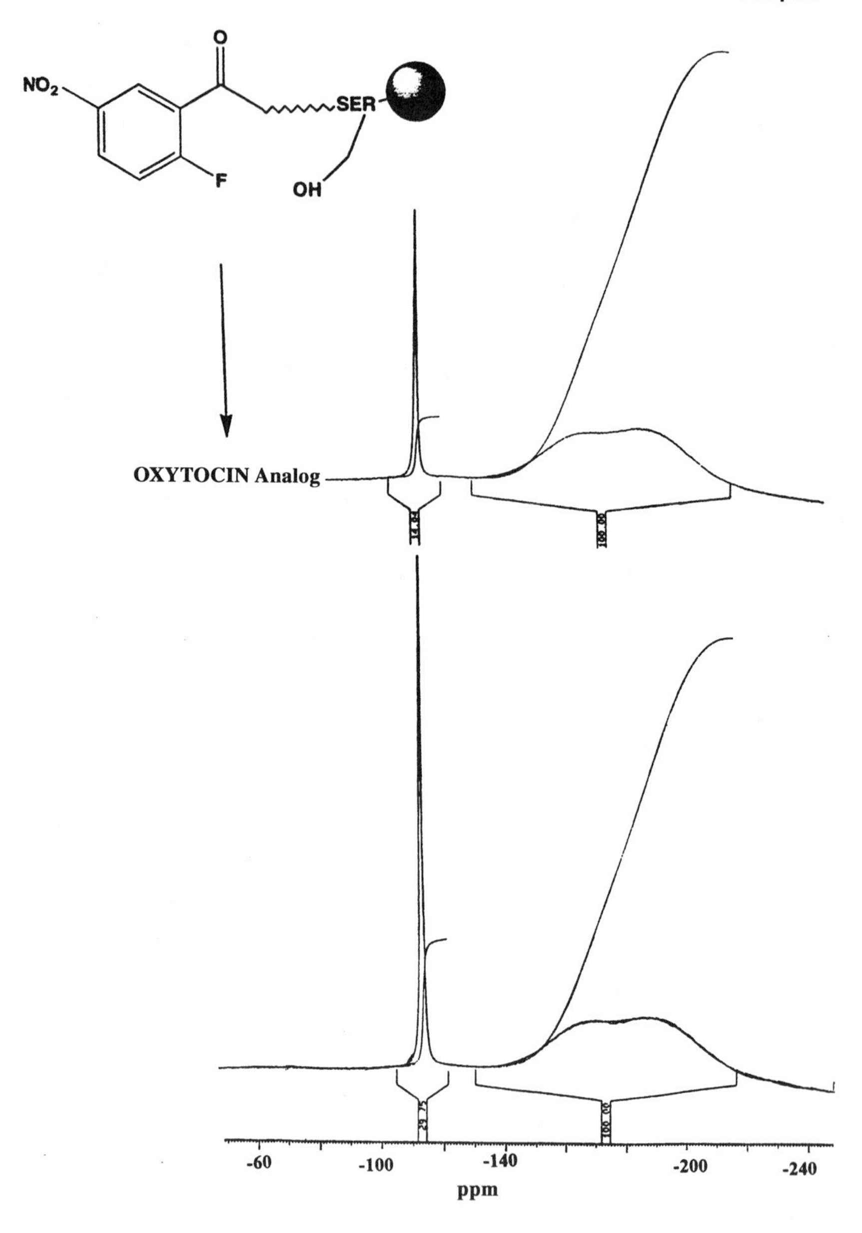

Figure 2 ^{19}F gel phase NMR spectrum for macrocyclization SNAr reaction for an oxytocin analog on Wang resin. Bottom spectrum is starting material and top spectrum is after a reaction time of 1 h. The broad ^{19}F signal arises from the probe components and can be used as an internal standard.

obtained during the course of the reaction. The broad peak in the spectrum arises from fluorine-containing components of the Bruker quadranuclear probe and was conveniently used as an external standard. Using TentaGel resin linked compounds, ^{19}F gel phase NMR data could be obtained in less than 10 min for the reactions shown in Fig. 3 (20). Since a wide variety of fluorine containing compounds are readily available, ^{19}F gel phase NMR is expected to increase in utility.

RO, F, OPfp, O

n-Butylamine. HOBt, r.t.,

RO, F, N, H, O

10a (-132.3)
10b (-132.4)

11a (-134.1)
11b (-134.1)

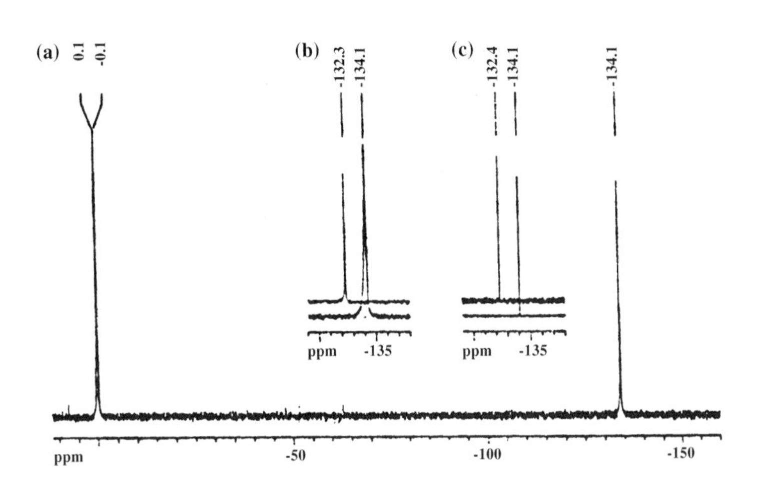

Figure 3 375 MHz ^{19}F gel phase NMR spectrum for resin linked butyl amide. Conversion to product is monitored (a) on TentaGel resin and (b) in solution.

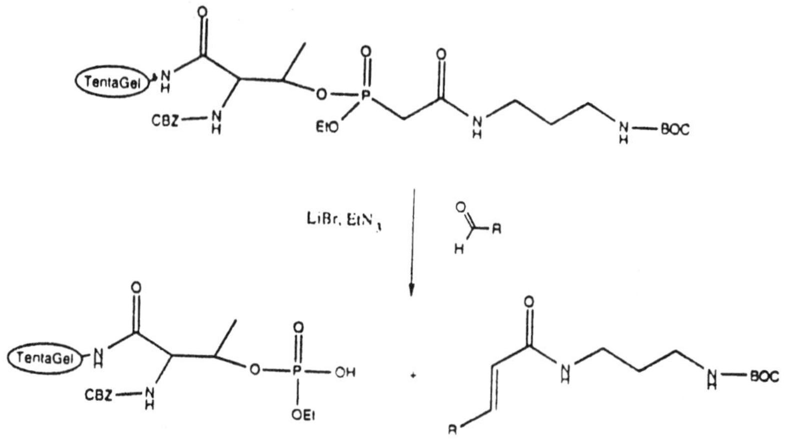

Figure 4 Horner-Wadsworth-Emmons reaction that was monitored by ^{31}P gel phase NMR.

As with ^{19}F NMR gel phase, ^{31}P NMR is useful because the resins currently used do not interfere with the signals produced. Spectra can be obtained in just 10 min with 50 mg of resin. The first use of ^{31}P gel phase data was described by Geralt et al. in conjunction with oligonucleotide synthesis (21). An especially attractive application of ^{31}P gel phase NMR was applied to the Horner-Wadsworth-Emmons olefination reaction depicted in Fig. 4 (22). The use of ^{31}P NMR proved to be a highly sensitive method to follow the reaction on solid phase and the NMR spectra were used to rapidly identify a competing side reaction that would have been difficult to show using other analytical methodologies. Van Etten used gel phase ^{31}P NMR to follow protection and deprotection of a phosphorylated benzyl group in peptide synthesis (23). Spectral resolution was sufficient to distinguish dibenzyl, monobenzyl, and nonbenzylated forms of a tyrosine phosphate group.

III. MAGIC ANGLE SPINNING NMR

A. ^{1}H Magic Angle Spinning NMR

It would be of great utility if a better methodology could be provided to monitor polymer-bound reactions than gel phase heteronuclear NMR. However, as previously mentioned, the utility of gel phase proton NMR in solid phase synthesis monitoring is of little value, as the spectra obtained are generally quite broad.

Spinning the sample at the magic angle (54.7°), MAS overcomes the limitations of using ^{1}H NMR for solvent-swollen resin samples. The broadening due to magnetic susceptibility and dipolar coupling are substantially removed. ^{1}H MAS NMR have been applied to study properties of crosslinked polystyrene gels for some time (24). However, the application of MAS ^{1}H NMR in combinatorial chemistry appeared only recently, where it has been demonstrated that ''high-resolution'' NMR data for resin-supported molecules can be obtained (25,26).

MAS NMR requires special equipment to obtain the data. High quality MAS NMR spectra can be obtained using specially designed probes that are now commercially available. While any standard MAS probe can also be used to collect NMR data, the resolution obtained will probably suffer due to mismatches in the magnetic susceptibility in the probes (27).

The quality of the NMR data is also dependent on the conditions by which the sample was prepared. All solvent/resin combinations do not yield high-quality ^{1}H MAS NMR spectra (28,29). The most important factor in obtaining high-quality ^{1}H MAS NMR data is the choice of resin, with TentaGel resins (TGT) generally giving rise to the best-quality data. The choice of swelling solvent can also play a critical role. For the most part, good solvents include CD_2Cl_2, $CDCl_3$, C_6D_6, DMF-d_7, and DMSO-d_6. For example, spectra obtained from Fmoc-Asp(OtBu)-NovaSyn TGT gave good line widths, less than 6 Hz, for CD_2Cl_2, DMF-d_7, and DMSO-d_6 where as using benzene-d_6 yielded 19.4 Hz. MAS data obtained on Fmoc-Asp(OtBu)-NovaSyn Wang yielded similar line widths for all the solvents with line widths ranging from 9 to 14 Hz. The compound attached to the resin also plays a role on the ''high-resolution-like'' appearance of the data (30).

The quality of the spectra can be enhanced by using presaturation of the resin signal to reduce the intensity of the aromatic resonances from the polystyrene or polyethylene glycol resonances in TentaGel-like resins. The spin-echo experiment distinguishes between the narrow lines of the substrates and the much broader lines of the polymer and can be used to remove the peaks from the polymer as well (31–33).

B. ^{13}C MAS NMR

In the late 1980s Frechet showed that high quality ^{13}C NMR data of swollen gels could be readily obtained by MAS using a standard MAS probe (34–36). The first use of ^{13}C MAS NMR data for combinatorial chemistry demonstrated that a potential reaction complication, the production of two similar compounds, could be monitored (37). The reaction of norbornane carboxylic acid

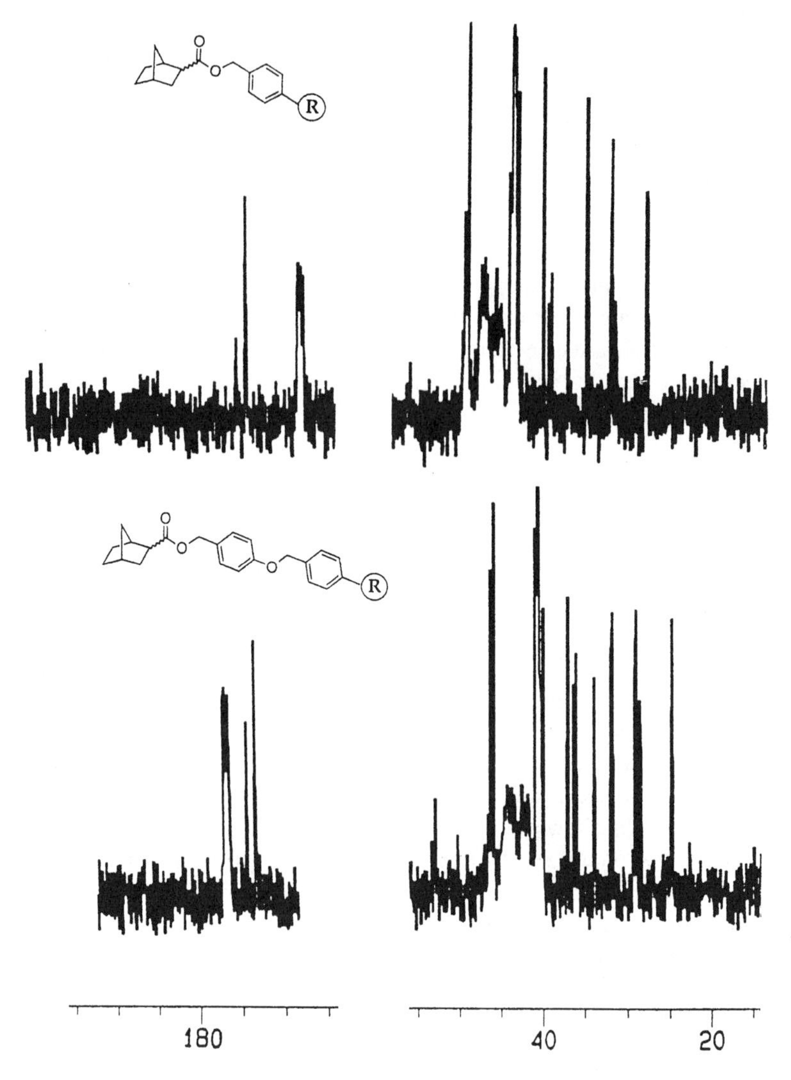

Figure 5 Utilization of MAS ^{13}C NMR for comparison of esterification reaction products for exo and endo norbornane-2-carboxylic acid mixture on Merrifield and Wang resins.

on Merrifield and Wang resins to produce the ester were compared to evaluate the potential usability of the particular resin in a library synthesis (Fig. 5). It was found that the stereochemistry of addition was different for the two resins. In the case of the Wang resin a 60:40 ratio of exo/endo product was obtained while the Merrifield resin yielded an 80:20 ratio. The spectral quality of the ^{13}C MAS data was virtually indistinguishable from that of the solution ^{13}C NMR data. The MAS ^{13}C NMR data can be collected in a much shorter time frame (20 min) than for gel phase ^{13}C NMR data which generally take several hours.

C. Other Nuclei

MAS ^{19}F NMR can also be used to follow appropriate reactions (19). The MAS data were seen to be an order of magnitude faster to collect than gel phase NMR for the same sample. While no reports of using MAS ^{31}P NMR have appeared at this writing, there is no reason why this nucleus and others could not prove valuable in following solid phase chemical reactions.

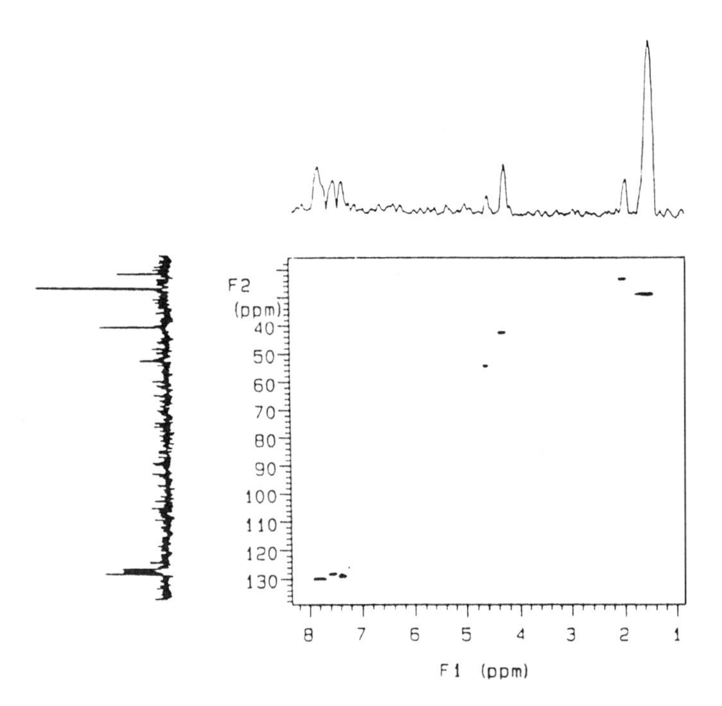

Figure 6 Two-dimensional CH-correlated gel phase NMR spectrum for chloroform-swollen Boc-Cys(Acm)-OCH_2-Pab-copoly (styrene 1% divinylbenzene).

IV. TWO-DIMENSIONAL NMR

Two-dimensional NMR spectroscopy has proven of great value for product analysis in solution phase chemistry and by its nature should provide a platform by which enhanced data quality and ease of interpretation is obtained. Geralt demonstrated that the use of 2D CH correlated NMR spectra (Fig. 6) enhances the utility of gel phase NMR even though the proton NMR spectrum was broad (36). Since MAS NMR spectra of the swollen resins are of higher quality than gel phase NMR, it is possible to treat these samples as if they

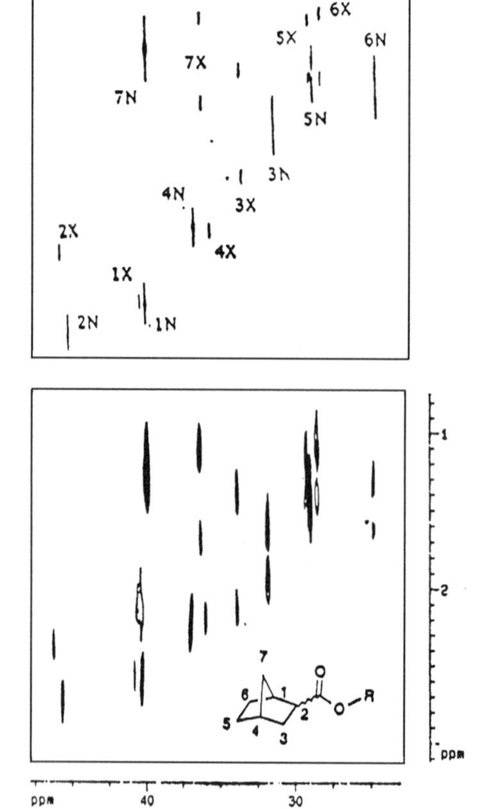

Figure 7 Two-dimensional CH-correlated NMR spectrum for a mixture of endo (N) and exo (X) norbornane epimers. (Top) High-resolution solution data for the mixture of norbornane-2-carboxylic acid. (Bottom) MAS NMR on benzene-swollen Wang resin.

were solutions and therefore use the same techniques that would be used in ''high-resolution'' NMR studies. The application of 2D NMR techniques to determine the structure on resin exemplifies this notion (26). The direct-observe MAS CH-correlated data for the exo/endo norbornane mixture on resin shown in Fig. 7 illustrates the resolution obtainable. Assignments of both proton and carbon resonances in both major and minor isomers can be readily made. The combination of heteronuclear multiple quantum correlation (HMQC) and total correlation spectroscopy (TOCSY) data, as shown in Fig. 8 for Wang-Fmoc-lys-Boc, allows complete assignment of the proton and carbon-13 spectra confirming the structure on resin (37). The step-by-step analysis of a multistep solid phase synthesis involving a Heck reaction was conveniently followed using MAS 2D NMR techniques (38). The use of 2D NMR

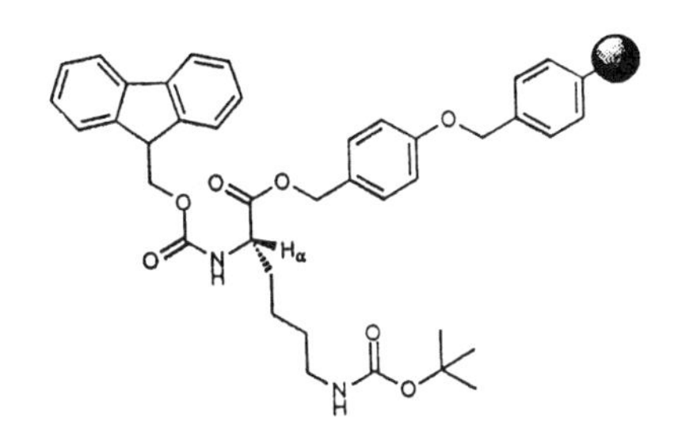

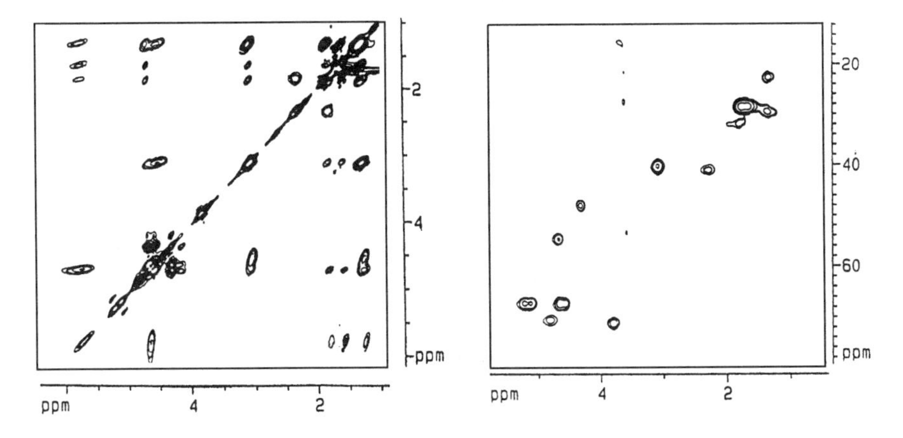

Figure 8 MAS HMQC and TOCSY NMR spectra for Fmoc-lys-tboc on benzene-swollen Wang resin.

TOCSY, HMQC, and nuclear Overhouser enhanced spectroscopy (NOESY) (39) data as shown in Fig. 9 was instrumental in the evaluation of products.

The main problem for NMR analysis of on-resin products, with the exception of TentaGel, is that the ^{1}H NMR spectra are generally broad with featureless line widths around 10 Hz or more (28). The additional complication of large polymer resonances can be attenuated by using a spin-echo sequence but the residual peaks can still be problematic. Peak assignment can be difficult due to the loss of coupling information relegating the interpretation of the

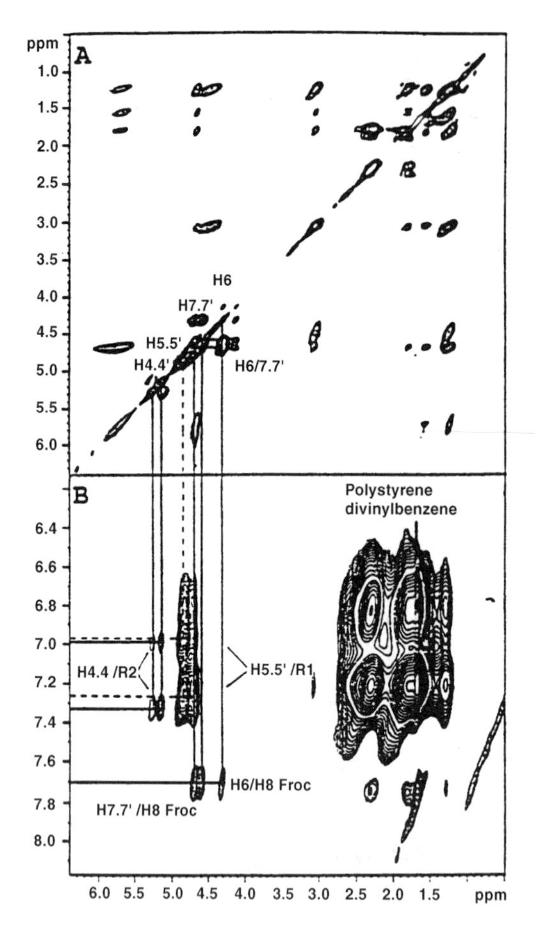

Figure 9 MAS (A) TOCSY and (B) NOESY NMR spectra for Fmoc-lys-tboc on benzene-swollen Wang resin.

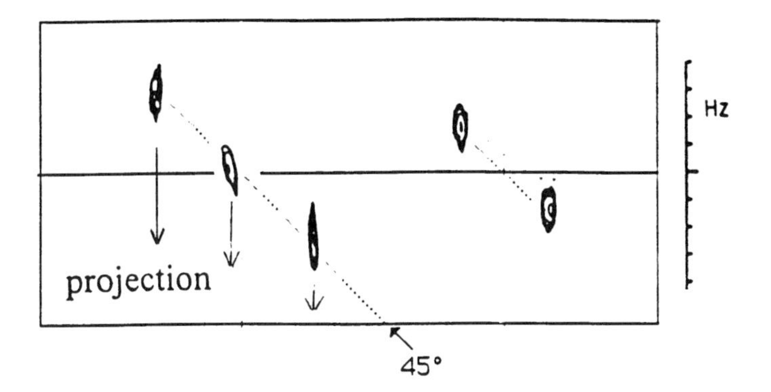

Figure 10 Schematic representation of nontilted 2D J-resolved experiment. Projection onto the chemical shift axis recovers the high resolution 1D NMR spectrum.

molecular structure to chemical shift information criteria only. Proton-proton spin coupling values can be obtained by performing a 2D J-resolved experiment (40,41). It was found that by using a nontilted projection from the 2D J-resolved experiment high quality as indicated in Figure 10, 1D proton NMR spectra of resin-bound molecules can be obtained. A comparison of the ^1H MAS NMR spectrum, obtained under spin echo and by the projection of the nontilted 2-D J-resolved data, for the two methyl groups of DMF-swollen isoleucine on Wang resin is shown in Fig. 11. The methyl resonances can be assigned by observation of the coupling patterns. This increase in resolution is further exemplified for Alloc-Asp derivatized oxazolidinone attached via its side chain carboxyl to SCAL-linked aminomethylpolystyrene as seen in Fig. 12 (42).

In the J-resolved projection, the aromatic rings of the SCAL are clearly present where as the polymer resonances have ''dropped out'' of the spectrum. In addition, the proton coupling constants arising from alloc group are readily measured as seen in the expanded spectrum (Fig. 13). If a more detailed measure of the couplings is desired, then the full 2D J-resolved spectrum can be evaluated in the normal manner as exemplified in Fig. 14. A similar method to obtain accurate proton-proton coupling constants based on E.COSY spectra has also appeared recently (43).

It would be advantageous if we could retain the enhanced resolution afforded by the projection of the 2D J-resolved data and also obtain spin system connectivities. MAS spin-echo correlated spectroscopy (SECSY) (44) allows both spin connectivities and enhanced resolution to be obtained. The

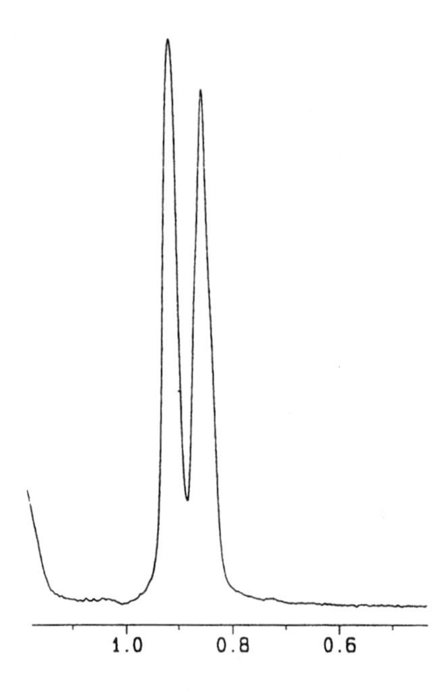

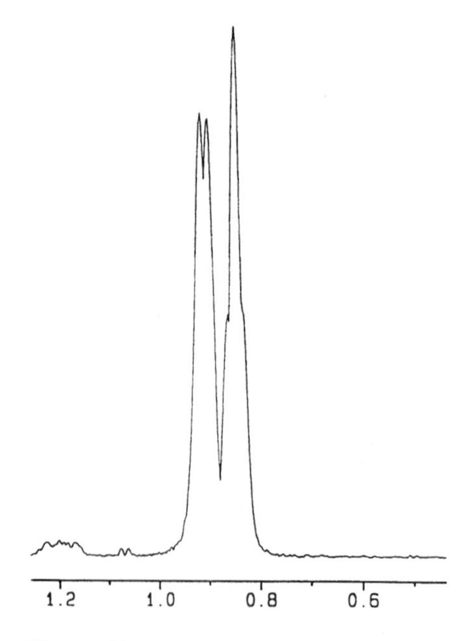

Figure 11 Projection of nontilted 400 MHz MAS J-resolved NMR spectrum for methyl region for Ile on DMF-d_6-swollen Wang resin.

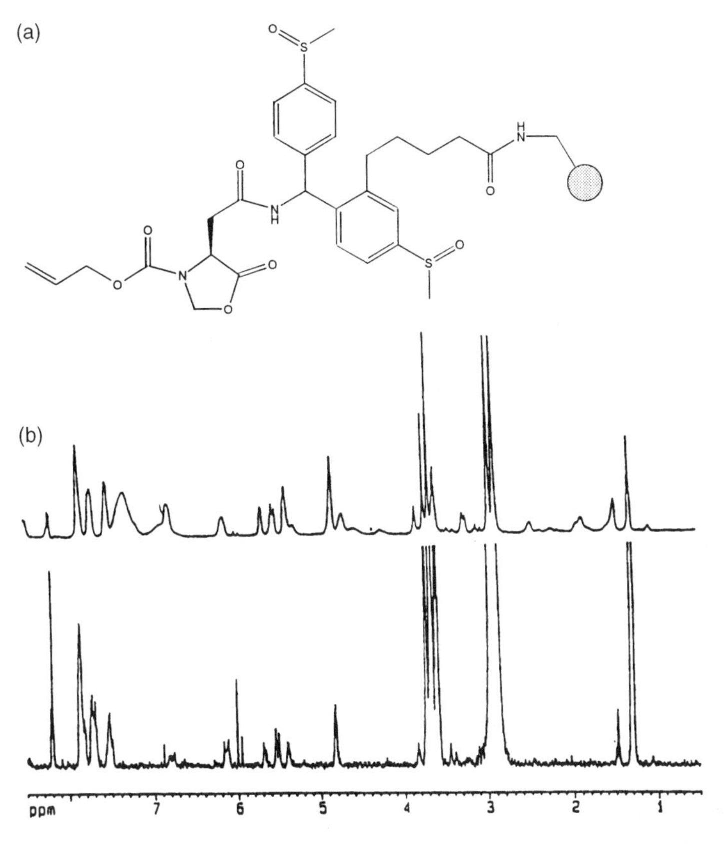

Figure 12 400 MHz MAS ^{1}H NMR spectrum for Alloc-Asp-derivatized oxazolidinone on SCAL-linked aminomethylpolystyrene. (a) Spin-echo ^{1}H NMR spectrum. (b) Nontilted projection from 2D J-resolved spectrum.

projection of the MAS SECSY data for isoleucine on Wang-1 resin shown in Fig. 15 illustrates the apparent increase in resolution. The apparent coupling constants observed in the 1D spectrum cannot be used (45).

The recent development of a gradient high-resolution MAS probe will extend the utility of 2D experiments by removing artifacts that generally accompany MAS 2D NMR data on resin samples (46). The lack of artifacts is illustrated by the high quality SECSY spectrum shown in Fig. 16. SECSY data contain the same information as a COSY spectrum, but the appearance of the spectrum is different. The diagonal lies along the F1 = 0 and the off-

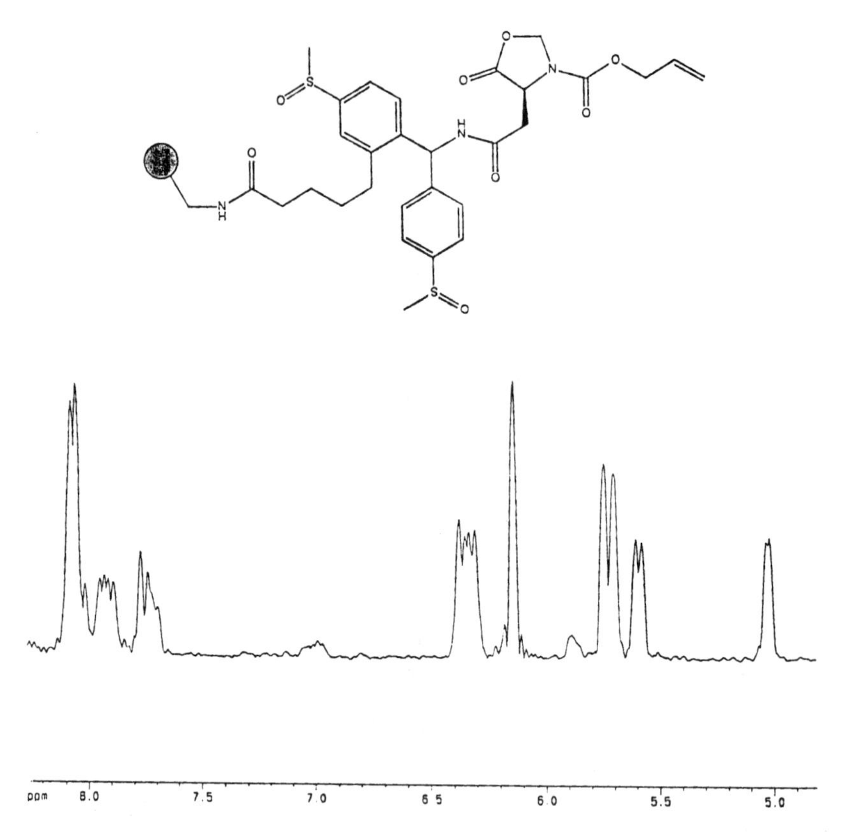

Figure 13 Partial NMR spectrum of projection of nontilted J-resolved NMR spectrum for Alloc-Asp-derivatized oxazolidinone attached via its side chain carboxyl to SCAL-linked aminomethyl polystyrene swollen with DMF-d_7.

diagonal peaks occur in pairs along lines that make an angle of 135° with the diagonal (47). The ability to obtain structural data from a single resin bead represents the ultimate level of sensitivity for MAS NMR (48). Reasonable ^1H NMR data can now be realized in about 1 h. This result, in principle, affords identification of the molecule on the resin directly without resorting to tagging methodologies (49). The incorporation of site specific ^{13}C labels aids the ability to obtain heteronuclear NMR data.

Robotic sample changing and automated data collection is an increasingly important aspect in the efficient utilization of MAS NMR for combinato-

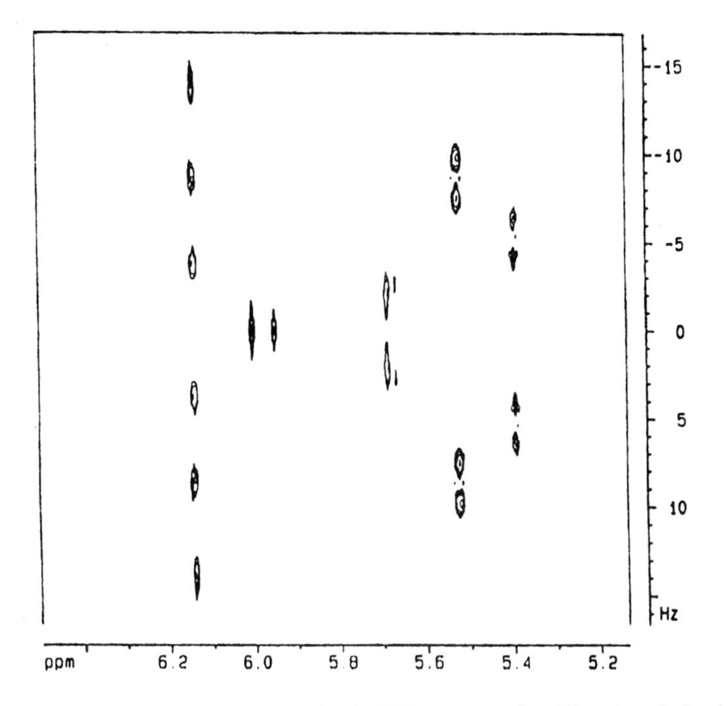

Figure 14 Partial 2D J-resolved NMR spectrum for Alloc-Asp-derivatized oxazolidinone attached via its side chain carboxyl to SCAL-linked aminomethylpolystyrene swollen with DMF-d_7.

rial chemistry. Sample changers for both MAS rotors and for nanoprobe samples are currently available from the instrument manufacturer.

V. MULTIPIN CROWNS

Combinatorial chemistry has developed using resin beads as a primary format for the synthesis of nonpeptide targets. An alternate format is the crown/pin system where resin is grafted onto a base polymer (50). FTIR, which is very useful for analysis of bead resin chemistry, as a general tool for analysis of crown chemistry is somewhat limited (51). The IR spectrum is hampered by interfering peaks that arise from the base plastic, the grafted resin, as well as from the linker molecule.

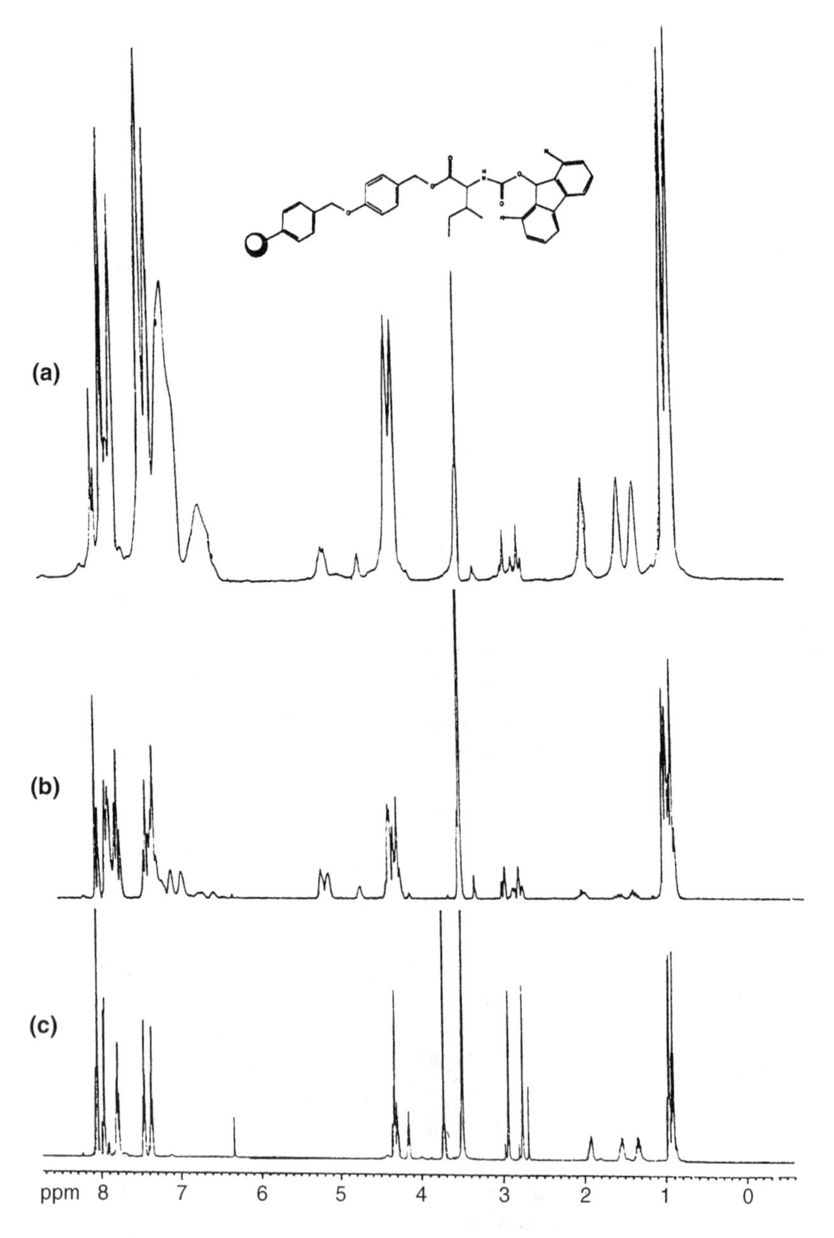

Figure 15 (a) Spin echo (b) projection of the SECSY data MAS ^1H NMR spectrum for DMF-d_7-swollen Fmoc-isoleucine on Wang resin **1** and (c) solution NMR spectrum for Fmoc-isoleucine methyl ester.

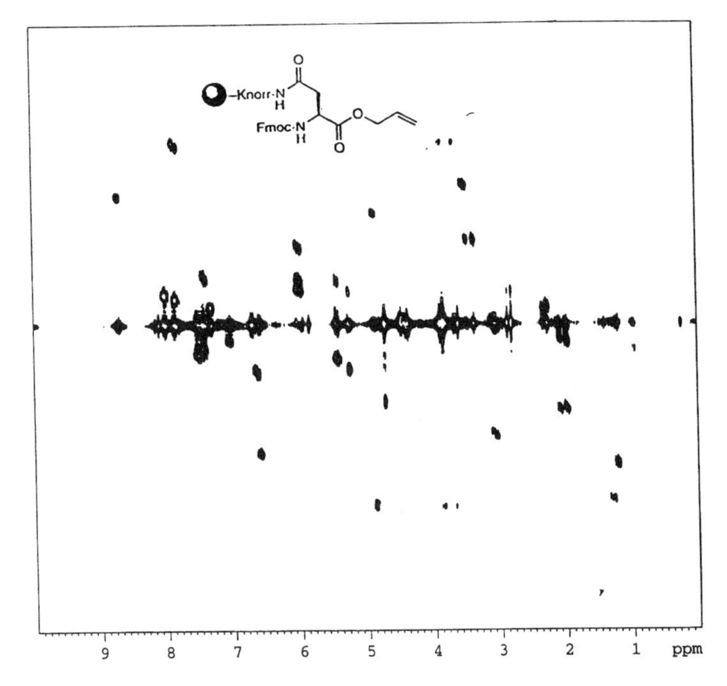

Figure 16 Gradient 2D SECSY NMR spectrum for Alloc-Asp-derivatized oxazolidinone on SCAL-linked aminomethylpolystyrene swollen with DMF-d_7 obtained with a 4-mm high-resolution gradient MAS probe.

One dimensional MAS NMR is not affected to the same extent as is IR. Reasonable NMR data were obtained using a 7-mm high-resolution MAS probe utilizing the spin-echo pulse sequence, allowing the removal of the broader resonances due to the plastic as well as the signals from the slower moving components of the resin. Use of MAS NMR allows the same crown to be sequentially monitored and returned to the reaction conditions until the desired transformation is complete (Scheme 1). The ability to follow the transformation of alcohol to aldehyde and then elaboration to olefin on the same crown exemplifies this technique (Fig. 17). The stereochemistry of the olefin was found to be trans as determined from the measured coupling constant of 16 Hz. Confirmation of the olefinic peaks was obtained by COSY (52). MAS NMR, following cleavage, on the same crown, showed complete absence of product peaks.

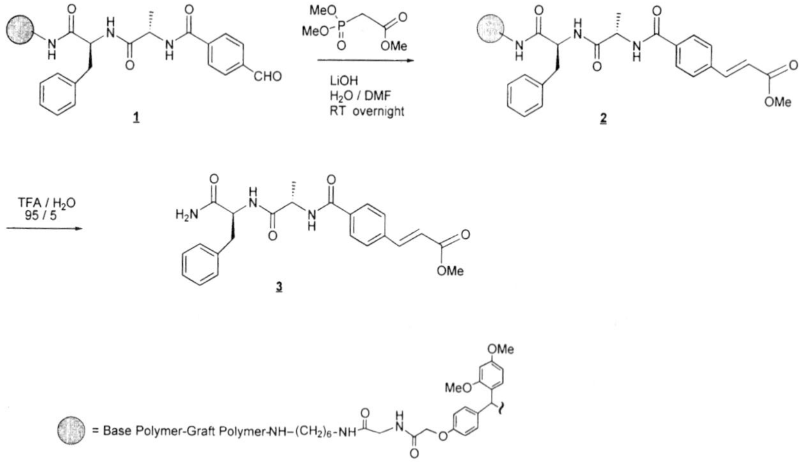

Scheme 1

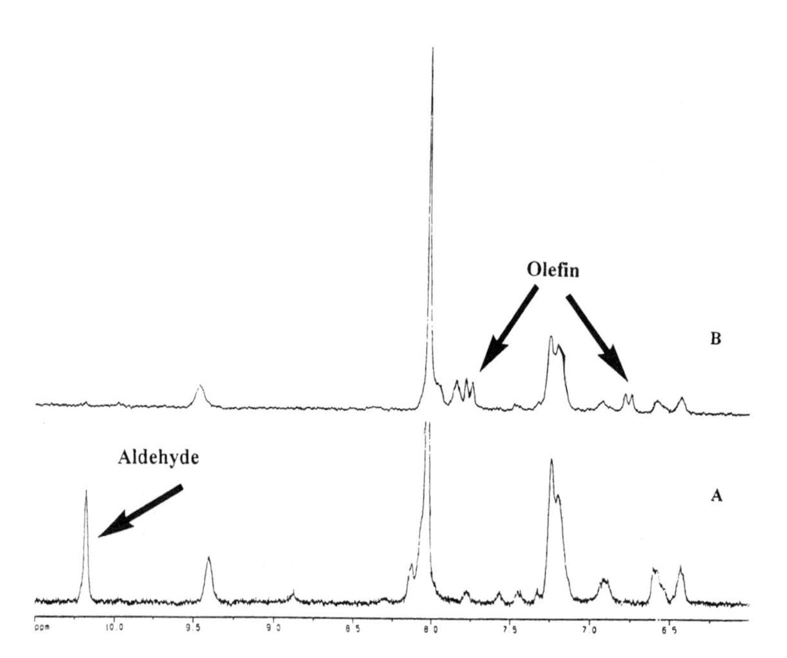

Figure 17 (A) 400 MHz spin-echo MAS ^{1}H NMR spectrum of a crown with aldehyde product attached to the grafted polymer and (B) with olefin attached, swelled in DMF-d_7.

VI. SOLUTION TECHNIQUES

A. HPLC/NMR

While the majority of attention in combinatorial syntheses has been on solid phase analysis, the use of traditional solution phase organic chemistry to form compound collections should not be disregarded. MS techniques are widely used for evaluation of mixtures produced by combinatorial chemistry (4). However, a potential problem with the MS methodology involves a situation when isomolecular weight compounds are present. The compounds can be stereoisomers, positional isomers or by chance identical molecular weight materials. In these cases, the identity of the substance as determined by MS can be ambiguous.

The recent redevelopment of HPLC-NMR allows for complete identification of individual compounds in complex mixtures (53). While the majority of reports using HPLC-NMR have been in the drug metabolism area, the utility for organic compounds and for peptides has been recently demonstrated (54). It is presently possible to obtain routine high-quality NMR data using this technique with as little as 5 mg of compound in the chromatogram peak, with the detection limit using this technology presently being on the order of 100 ng (55).

The utility of HPLC-NMR for analysis of isomolecular weight compounds is illustrated by the analysis of a mixture of dimethoxybenzoylglycines prepared by split-mix synthesis having aromatic ring positional isomers. The results from the dimethoxybenzoylglycines was analyzed by stop-flow HPLC-NMR where individual components are evaluated as each peak enters the NMR probe while the separation is paused. The assignment of structure from the separated components is straightforward by consideration of the shifts and the coupling patterns for the region of interest shown in the stacked plot NMR spectrum shown in Fig. 18.

On-flow HPLC-NMR analysis can also be performed when sufficient material is available. It involves collecting the NMR data continuously as the sample passes through the probe. This is the most efficient method for structure evaluation by HPLC-NMR. The NMR data are represented in a 2-D plot where the x direction contains chemical shift information and the y direction is representative of the LC retention time. The individual spectra can be extracted from the 1D slices along the x axis if so desired. The resolutions in the individual spectra are of somewhat lower quality than in the stop-flow method; however, the introduction of the second dimension allows for easy structure assignment even for overlapping peaks in the LC separation. As seen in Fig. 19, the on-flow HPLC-NMR characterization shows four distinct sets of resonances.

The HPLC/NMR for peptides is illustrated by the data obtained for pen-

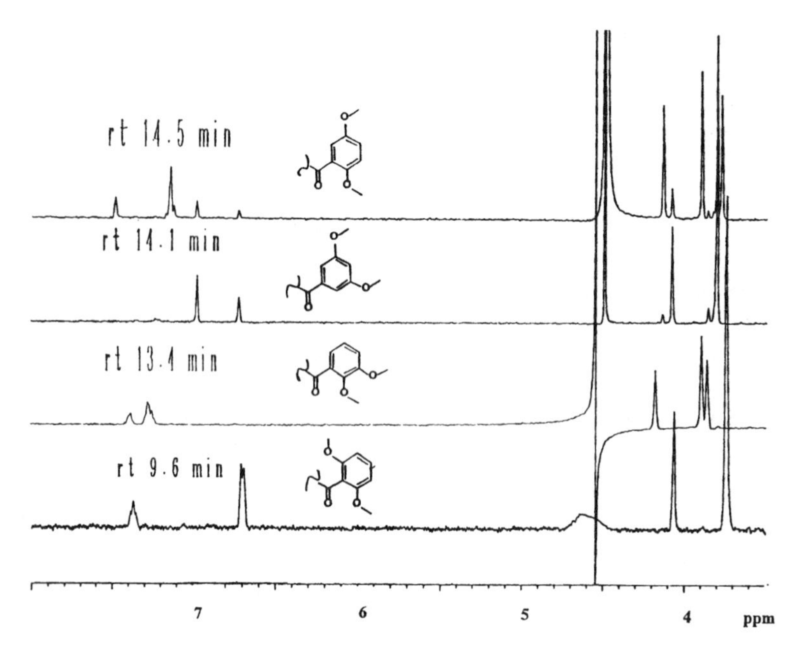

Figure 18 500 MHz ^{1}H HPLC/NMR stack plot of chemical shift vs. retention time for a mixture of isomolecular weight dimethoxybenzoylglycines.

tapeptide mixtures FNXEF-OH where X = D, Q, I, K, or T. In each of these five component mixtures there are two compounds having identical molecular weights; therefore, it would be difficult to unambiguously assign structures by MS without resorting to special techniques. Stop-flow HPLC-NMR allowed all of the compounds to be unambiguously assigned utilizing 2D TOCSY data as exemplified in Fig. 20 for the compound having a retention time of 10.8 min. The TOCSY spectrum showed resonances at δ = 1.3, 1.6, 2.9, and 4.1, indicative of a lysine residue. The additional resonances are consistent with the amino acids; phenylalanine δ = 4.2 and 3.1, and 4.6, 3.18, and 2.98; asparagine δ = 4.7, 2.7, 2.55; and glutamate δ = 4.27, 2.34–2.05. This clearly identifies the peptide as FNKEF-OH. The isomolecular weight complication of lysine vs. glutamine was readily resolved. In another example, using standard HPLC conditions, assignment of a majority of compounds from a library of 27 tripeptides (AYM) was achieved (56). This task, which was accomplished in a single pass using on-flow HPLC-NMR, would have been difficult if not impossible by any other technique.

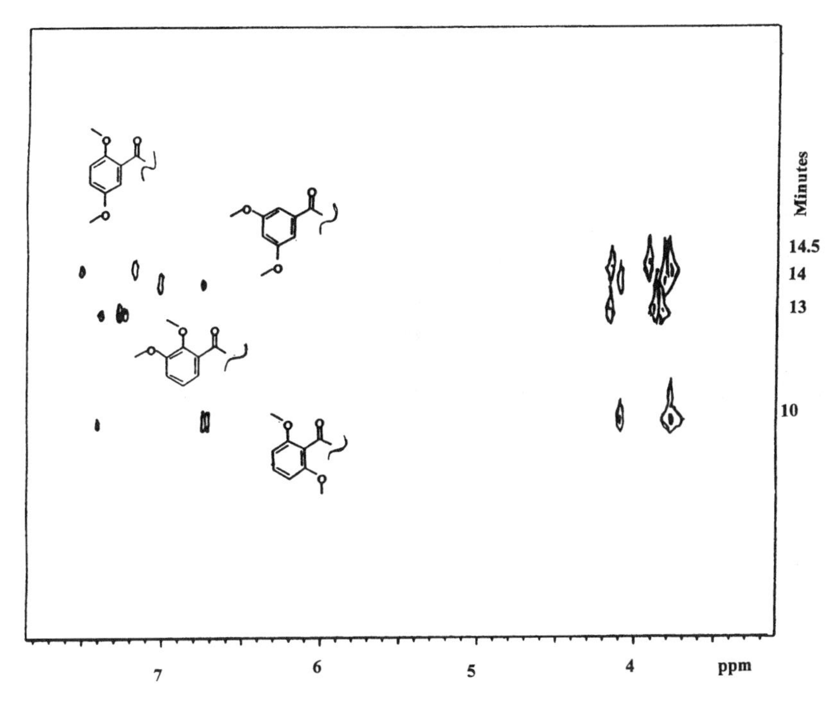

Figure 19 On-flow 500 MHz HPLC-NMR contour plot spectrum of chemical shift vs. retention time for a mixture of isomolecular weight dimethoxybenzoylglycines.

The simultaneous determination of structure by the combination of HPLC-NMR-MS should prove to be an extremely powerful tool for combinatorial chemistry.

B. Flow NMR

The ability to obtain routine NMR data in a reasonable time frame is greatly enhanced by the use of robotic sample changers. Currently it is possible to obtain data from over 100 samples in an unattended manner. The inefficiency in this procedure is that the samples generally are made by manual methods restricting the potential throughput and capacity. Mass spectrometry techniques have been developed to obtain spectra in about a minute or so each, using an autosampler that can take samples directly from 96 welled plates. These same plates are also used in the high-throughput screening efforts, thereby minimizing the effort to prepare samples. It would be convenient if

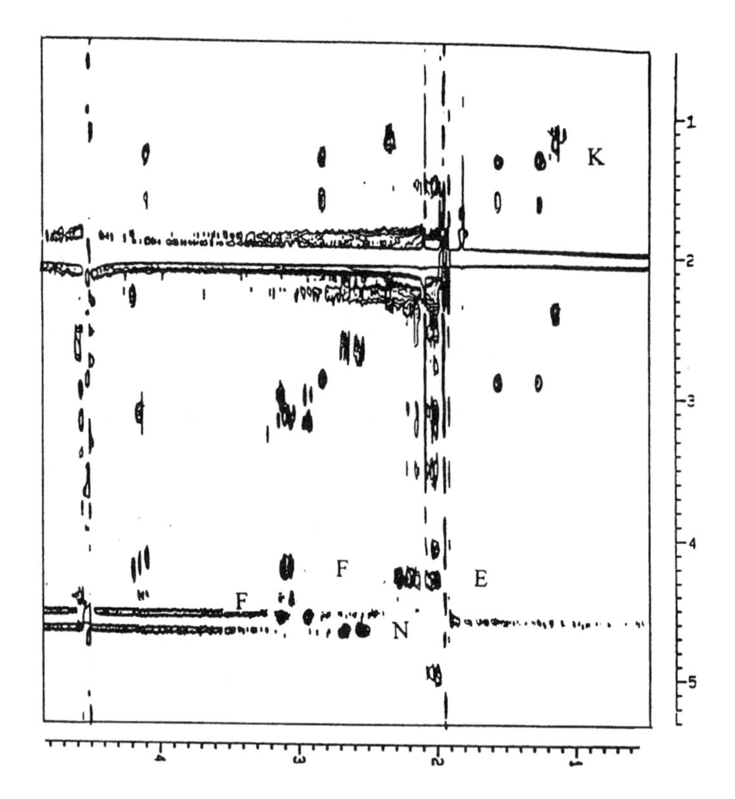

Figure 20 2D TOCSY spectrum for FNKEF-OH pentapeptide derivative obtained using stop-flow HPLC-NMR. Amino acid assignments are indicated by the letter code.

the same format could be used to obtain NMR spectra. Flow NMR has been recently introduced using a similar format MS and is commercially available. NMR most likely won't be able to obtain data in the same time scale as MS; however, high-quality spectra can most probably be obtained in around 3 min. The robotics used to obtain these spectra will allow samples to be spot-checked or to permit one to obtain data on the identified active well as suggested in the cartoon in Fig. 21.

Since the solvent used in these wells is generally DMSO/water, this makes the flow NMR experiment that much more difficult. Techniques have been described to efficiently suppress the protons from multiple solvents and obtain NMR data on small samples in a relatively short time (57,58). High-quality NMR data on a 26-μg sample can be obtained in just over 2 min using

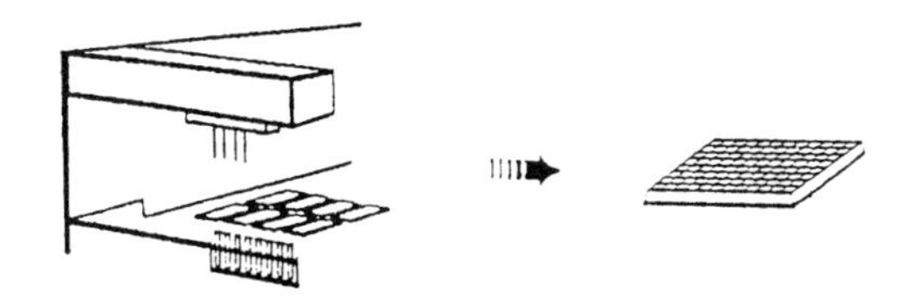

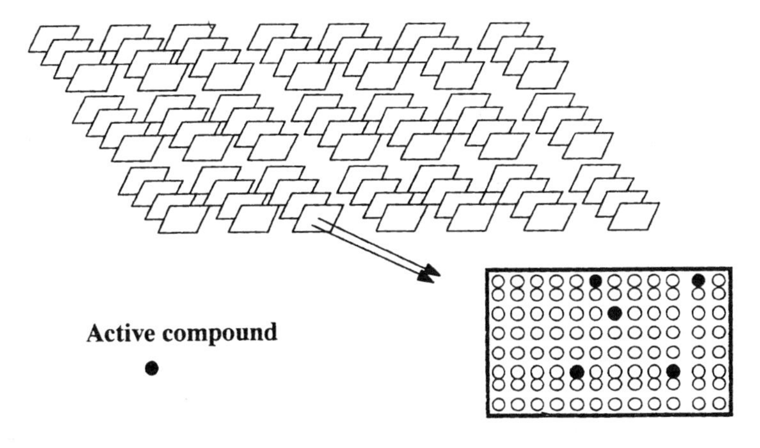

Figure 21 Schematic representation of flow NMR using 96 welled plates.

either a microprobe or an HPLC-NMR probe. The HPLC probe in principle can be adapted to handle streams of samples using a modified stop-flow procedure. The perfection of this technique will replace the automation systems as we presently know them and may render NMR tubes superfluous.

One aspect of combinatorial chemistry is the synthesis of mixtures of structurally related compounds. This facilitates a high throughput in both synthesis and screening. The split-and-mix synthesis technique produces a mixture of compounds as the final product (59,60). In developing the synthesis of large libraries, smaller test systems are studied to optimize synthetic strategies. The utility of NMR for studying intact mixtures has not been extensively demonstrated. 2D NMR methods such as TOCSY have been used in relatively simple mixture analysis (61). However, for compounds that have their spin systems insulated by groups such as esters or ethers, TOCSY methodologies are not sufficient for complete analysis.

Recently, the use of pulsed-field gradient (PFG) technology to obtain diffusion coefficients of molecules has been demonstrated as a useful technique for mixture analysis (53). Unlike any other 2D experiment, size-resolved or diffusion-resolved NMR assigns the resonances based on the diffusion coefficient for each proton (or other spin) in the molecule and therefore can be used to distinguish resonances arising from different molecules (63–70) (Fig. 22). A method that involves the use of PFG and TOCSY, called diffusion-encoded spectroscopy (DECODES), simplifies mixture analysis by NMR (71). The combination of PFG and TOCSY ''decodes'' the spin systems, allowing individual components in complicated mixtures to be assigned. A typical DECODES spectrum obtained in this manner is shown in Fig. 23. The use of TOCSY aids the calculation of the diffusion coefficient and determination of molecular identity.

The synthesis and screening of mixtures of compounds offers increased efficiency and throughput compared to making and testing individual compounds. However, utilization of mixtures of compounds requires a method to determine which molecule in the mixture is responsible for the desired effect. Typically, mixtures of compounds are prepared by design and these mixtures are tested without separation. When there is evidence of sufficient activity,

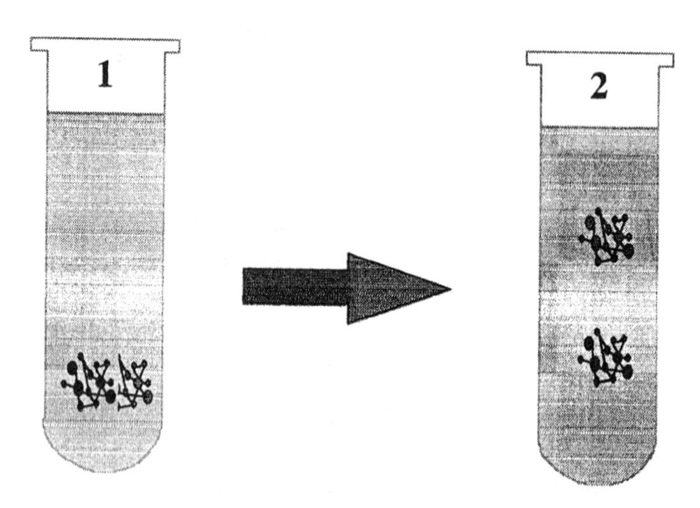

Figure 22 Schematic representation of molecular translation in solution. The DOSY experiment spatially encodes the molecule in 1 and then DECODES the movement in 2. The spectrum indicates the translational movement by a decreased observed signal intensity.

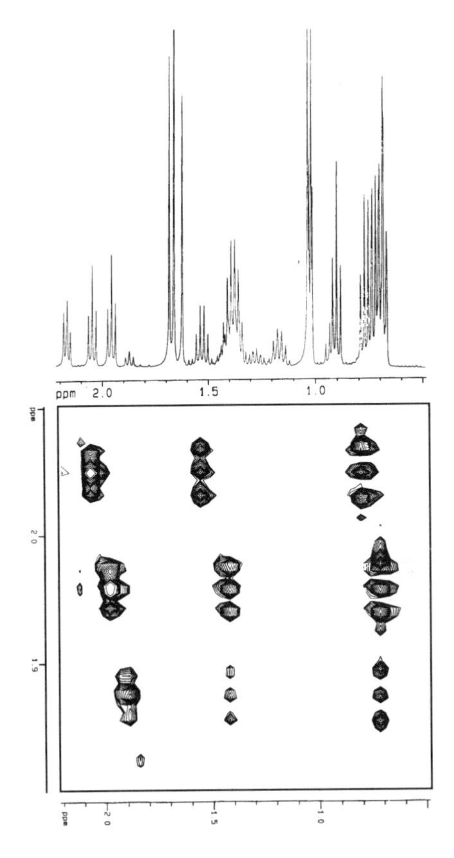

Figure 23 Selected region of TOCSY spectrum obtained via DECODES sequence of an ester mixture in benzene. Peaks arise from propyl acetate, ethyl butyryl acetate, and butyl levulinate.

the mixture is deconvoluted to identify the active component. Several approaches to identify interesting components in a mixture have been described (72–77). Methods that identify active components of mixtures without the need for deconvolution could eliminate ''false positives'' and greatly reduce the effort required to analyze mixtures. One such method under investigation is affinity MS (78–80).

A somewhat similar method, termed affinity NMR, has shown that the diffusion coefficient of a small molecule binding with a ''receptor'' in solution is significantly different from the small compound alone (81). Thus the inter-

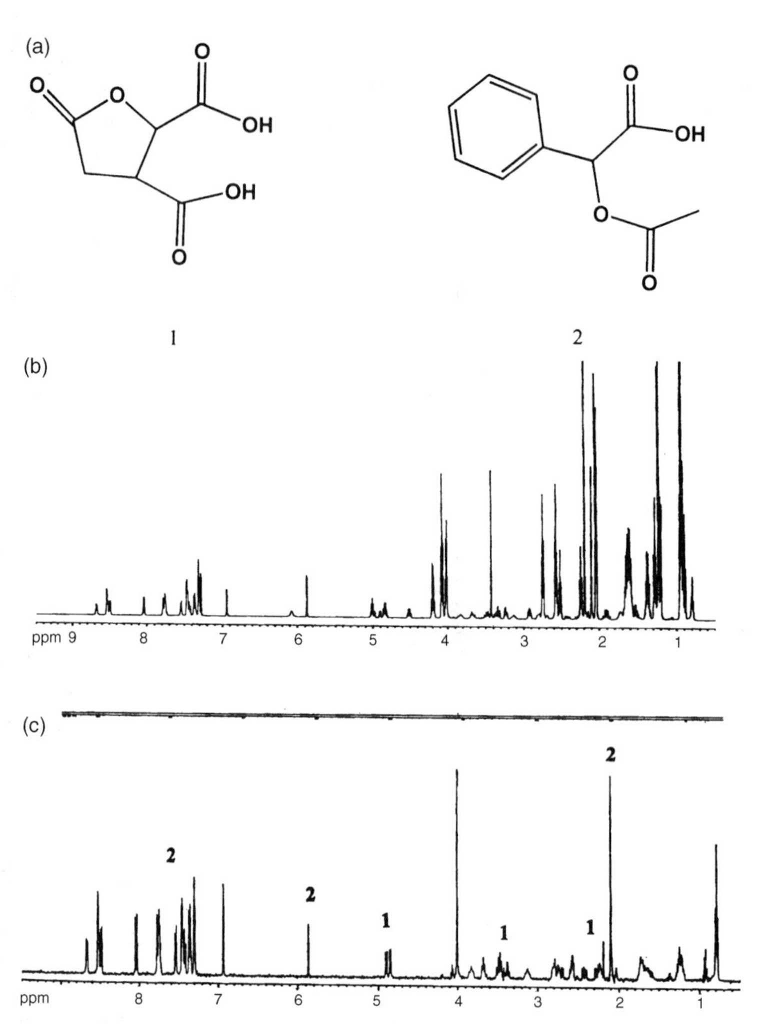

Figure 24 (a) 1D 400 MHz ^{1}H NMR spectrum of the nine-component mixture in $CDCl_3$. (1) DL-isocitric lactone, (2) (S)-(+)-*O*-acetylmandelic acid, (3) DL-*N*-acetylhomocysteine thiolactone, (4) (+)-*sec*-butyl acetate, (5) propyl acetate; (6) isopropyl butyrate, (7) ethyl butyryl acetate, (8) butyl levulinate, (9) hydroquinine-9-phenanthryl ether. (b) PFG 1D ^{1}H NMR spectrum of the mixture without hydroquinine-9-phenanthryl ether. (c) PFG 1D ^{1}H NMR spectrum of the nine component mixture using LED sequence. Chemical shifts arising from compounds 1 and 2 are shown. All other shifts are from compound 9. The PFG conditions were the same as in (b).

acting molecules can be distinguished from inert molecules in a manner reminiscent of physical separation of molecules by affinity chromatography. Affinity NMR, using hydroquinine 9-phenanthryl ether as a model receptor, was successfully applied to a nine-component mixture containing seven inert materials and the two carboxylic acids shown.

Figure 24 shows the normal 1D ^{1}H NMR spectrum for the nine component mixture; without PFG, a control experiment performed on the mixture in the absence of hydroquinine 9-phenanthryl ether under identical PFG conditions and the 1D ^{1}H NMR spectrum of the same mixture under the PFG conditions. Only signals from hydoquinine 9-phenanthryl ether and compounds 1 and 2 are observed in the bottom spectrum. As expected no NMR signals are present in the absence of molecular interactions. The structures of the compounds that interacted with hydroquinine 9-phenanthryl ether were identified directly in the mixture using the DECODES method.

Since the relatively high concentration of each component required by NMR adds up to a high total concentration of compounds for the mixture, the application of this methodology to screen combinatorial chemistry mixtures for biological activity will likely be limited by the total compound concentration tolerated by the biological target. Nevertheless, this NMR method, when applied to suitable systems, should add a powerful tool for mixture analysis.

C. SAR by NMR

A method for identifying active compounds from a library of low molecular weight ligands using ^{15}N-labeled proteins has been recently reported (82). The binding is determined by the ^{15}N or ^{1}H chemical shift changes in the protein upon the addition of the ligand. The method, which at present is limited to small biomolecular receptors, promises to play an integral part of the drug discovery process.

VII. FUTURE

''I am not aware of any other field of science outside of NMR that offers so much creative freedom and opportunity for a creative mind to invent and explore new experimental schemes that can be fruitfully applied in a variety of disciplines.''—Richard Ernst, Nobel Prize, 1993

REFERENCES

1. CA Fyfe. Solid State NMR for Chemists; CFC Press: Guelph, Ontario, 1983
2. B Yan, G Kumaravel, H Anjaria, A Wu, R Petter, C Jewell, J Wareing. Infrared spectrum of a single resin bead for real-time monitoring of solid-phase reactions. J Org Chem 60:5736–5738, 1995.
3. B Yan, G Kumaravel. Probing solid-phase reactions by monitoring the IR bands of compounds on a single ''flattened'' resin bead. Tetrahedron 52:843–848, 1996.
4. JA Boutin, P Hennig, PS Bertin, L Petit, J-P Mahieu, B Serkiz, J-P Volland, J-L Fauchere. Combinatorial peptide libraries: robotic synthesis and analysis by nuclear magnetic resonance, mass spectrometry, tandem mass spectrometry, and high performance capillary electrophoresis techniques. Anal Biochem 234:126–141, 1996.
5. BJ Egner, GJ Langley, M. Bradley. Solid phase chemistry: direct monitoring by matrix-assisted laser desorption/ionization time of flight mass spectrometry. A Tool for Combinatorial Chemistry. J Org Chem 60:2652–2653, 1995.
6. K McMellop, W Davidson, G Hansen, D Freeman, N Pallai. The characterization of crude products from solid-phase peptide synthesis by ν-HPLC/fast atom bombardment mass spectrometry. Peptide research 4:40–46, 1991.
7. H Sterlicht, GL Kenyon, EL Packer, J Sinclair. J Am Chem Soc 93:199–208, 1971.
8. R Epton, P Goddard, KJ Ivin. Gel phase 13C NMR spectroscopy as an analytical method in solid (gel) phase peptide synthesis. Polymer 21:1367–1371, 1980.
9. FGW Butwell, R Epton, EJ Mole, N Muzaffar, S Phillips. Deprotection studies in ultra-high load solid (gel) phase peptide synthesis. 13C NMR investigation of the efficacy of boron trifluoride-based side-chain deprotection cocktails. Innov Persp Solid Phase Synth Collect Pap Int Symp 121–132, 1990.
10. E Geralt, J Rizo, E Pedroso. Application of gel-phase 13C-NMR to solid-phase peptide Synthesis. Tetrahedron 40:4141–4152, 1984.
11. D Liebfritz, W Mayr, R Oekonomopulos, J Jung. 13C NMR spectroscopic studies on the conformation during stepwise synthesis of peptides bound to solubilizing polymer supports. Tetrahedron 34:2045-2050 1978.
12. SL Mannatt, D Horowitz, R Horowitz, RP Pinnell. Solvent swelling for enhancement of carbon-13 nuclear magnetic resonance spectral information from insoluble polystyrenes. Anal Chem 52:1529–1534, 1980.
13. AJ Jones, CC Leznoff, PI Svirskaya. Characterization of organic substrates bound to cross-linked polystyrenes by 13C NMR spectroscopy. Org Magn Reson 18:236–240, 1982.
14. BE C Lossey, RG Cannon, WT Ford, M Periyasamy, S Mohanraj. Synthesis, reactions, and 13C FT NMR spectroscopy of polymer-bound steroids. J Org Chem 55:4664–4668, 1990.

15. MM Azure, B Calas, A Cave, J Parello. Synthese peptidique en phase solide. Caracterisation par RMN du proton d'un peptide immobilise sur un support polyacrylique. C R Acad Sc Paris 303:553–556, 1986.
16. GC Look, CP Holmes, JP Chinn, MA Gallop. Methods for combinatorial organic synthesis: the use of fast 13C NMR analysis for gel phase reaction monitoring. J Org Chem 59:7588–7590, 1994.
17. GC Look, MM Murphy, DA Campbell, MA Gallop. Trimethylorthoformate: a mild and effective dehydrating reagent for solution and solid phase imine formation. Tetrahedron Lett 36:2937–2940, 1995.
18. SL Manatt, SF Amsden, CA Bettison, WT Frazer, JT Gudman, BE Lenk, JF Lubetich, EA McNelly, SC Smith, DJ Templeton, RP Pinnell. A fluorine-19 NMR approach for studying Merrifield solid-phase peptide synthesis. Tetrahedron Lett 21:1397–1400, 1980.
19. MJ Shapiro, G Kumaravel, RC Petter, R Beveridge. 19F NMR monitoring of a SNAr reaction on solid support. Tetrahedron Lett 37:4671–4674, 1996.
20. A Svensson, T Fex, J Kihlberg. Use of 19F NMR spectroscopy to evaluate reactions in solid phase organic synthesis. Tetrahedron Lett 37:7649–7652, 1996.
21. F Bardella, R Eritja, E Pedroso, E Geralt. Gel-phase 31P-NMR. A new analytical tool to evaluate solid phase oligonucleotide synthesis. Bioorg Med Chem Lett 3:2793–2796, 1993.
22. CR Johnson, B Zhang. Solid phase synthesis of alkenes using the Horner-Wadsworth-Emmons reaction and monitoring by gel phase 31P NMR. Tetrahedron Lett 36:9253–9256, 1995.
23. Z Tian, C Gu, RRW Roeske, M Zhou, RL Van Etten. Synthesis of phosphotyrosine-containing peptides by the solid-phase method. Int J Peptide Protein Res 42:155–158, 1993.
24. B Schneider, D Doskocilova, JJ Dybal. Motional restrictions and chain conformation in various swollen crosslinked polystyrene gels from 1H NMR line-shape analysis. J Polymer 26:253–259, 1985.
25. WL Fitch, G Detre, CP Holmes, JN Schoolery, P Keifer. High resolutions 1H NMR in solid phase organic synthesis. J Org Chem 59:7955–7956, 1994.
26. RC Anderson, MA Jarema, MJ Shapiro, JP Stokes, M Ziliox. Analytical techniques in combinatorial chemistry: MAS CH correlation in solvent-swollen resin. J Org Chem 60:2560–2651, 1995.
27. PA Keifer, L Baltosis, DM Rice, AA Tymiak and JN Shoolery. A comparison of NMR spectra obtained for solid-phase-synthesis resins using conventional high-resolution, magic-angle-spinning, and high-resolution magic-angle-spinning probes. J Magn Reson A 119:65–75, 1996.
28. P Keifer. Influence of resin structure, tether length, and solvent upon the high-resolution 1H NMR spectra of solid-phase-synthesis resins. J Org Chem 61: 1558–1559, 1996.
29. IE Pop, CF Dhalluin, BP Deprez, PC Melnyk, GM Lippens, AL Tartar. Monitoring of a three-step solid phase synthesis involving a heck reaction using magic angle spinning NMR spectroscopy. Tetrahedron 52: 12209–12222, 1996.

30. MJ Shapiro, unpublished work.
31. EL Hahn, Spin echoes Phys Rev 80:580 1950.
32. DL Rabinstein, KK Mills, EJ Strauss. Proton NMR spectroscopy of human blood plasma and red blood cells. Anal Chem 60:1380A–1391A, 1988.
33. T Wehler, J Westman. Magic angle spinning NMR: a valuable tool for monitoring the progress of reactions in solid phase synthesis. Tetrahedron Lett 37:4771–4774, 1996.
34. HDH Stover, JM Frechet. Direct polarization 13C and 1H magic angle spinning NMR in the characterization of solvent-swollen gels. J Macromolecules 22: 1574–1576, 1989.
35. HDH Stover, JM Frechet. NMR characterization of cross-linked polystyrene gels. J Macromolecules 24:883–888, 1991.
36. E Giralt, F Alberico, F Bardella, R Eritja, M Feliz, E Pedroso, M Pons, J Rizo. Gel-phase NMR spectroscopy as a useful tool in solid phase synthesis. Innov Persp Solid Phase Synthesis Collect. Pap., Int Symp. R. Epton (ed.) 111–120, 1990.
37. RC Anderson, MJ Shapiro, JP Stokes. Structure determination in combinatorial chemistry: utilization of magic angle spinning HMQC and TOCSY NMR spectra in the structure determination of Wang-bound lysine. Tetrahedron Lett 36:5311–5314, 1995.
38. IE Pop, CF Dhalluin, BP Deprez, PC Melnyk, GM Lippens, AL Tartar Monitoring of a three-step solid phase synthesis involving a heck reaction using magic angle spinning NMR spectroscopy. Tetrahedron 52:12209–12222, 1996.
39. A Kumar, RR Ernst, K Wuthrich. A two-dimensional nuclear overhauser enhancement (2D NOE) experiment for the elucidation of complete proton-proton cross-relaxation networks in biological macromolecules. Biochem Biophys Res Commun 95:1–6, 1980.
40. WP Aue, E Bartholdi, RR Ernst. Two-dimensional spectroscopy. Application to nuclear magnetic Resonance. J Chem Phys 64:2229–2246, 1976.
41. MJ Shapiro, J Chin, RE Marti, MA Jarosinski. Enhanced resolution in MAS NMR for combinatorial chemistry. Tetrahedron Lett 38:1333–1336, 1997.
42. M Patek, M Lehl. Safety-catch anchoring linkage for synthesis of peptide amides by Boc/Fmoc Strategy. Tetrahedron Lett 32:3891–3894, 1991.
43. A Meissner, P Bloch, E Humpfer, M Spraul, OW Sorensen. Reduction of inhomogeneous line broadening in two dimensional high-resolution MAS NMR spectra of molecules attached to swelled resins in solid-phase synthesis. J Am Chem Soc 119:1787–1788, 1997.
44. K Nagayama, K Wuthrich, RR Ernst. two-dimensional spin echo correlated spectroscopy (SECSY) for 1H NMR studies of biological macrcomolecules. Biochem Biophys Res Commun 90:305–311, 1979.
45. Y Kim, JH Prestegard. Measurement of vicinal couplings from cross peaks in COSY spectra, J Magn Reson 84:9–13, 1989.
46. WE Mass, FH Laukien and DG Cory. Gradient, high resolution, magic angle sample spinning NMR. J Am Chem Soc 118:13085–13086, 1996.

47. The diagonal peaks lie along the F2 at F1 = 0 axis and represent the chemical shift as in a COSY spectrum. The F1 axis represents one-half of the chemical shift difference between the coupled spins.

48. S Sarkar, RS Garigipati, JL Adams, PA Keifer. An NMR method to identify nondestructively chemical compounds bound to a single solid-phase-synthesis bead for combinatorial chemistry applications. J Am Chem Soc. 118:2305–2306, 1996.
49. A Borchard, WC Still. Synthetic receptor binding elucidated with an encoded combinatorial library. J Am Chem Soc 116:373–374, 1994.
50. RM Valerio, AM Bray, RA Campbell, A Dipasquale, C Margellis, S. J. Rodda, HM Geysen, NJ Maeji. Multipin peptide synthesis at the micromole scale using 2-hydroxyethyl methacrylate grafted polyethylene supports. Int J Peptide. Protein Res 42:1–9, 1993.
51. HU Gremlich, SL Berets. Use of FT-IR internal reflection spectroscopy in combinatorial chemistry applied spectroscopy 50:532–536, 1996.
52. WP Aue, E Bartholdi, RR Ernst. Two dimensional spectroscopy. Application to nuclear magnetic resonance. J Chem Phys 64:2229, 1976.
53. M Spraul, M Hoffman, P Dvortsak, JK Nicholson, ID Wilson. High-performance liquid chromatography coupled to high-field proton nuclear magnetic resonance spectroscopy: application to the urinary metabolites of ibuprofen. Anal Chem 65:327–330, 1993.
54. K Albert Angew. Direct on-line coupling of capillary electrophoresis and 1H NMR spectroscopy. Chem Int Ed Engl 34:641–642, 1995.
55. DL Olson, TL Peck, AG Webb and JV Sweedler. On-line NMR detection for capillary electrophoresis applied to peptide analysis. Peptides: Chem Struct Biol PTP Pravin, RS Hedges, (eds.). Mayflower Scientific, Ltd. 730–731, 1996.
56. JC Lindon, RD Farrant, PN Sanderson, PM Doyle, SL Goughg, M Spraul, M Hoffman, JK Nicholson. Separation and characterization of components of peptide libraries using on-flow coupled HPLC-NMR spectroscopy. Magn Reson Chem 33:857–863, 1995.
57. S Smallcombe, SL Patt, PA Keifer. WET solvent suppression and its applications to LC NMR and high-resolution NMR spectroscopy. J Magn Reson A 117:295–303, 1995.
58. C Dalvit, JM Bohlen. Multiple-solvent suppression in double-quantum NMR experiments with magic angle pulsed field gradients. Magn Reson Chem 34:829–833. 1996.
59. A Furka, F Sebestyan, M Asgedom, G Dibo. General method for rapid synthesis of multicomponent peptide mixtures. Int J Peptide Protein Res 37:487–493, 1991.
60. F Sebestyen, G Dibo, A Kovacs, A Furka. Chemical synthesis of peptide libraries. Bioorg Med Chem Lett 3:413–418, 1993.
61. K Johnson, LG Barrientos, PPN Murthy. Application of two-dimensional total correlation spectroscopy for structure determination of individual inositol phosphates in a Mixture. Anal Biochem 231:421–431, 1995.

62. P Stilbs, K Paulsen, PC Griffiths. Global least squares analysis of large, correlated spectra data sets. Application to component-resolved FT-PSGE NMR spectroscopy. J Chem Phys 100:8180–8189, 1996.
63. SJ Gibbs, CS Johnson Jr. A PFG NMR experiment for accurate diffusion and flow studies in the presence of eddy currents. J Magn Reson 93:395–402, 1991.
64. KF Morris, CS Johnson Jr. Diffusion-ordered two-dimensional nuclear magnetic resonance spectroscopy. J Am Chem Soc 114:3139–3141, 1992.
65. KF Morris, CS Johnson Jr. Resolution of discrete and continuous molecular size distributions by means of diffusion-ordered 2D NMR spectroscopy. J Am Chem Soc 115:4291–4299, 1993.
66. KF Morris, P Stilbs, CS Johnson Jr. Analysis of mixtures based on molecular size and hydrophobicity by means of diffusion-ordered 2D NMR. Anal Chem 66:211–215, 1994.
67. M Lin, DA Jayawickrama, RA Rose, JA Delvisio, CK Larive. Nuclear magnetic resonance spectroscopy analysis of the selective complexation of the cis and trans isomers of phenylalanylproline by B-cyclodextrin. Anal Chim Acta 307: 449–457, 1995.
68. M Liu, JK Nicholson, JC Lindon. High-resolution diffusion and relaxation edited one- and two-dimensional 1H NMR spectroscopy of biological fluids. Anal Chem 68 3370–3376, 1996.
69. N Birlirakis, E Guittet. A new approach in the use of gradients for size-resolved 2D-NMR experiments. J Am Chem Soc 118:13083–13084, 1996.
70. H Barjat, GA Morris, S Smart, AG Swanson, SC Williams. High-resolution diffusion-ordered 2D spectroscopy (HR-DOSY)—A new tool for the analysis of complex mixtures. J Magn Reson Series B 108:170–172, 1995.
71. M Shapiro, M Lin. Mixture analysis in combinatorial chemistry. Application of diffusion-resolved NMR spectroscopy. J Org Chem 61:7617–7619 1996.
72. CT Dooey, NN Chung, BC Wilkes, PW Schiller, JM Bidlack, GW Pasternak, RA Houghten. An all D-amino acid opioid peptide with central analgesic activity from a combinatorial library. Science 266:2019–2022, 1994.
73. K Burgess. Combinatorial technologies involving reiterative division/coupling/ recombinations: statistical considerations. J Med Chem 37:2985–2987, 1994.
74. E Erb, KD Janda, S Brenner. Recursive deconvolution of combinatorial chemical libraries. Proc Nat Acad Sci USA 91:11422–11426, 1994.
75. DJ Ecker, TA Vickers, R Hanecak, V Driver, K Anderson. Rational screening of oligonucleotide combinatorial libraries for drug discovery. Nucleic Acids Res 21:1853–1856, 1993.
76. R Fathi, MJ Rudolph, RG Gentles, R Patel, EW MacMillan, MS Reitman, D Pelham, AF Cook. Synthesis and properties of combinatorial libraries of phosphoramidates. J Org Chem 61:5600–5609, 1996.
77. S Borman, C&E N, 29–54, 1996.
78. RS Youngquist, GR Fuentes, MP Lacey, T Keough. Generation and screening of combinatorial peptide libraries designed for rapid sequencing by mass spectrometry. J Am Chem Soc 117:3900–3906, 1995.

79. X Cheng, R Chen, JE Bruce, BL Schwartz, GA Anderson, HSA Ofstadler, DC Gale, RD Smith, J Gao, GB Sigal, M Mammen, GM Whitesides. Using electrospray ionization FTICR mass spectrometry to study competitive binding of inhibitors to carbonic anhydrase. J Am Chem Soc 117:8859–8860, 1995.
80. Y-H Chu, YM Dunayevskiy, DP Kirby, P Vouros, BL Karger. Affinity capillary electrophoresis–mass spectrometry for screening combinatorial libraries. J Am Chem Soc 118:7827–7835, 1996.
81. M Lin, MJ Shapiro, JR Wareing. Diffusion-edited NMR—affinity NMR for direct observation of molecular interactions. J Am Chem Soc 119:5249–5250, 1997.
82. SB Shuker, PJ Hajduk, RP Meadows, SW Fesik. Discovering high-affinity ligands for proteins: SAR by NMR. Science 274:1531–1534, 1996.

5
The Role of Liquid Chromatography

Michael E. Swartz
Waters Corporation
Milford, Massachusetts

I. INTRODUCTION

Chromatography alone, or in combination with other analytical techniques, has been used for a number of years in the drug discovery process in support of traditional organic synthesis for compound identification, compound purity and stability determinations, from lead discovery to final lead optimization, testing, and candidate selection. However, in response to increasing demands in the pharmaceutical industry to accelerate the drug discovery process and identify lead compounds in increasing numbers, new avenues of approach, such as combinatorial chemistry, must be investigated.

Combinatorial chemistry synthesis techniques have presented new challenges to the analytical chemist. During lead discovery, libraries of large numbers of compounds, numbering from 10–20, to hundreds, thousands, ten of thousands, or even millions of compounds are generated. Therefore, due to the sheer numbers of compounds, assays must be rapid, as well as capable of determining quantity, purity, and whether or not the proper compound was synthesized. Further along the drug discovery path, during lead optimization and testing leading to candidate selection, compounds are required in larger quantities. Analytical techniques used in lead discovery now give way to preparative, mass-directed autopurification techniques, capable of isolating and purifying 10–20 mg of the compound of interest during a single chromatographic analysis. Furthermore, all of the assays, from the analytical to preparative scale, must be accessible to everyone in the drug discovery process, in what has become to be known as an ''open access'' environment. In response to these challenges, chromatographers have had to rethink their strategy in order to provide timely and complete information and feedback.

In order to be successful, a proper method, like any other type of assay, run on the appropriate instrument, must be employed. In addition, software, both for systems operation/integration and data analysis, also plays a central role. This chapter will address how chemists, using chromatography, have adapted in answering the challenges presented by a combinatorial chemistry program of drug discovery. While not intended to be an exhaustive review, this chapter focuses predominantly on liquid chromatography (LC), discussing various applications and techniques that highlight its use in the drug discovery process.

Since application of fast, generic LC methods with mass spectrometry (LC/MS) is emerging as the technique of choice for assessing the progress and final quality of large combinatorial arrays in drug discovery, it will be discussed in some detail, along with other detection techniques. Mass-directed purification and characterization on the preparative scale will also be addressed.

II. BASIC HIGH-THROUGHPUT ANALYSIS AND CHARACTERIZATION

A. Chromatographic Optimization and Injection Overhead Reduction

The first step in drug discovery is lead identification before high-throughput screening. This step is characterized by use of combinatorial libraries varying in size from a few hundred to tens of thousands (or more) compounds. Assays should provide simple, basic information such as compound identification on a molecular weight basis, and the synthesis yield (purity). In order to answer these questions, methods must satisfy certain requirements. Gradient reverse phase liquid chromatography (RP-LC) with various detection modes can satisfy these requirements and is currently the method of choice for the analysis of combinatorial libraries and synthesis-related products (1–5).

The sheer number of samples, their diversity, and a lack of suitable standards for quantitation can at first seem an insurmountable challenge. Since method development for individual samples is not feasible due to time and economic factors, a broadly applicable ''generic'' method must be developed, without sacrificing information content. Also, an inject-to-inject cycle time of

2–5 min (or less!) is desired to accomplish the required sample throughput. Methods and instruments also need to be compatible with various detection methods in addition to MS, such as photodiode array (PDA), evaporative light scattering, and nitrogen- and sulfur-specific chemiluminescence detection. Finally, it is important to keep in mind that once a lead has been identified, it may be necessary to scale up the separation (with instrument and chemistry implications) to isolate larger amounts of pure material for additional studies.

Chromatographic instruments used in support of combinatorial chemistry require features somewhat different from those of traditional systems (5–8). The basic components of a system are the same: a solvent manager (pump), a sample manager (autosampler or injector), a detector(s), a data system, and, in some instances, a fraction collector. However, some important differences exist. The most significant difference is the sample manager. Samples can be presented in standard vials, tubes, or microtiter plates of various sizes and capacities, either singly or in various combinations. Therefore some type of ''XYZ'' sample management device is dictated. Other important differences include software instrument access and control, and data reduction and reporting.

Let's examine a typical hypothetical situation. Assume an analyst has just received a combinatorial library on six high-density 384-well microtiter plates, for a total of 2304 samples of which all are unknown, all are different, and all differ in the degree of purity, for analysis in the lab. The traditional approach would be to use a gradient RP-LC method similar to that presented in Fig. 1. Using a long column and a shallow gradient is typically the first step in developing and optimizing a method. However, given the number of compounds involved, individual method development is highly impractical. Under the conditions used in Fig. 1, including 20 min of postrun reequilibration time, analysis of the 2304 samples would take more than 96 days! Although multiple systems could be used to analyze the samples, additional steps must be taken to improve sample throughput and to maintain final method detector compatibility in a full-time combinatorial chemistry support laboratory.

Several options, either alone or in combination, can be employed to improve sample throughput. Obvious options include using shorter columns, higher flow rates and temperatures, and/or sacrificing resolution. Figure 2A shows a separation of the same sample illustrated in Fig. 1, but with a smaller column at higher flow rates. The chromatographic test sample used spans a wide elution range; however, the peak capacity of the method remains high. Under these conditions, the total time for the analysis of the 2304 samples was decreased by an order of magnitude to less than 8 days. The column used

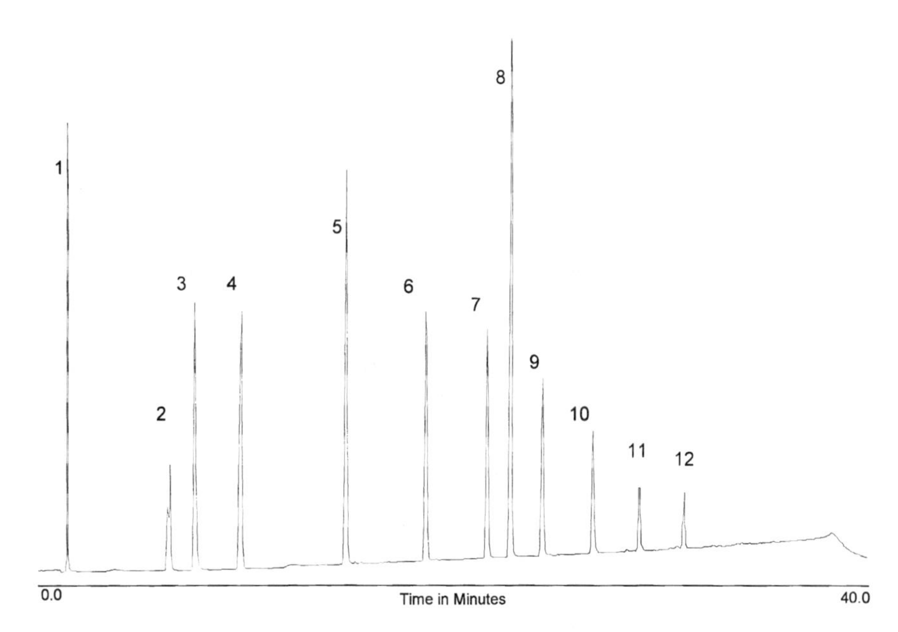

Figure 1 Separation of a chromatographic test mixture by a traditional gradient method. Separation was performed on Waters Alliance® HPLC System (Waters Corporation, Milford, MA) and a 3.9 by 150mm 5 micron particle size Symmetry® C18 column at 30°C. The mobile phase consisted of 0.1% phosphoric acid as the A solvent, and acetonitrile as the B solvent, run as a linear gradient from 0–80% B over 40 minutes at 1.0 mL/min. Total analysis time does not include twenty minutes of post run re-equilibration of the column and system. UV detection at 254nm, and a 20 μL injection was used. Peaks 1–12 (0.1mg/mL each in 50/50 methanol/water) are uracil, theophyline, acetylfuran, acetanilide, acetyl-, propio-, butyro-, benzo-, valero-, hexano-, heptano-, and octano-phenone, respectively.

to generate the example chromatogram shown in Fig. 2A had a smaller particle size (3.5 μm) and an increased internal diameter was used (4.6 mm) to accommodate the increased flow rate of 4 mL/min. A formic acid instead of a phosphate-buffered mobile phase was used for better MS compatibility.

Besides using a lot of mobile phase, analyses run at these high flow rates are not compatible with an MS detector without some sort of split to divert flow. For many LC analyses, flow splitters are a fact of life when using MS detection. However, Fig. 2B illustrates that separation of the chromatographic test sample can be further scaled to a 2.1-mm i.d. column, now running at an equivalent linear velocity (1.0 mL/min). Under the conditions used in

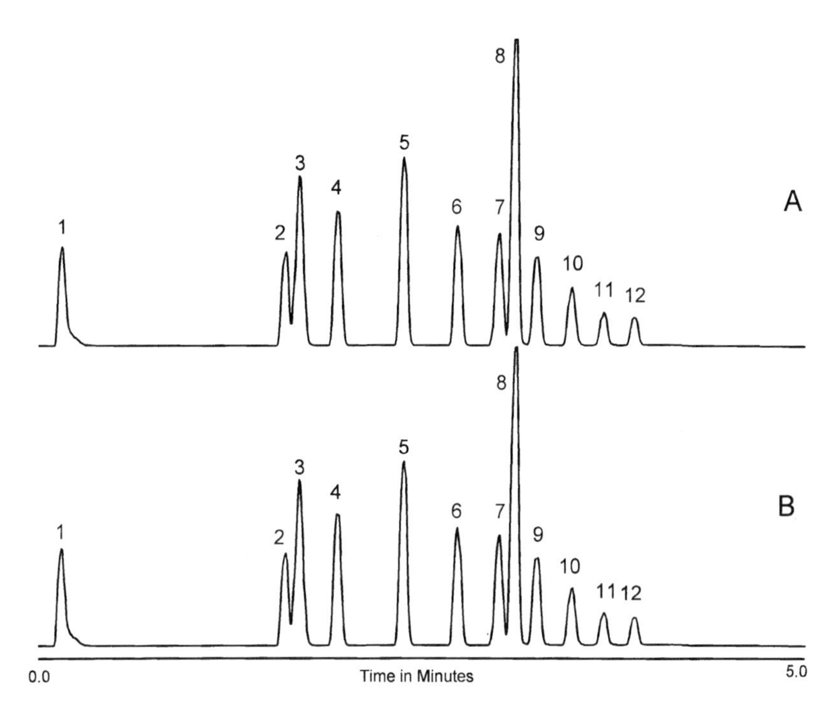

Figure 2 (A) Generic method optimized for high-throughput. Separation was performed on a 4.6 by 50mm 3.5 micron particle size Symmetry® C18 column (Waters Corporation, Milford, MA) at 30°C. The mobile phase consisted of 0.1% formic acid as the A solvent, and acetonitrile as the B solvent, run as a linear gradient from 0–100% B over 3 minutes at 4.0 mL/min. All other conditions (injection, detection, and sample) were identical to those listed in Figure 1. (B) Chromatography optimized by APCI/MS detection. Separation was performed on a 2.1 by 50mm 3.5 micron particle size Symmetry® C18 column (Waters Corporation, Milford, MA) at 1.0 mL/min. and at 30°C. Mobile phase and gradient conditions were identical to Figure 2A. All other conditions (injection, detection, and sample) were identical to those listed in Figure 1.

Fig. 2B, atmospheric pressure ionization MS can be used directly without splitting the flow. However, depending on the compound class or type in the library being analyzed, electrospray ionization might be preferred and in some cases might still necessitate a split flow, depending on the flow rate and mobile phase composition.

Figures 3 and 4 illustrate the diversity of the method. In Fig. 3, six penicillin-type antibiotics representing a somewhat more realistic sample are separated. Peak 1 is the synthetic precursor for the rest of the compounds in the

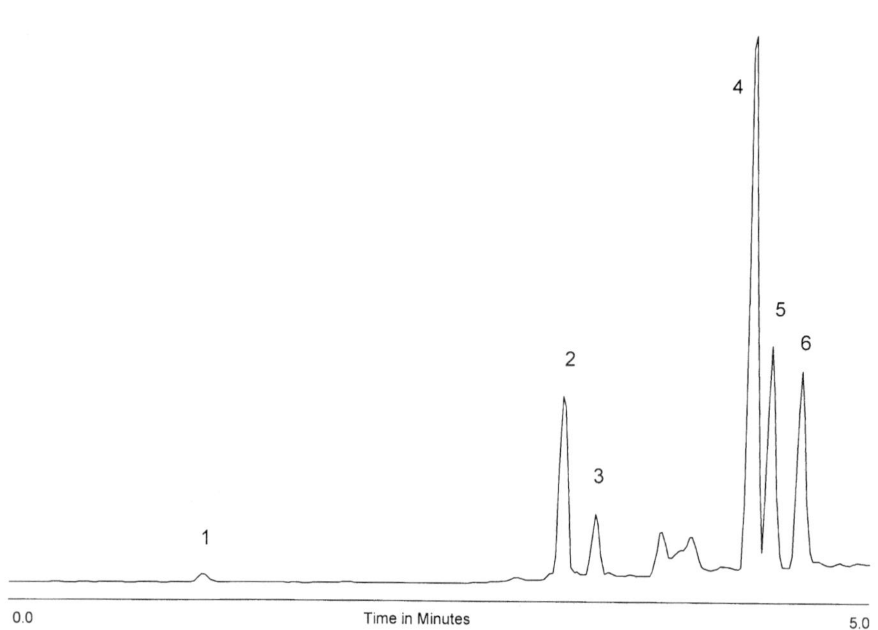

Figure 3 Separation of penicillin type antibiotic homologous series. Separation conditions are identical to those reported in Figure 2B. Peaks 1–6 are: 6-aminipenicillanic acid, amoxicillin, ampicillin, oxicillin, cloxicillin, and dicloxicillin, respectively.

mixture, and peaks 4–6 differ only by a chlorine atom, representing possible synthesis byproducts. The generic method readily resolves all of the components in this series of homologous compounds, in spite of the wide polarity range. Figure 4 shows the separation of another test mix of ''drug-like'' compounds reported in the literature for use as a chromatographic test mixture (8).

Although the improvements illustrated in Figs. 2–4 are significant, it is possible to improve sample throughput even more dramatically. When determining sample throughput capabilities for short run times, additional system timing issues must be taken into account. That is, to determine actual sample throughput, cycle-to-cycle inject times must be determined that take into account all other aspects of each individual sample analysis. Run time, while important, is only one factor in determining overall cycle-to-cycle inject times. The amount of time it takes to position the injector, aspirate a sample, load the sample loop, inject the sample, and rinse the injector apparatus is referred to as ''injection overhead.'' While injection overhead varies from instrument to instrument, times from 1 to 2 min per injection cycle are not uncommon. In

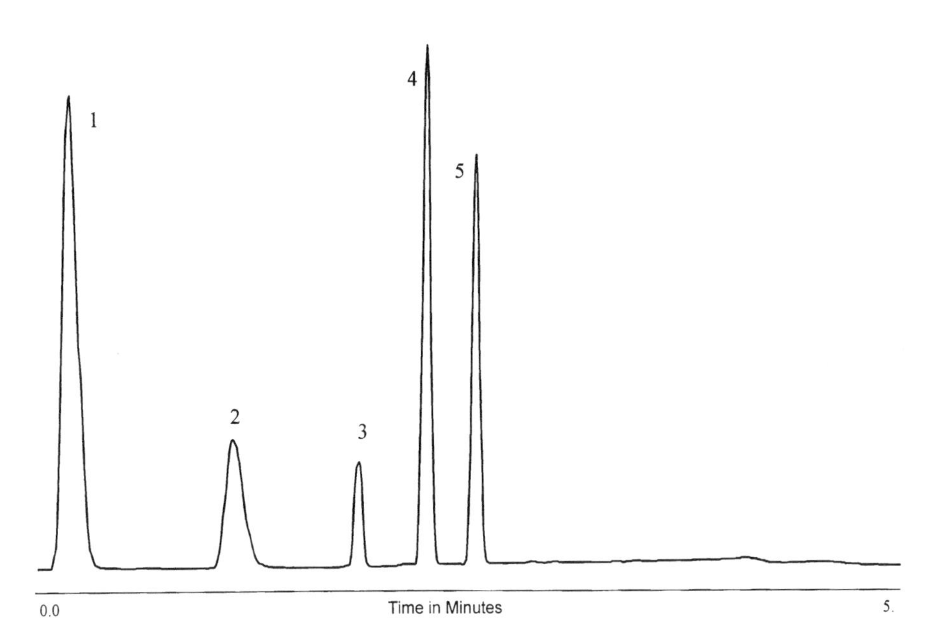

Figure 4 Separation of chromatographic test mix showing generic method diversity. Separation conditions are identical to those reported in Figure 2B. Peaks 1–5 are uracil, 1-hydroxy-7-azabenzotriazole, methoxybenzenesulfonamide, methyl-3-amino-2-thiophenecarboxylate, and 4-aminobenzophenone, respectively.

a 40-min run, with 15–20 min of post run time, 1–2 min is hardly significant. However, with run times as short as 5 min, the injection overhead can add 20% or more to the cycle-to-cycle inject time. Another parameter that must be considered in the cycle-to-cycle inject time is the time it takes to reequilibrate the column and the system following the gradient. Each of these functions is illustrated in Fig. 5 for an LC system operated in the traditional ''sequential'' mode. Obviously, anything that can be done to reduce the time it takes to accomplish these tasks will also improve sample throughput. As shown in Fig. 6, a separation of an 11-component mix in the sequential mode with a 5-min run time can consume an additional 1.2 min due to injection overhead, for a total cycle to cycle inject time of 6.2 min. If, however, some of these functions can be performed in a ''parallel'' mode, as illustrated in Fig. 7 (p. 122), additional time savings could be realized. In the parallel mode of operation, sample-manager needle and inject port washes, and aspiration of the next sample takes place during the actual run, saving considerable time.

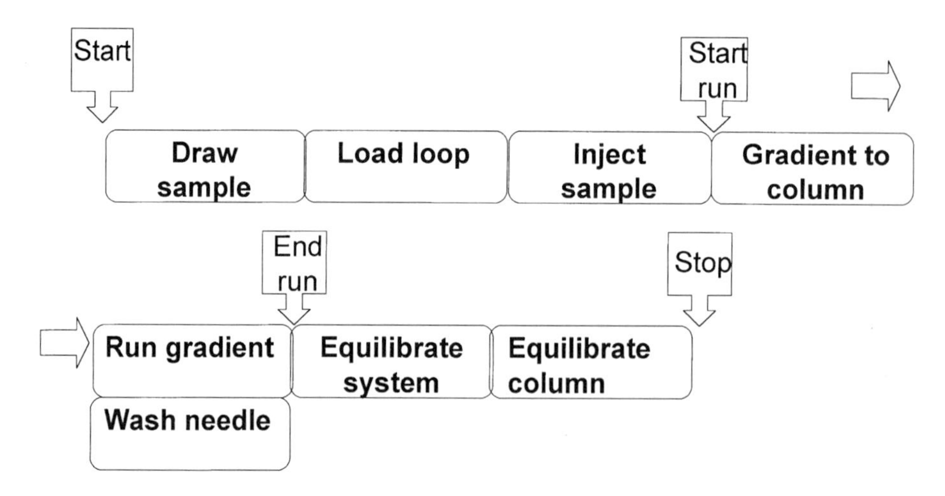

Figure 5 Schematic of traditional sequential mode injection to injection cycle time (injection overhead) operation of an LC system.

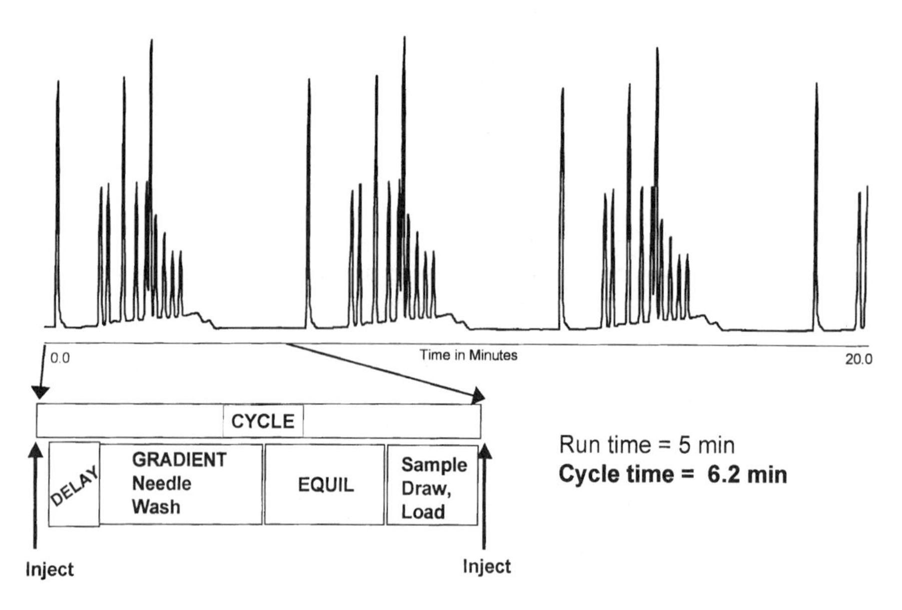

Figure 6 Traditional sequential mode injection showing cycle to cycle inject time. Separation conditions and peak identification are identical to those reported in Figure 2B. Data system run time was extended to 20 minutes to capture LC instrument timing.

Equilibrating the system separately from the column can also save additional time. Equilibration time is an essential part of gradient chromatography. Both the LC system and the column must be returned to initial mobile phase conditions prior to the next run to ensure repeatability. Traditionally, equilibration times of 5–10 column volumes have been used (9). However, with the use of smaller diameter and shorter LC columns, a 10-column volume is not always enough to equilibrate both the LC system and the column. To compensate for this situation, the reequilibration volume has been divided into two parts: the volume required for system equilibration and the volume required for column equilibration (10). Total equilibration time is then given by the formula:

$$T_r = (3V_T + 5V_C)/F$$

where T_r is the equilibration time (min), V_T is the total system volume, V_C is the column volume (mL), and F is the flow rate (mL/min). As the column gets smaller (with a correspondingly lower flow rate), more equilibration time is taken up by returning the system volume of the LC to initial conditions. In advanced LC systems with integrated fluidics and control, a purge step can be added post run at high flow rates (5–7 mL/min) with the column off-line to significantly reduce system equilibration times. By starting the gradient but holding off on the injection for the amount of time proportional to the system volume, the volume of the system can also be used to aid in column equilibration—a technique referred to as a ''just-in-time gradient.'' Loading the sample loop, as illustrated in Fig. 8 (p. 123), during the equilibration saves additional time by further reducing injection overhead. Comparing Figs. 6 and 8, the parallel mode with rapid equilibration increases throughput by about 30% without sacrificing chromatographic information or integrity.

Rapid equilibration techniques are particularly helpful to save time as the flow rate decreases, in, for example, microbore applications. Figure 9 (p. 124) shows a separation on a 1 × 50 mm column, at 0.3 mL/min, with a cycle-to-cycle inject time of 5.5 min using the parallel mode with rapid equilibration. Run in the traditional sequential manner, the corresponding cycle-to-cycle inject time would be in excess of 7.5 min. At lower flows, the time saving during reequilibration would be even more significant.

Some general notes on operating LCs in a high-throughput mode are as follows:

With the trend toward smaller columns, extra care should be taken to minimize extra column band broadening by using reduced diameter tubing, smaller volume detector cells, and properly fitting connections (9).

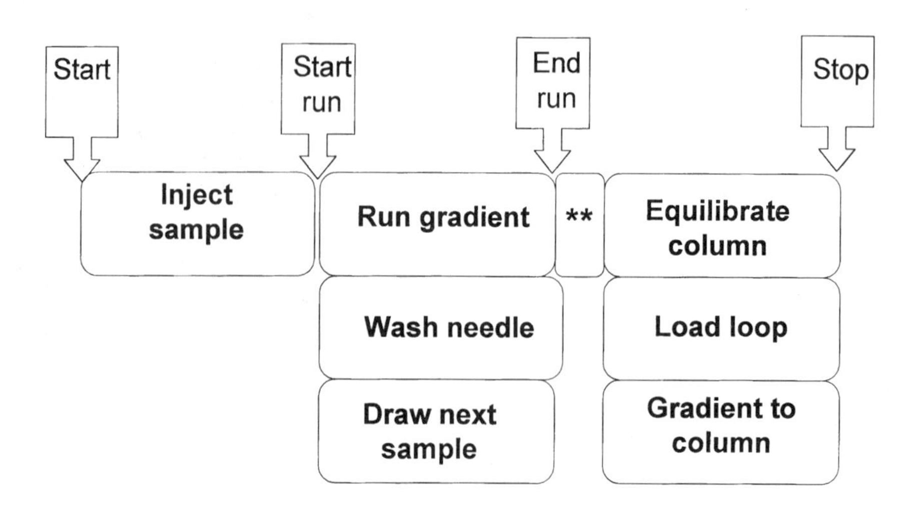

Figure 7 Schematic of parallel mode operation with rapid equilibration.

Photolabile or temperature-sensitive compounds may require special handling. LC systems are available that maintain samples in the dark and/or chilled if required.

The small peak volumes and resolution demands from high-throughput analyses require faster data collection rates for accurate results. Adequate detector time constants should also be used.

III. ADVANCED HIGH-THROUGHPUT ANALYSIS TECHNIQUES

A. Sample Pooling

Sample pooling can also be used to reduce analysis times. Pooling is commonly used in drug discovery and early development for pharmacokinetic screening in animals, intestinal permeability screening using Caco-2 cell–based assays, and combinatorial library analyses. Sample pooling is an injection technique that uses multiple sample aspirations and then coinjects the samples, decreasing analysis times by reducing the number of chromato-

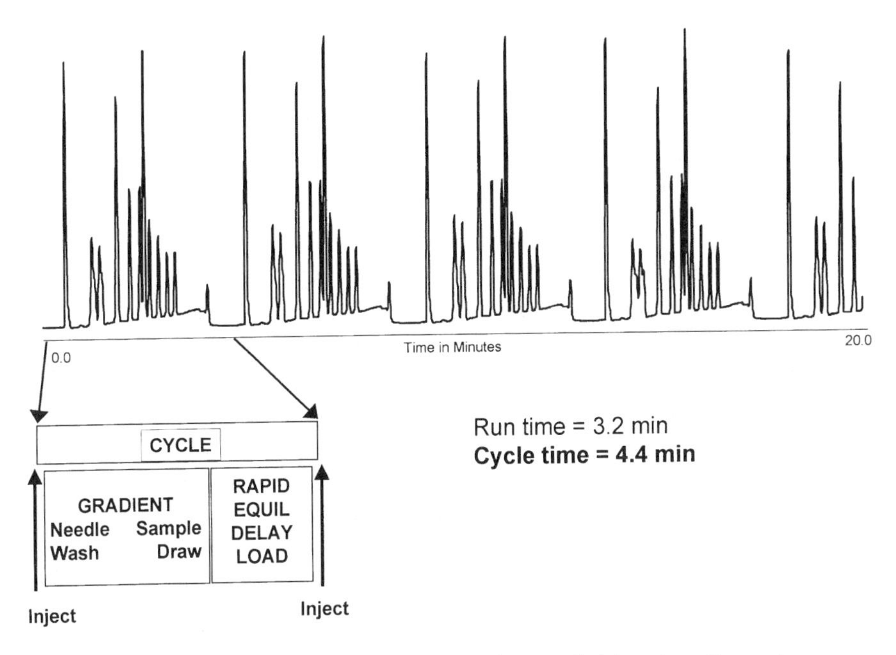

Figure 8 Parallel mode injection showing cycle to cycle inject time. Separation conditions and peak identification are identical to those reported in Figure 2B. Data system run time was extended to 20 minutes to capture LC instrument timing. Note differences in cycle to cycle timing relative to Figure 6.

graphic runs. It is performed as diagrammed in Fig. 10 (p. 126) either, for example, within a microtiter plate (intraplate) or between plate (interplate) pooling, with mixing or air or solvent segmentation. Figure 11 (p. 128) illustrates the results of an interplate pooling experiment using two compounds and two internal standards, run parallel with rapid equilibration mode conditions used in Fig. 6. The technique is quantitative (good accuracy and precision), as summarized by the data shown in Table 1. In this simple experiment, sample throughput is doubled without compromising inject-to-inject cycle time or injection overhead.

B. Two-Column Regeneration

A technique referred to as two-column regeneration has also been used to increase sample throughput. In this technique, while one LC column is running the gradient method, a second isocratic pump is used to equlibrate a second

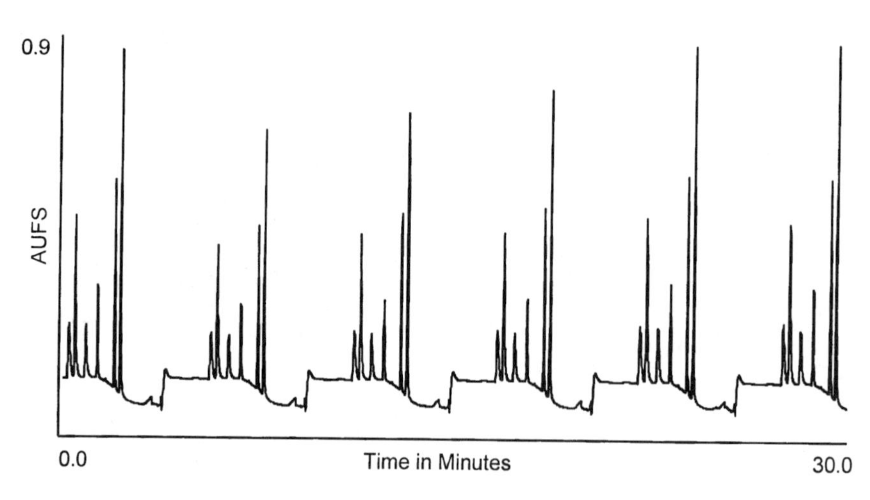

Figure 9 A μbore separation cycle time comparison. Separation was performed on a Waters Alliance® HT HPLC System (Waters Corporation, Milford, MA) using a 1.0 by 50mm 3.5 micron particle size Symmetry® C18 column (Waters Corporation, Milford, MA) at 60°C. The mobile phase consisted of 0.1% phosphoric acid as the A solvent, and acetonitrile as the B solvent, run as a linear gradient from 10–95% B over 2 minutes at 300 μL/min. UV detection at 220nm, and a 2 μL injection was used. Peaks 1–6 (0.1mg/mL each in 50/50 methanol/water) are uracil, caffeine, primidone, phenacetin, benzophenone and biphenyl, respectively.

Table 1 Quantitative Results of Interplate Pooling Experiment from Figure 9 (n = 6)

	A/IS	B/IS	A/IS Pooled	B/IS Pooled
Avg. ratio	1.129	2.632	1.123	2.618
%RSD	0.55	0.52	0.21	0.11
Avg. cycle time (min.)	4.68	4.69	4.69	
%RSD	0.30	0.08	0.10	

column to initial gradient conditions. After the analysis is complete, the columns are switched, and the first column is reequilibrated while the second column is run. Postrun equilibration time can be decreased using this technique on a traditional LC system; however, when modern LC systems are used that operate in the parallel rapid equilibration mode outlined above, any throughput advantages from the two-column regeneration technique may be minimal on the analytical scale. Throughput can, however, be increased by as much as 25% on the preparative scale (11).

C. Automated Multichannel Chromatography

Another way to save time is to use multiple separation and analysis channels on the same instrument. Recently, new technology was reported that allows multiple samples to be analyzed in parallel as illustrated in Fig. 12 (p. 129) (12). This technique allows for four LC columns running identical gradients in parallel on a single LC system to be multiplexed to one MS detector. In the LC/MS interface the traditional electrospray probe and outer source assembly are replaced by an array of four miniaturized, pneumatically assisted electrosprays. The interface has a sampling rotor that is monitored in real time enabling the four liquid inlets to be indexed. Four separate chromatograms are collected into four separate data files, allowing conventional data analysis, as presented in Fig. 13 (p. 130). Depending on LC conditions, analysis time for an entire 96-well microtiter plate can be reduced to as little as 1.1 h, resulting in sample throughput approaching 2000 samples per day.

D. Auxiliary Detection

As noted previously, fast, generic LC/MS methods are the techniques of choice for assessing the progress and final quality of large combinatorial arrays in drug discovery. However, ultraviolet (UV) or MS detection alone cannot always provide the type of quantitative data required when assessing compound purity. To obtain accurate quantitative data, UV detection requires the use of well-characterized reference standards due to the differences in molar absorptivity that may exist between members of a combinatorial library. Since reference standards may not be available for even a fraction of the members of any combinatorial library, detectors that respond to physical characteristics independent of compound-to-compound variations must be used, i.e., the ''universal'' detector. While no truly universal detectors exist, detectors do exist that provide similar responses to compounds in a particular class. These detectors include those used in evaporative light scattering detection (ELSD),

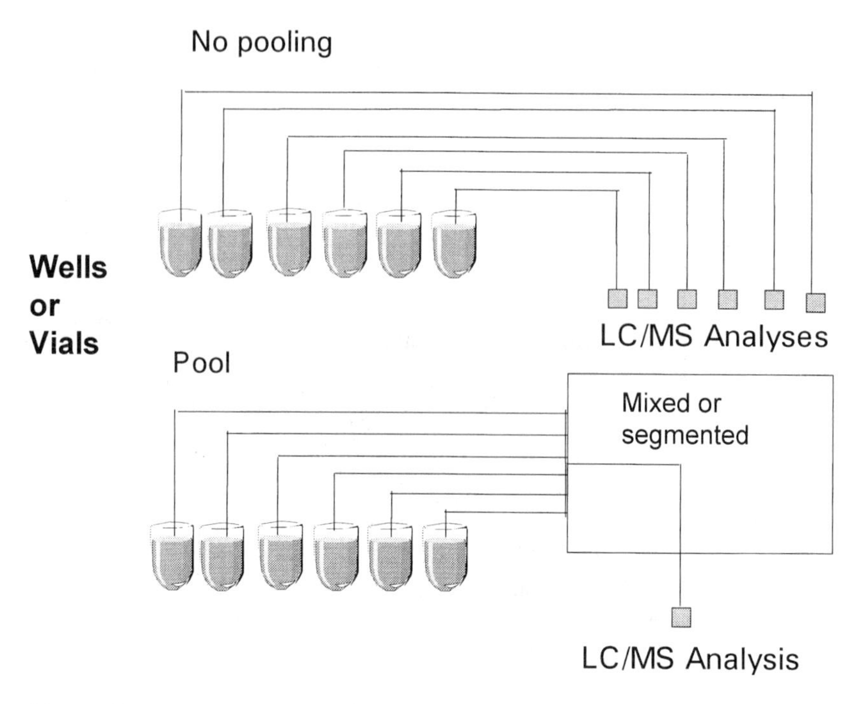

Figure 10 Injection pooling schematic.

chemiluminescent nitrogen detection (CLND) and its cousin, chemiluminescent sulfur detection (CLSD). Used alone or in combination with UV and MS, use of these auxiliary techniques allows a more accurate and complete assessment of purity to be obtained.

The principles of ELSD date back a number of years (13–15). More recently, ELSD has been applied specifically to pharmaceutical analyses and combinatorial uses (2,16–18). Detection by light scattering is based on the available mass and not absorptivity (UV) or ionization efficiency (MS), making it more accurate in some applications. In ELSD, nebulized column effluent enters a heated drift tube where rapid evaporation of the LC mobile phase takes place. A stream of nitrogen gas sweeps any nonvolatile solutes toward a detection region. Detection is accomplished by a laser and photodiode at an angle of 90°, perpendicular to the central axis of the drift tube. As the solute particles pass through the laser beam, the source is scattered. The intensity of the scattered light measured by the photodiode is proportional to the amount of solute in the column effluent.

Although most compounds respond well to ELSD, the volatility of some low molecular weight compounds may cause problems during the evaporation process. Varying response in different stages of the gradient can also cause problems. Use of ELSD in combination with UV detection minimizes these potential limitations.

CLND and, more recently, CLSD play a role similar to that of ELSD in quantitative assays of true unknowns. The CLND and CLSD detectors respond to the nitrogen and sulfur content of a compound, respectively. Both detectors operate under relatively the same principles and have been used with considerable success in drug discovery (19–21). In CLND, the analyte is oxidized at 1050°C, converting nitrogen-containing compounds to nitric oxide. The nitric oxide reacts with ozone to produce nitrogen dioxide in the excited state, which releases a photon when decaying to the ground state. The photons are measured by a photomultiplier tube and converted to an analog signal dependent only on the total mass injected.

Since a vast majority of drug compounds contain nitrogen, CLND is very useful for pharmaceutical analyses. The CLND response has been shown to be independent of gradient composition and, again, is often used in combination with other (UV and MS) detection systems (18). However, the presence of nitrogen-containing impurities in the sample or solvent will bias results. Mobile phases must of course be nitrogen-free, dictating the use of methanol or another alcohol rather than acetonitrile as mobile phase modifier.

IV. LC/MS INSTRUMENTS IN ROUTINE HIGH-THROUGHPUT USE

A. Open Access Instrument Operation

Traditionally, MS analyses have been performed in a centralized facility, often on highly specialized instruments that required constant operator intervention and maintenance. This situation is highly impractical when supporting a combinatorial program because it inhibits high throughput and general access to instrumentation and data. In response, instruments are now often operated in an ''open access'' environment. In such an environment, people not trained in LC or MS can submit samples on a continuous basis and get rapid turnaround. The use of MS and its wealth of information is promoted, and spectrometrists are freed from the mundane, tedious task of repetitive sample analysis of perhaps thousands of samples.

An open access setup usually consists of a workstation at the point of need. The walkup user has access to the workstation as well as to the sample

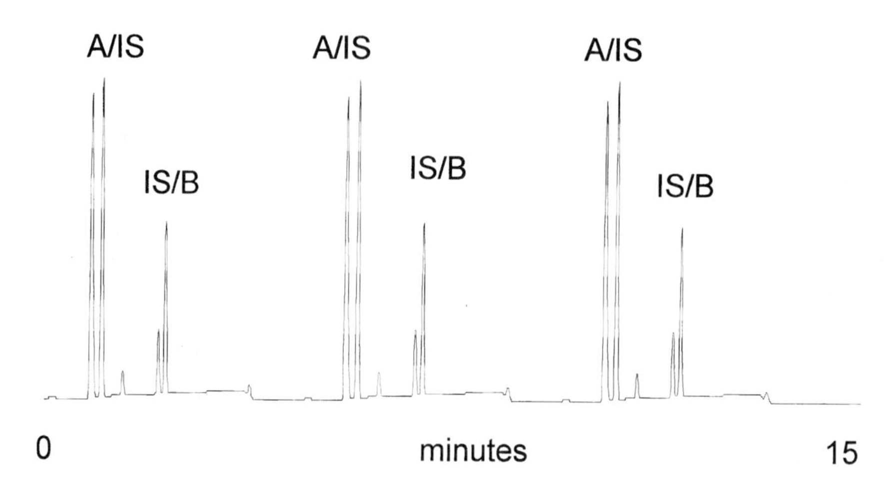

Figure 11 Interplate pooling with internal standards. Separation conditions are identical to those reported in Figure 2B. Peaks A and B are acetylfuran and valerophenone. The internal standards for peaks A and B are acetanilide and benzophenone, respectively. Data system run time was extended to 15 minutes to capture LC instrument timing. Cycle to cycle injection time is 4.7 minutes including interplate pooling.

manager or autosampler used for injection. A system administrator is responsible for setting up the system and has access to and control of all components. The walkup user, perhaps a medicinal or synthetic organic chemist, logs in sample(s) (or an entire plate) at the workstation, places them into the directed location, and, depending on the queue and system setup, gets a postanalysis report at the point of use or e-mailed to his desk. Results are typically displayed in an integrated browser format, similar to that illustrated in Fig. 13. The browser provides a graphical display for review of the results, a confirmation of molecular weight, and displays corresponding chromatograms and spectra.

B. Mass-Directed Autopurification

As mentioned previously, during lead optimization and testing leading to candidate drug selection, compounds are often required in larger quantities. Analytical techniques used in lead discovery eventually give way to semipreparative or preparative mass-directed autopurification techniques, capable of isolating and purifying 10–20 mg or more of the compound of interest during a single chromatographic analysis. Short, wide-diameter columns operated at

10–50 mL/min are the norm. Isolation and purification on the preparative scale involves fraction collection and reanalysis. Fractions are reanalyzed on the analytical scale to see how efficient the purification was. The analytical scale separation is also run on the system to check to develop the preparative method prior to actual use. Generic methods exactly like those outlined previously and scalable column chemistries are used.

Automated purification has evolved over the years from ''collect everything'' on a time basis to UV-based collection. In both instances, time-consuming secondary analyses must be carried out to correctly identify the correct or desired fraction. More recently, techniques have evolved using mass-directed and intelligent automated purification. Mass-directed fraction collection is defined as collecting a fraction of a certain specified mass only. Intelligent automated fraction collection takes fraction collection one step further by allowing fraction collection based on masses, substructures/fragments, multiple masses, adducts, etc. This results in higher quality data, single fractions, less time, fewer steps, and less chance for errors. A typical high-throughput LC/MS mass-directed autopurification system is highlighted in Fig. 14.

The system as diagrammed in Fig. 14 used two-column regeneration to improve throughput. The system has both preparative and analytical capability so that samples can be run on the same system for rapid screening of original samples, or an initial purity assessment, as well as fraction collection capability. In addition to the two 6-port column-switching valves, two flow splitters

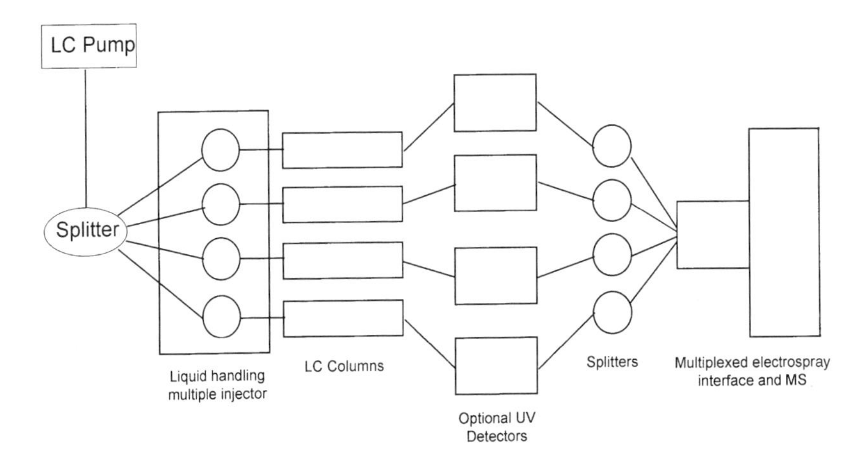

Figure 12 Schematic of an automated multichannel LC/MS system.

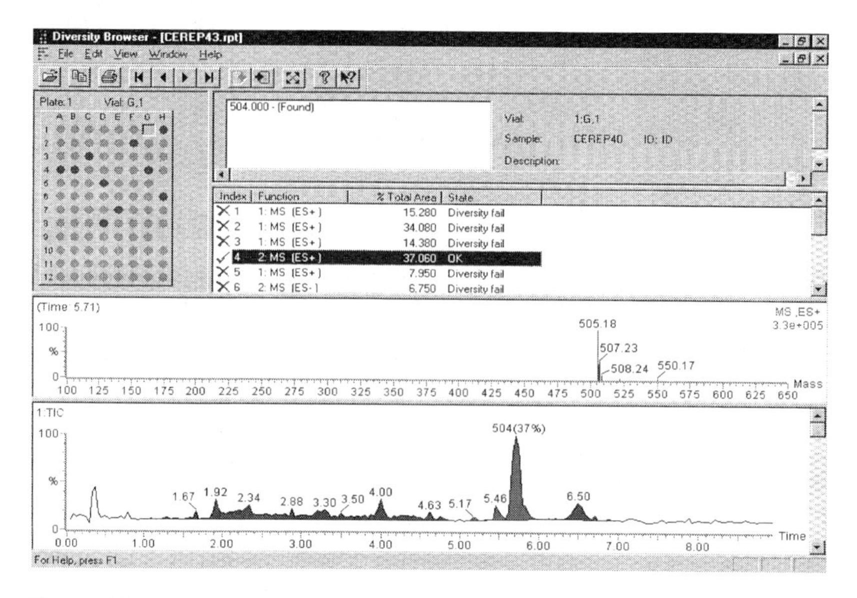

Figure 13 MassLynx™ (Micromass UK Limited, Manchester, UK) OpenLynx™ open access browser data report. Plate configuration is shown in upper left. MS and UV data is displayed for the selected well position. If the requested mass is found, the well position is highlighted in green. If it is not, it is highlighted in red. Browser's of this type are also used to track fractions in mass directed auto-purification systems.

are also used in this configuration. The upstream splitter divides the preparative flow and is an integral part of the system. The second splitter splits the analytical flow for parallel MS and PDA detection.

The use of an upstream splitter as outlined in Fig. 14 has several distinct advantages. Besides being easy to use and reproducible, this type of splitter provides constant delay times across a wide flow range, without the need for plumbing different tubing lengths or diameters. The use of a makeup flow allows high column loading without saturating the detector (by dilution) and allows flow to be split to multiple detectors without back streaming (explained below). The splitter essentially has two sides: a high-pressure side (column to collector) and a low-pressure side (makeup to detectors). As long as this balance is maintained the system works correctly. In operation, however, as the gradient composition changes from aqueous to organic, the back pressure

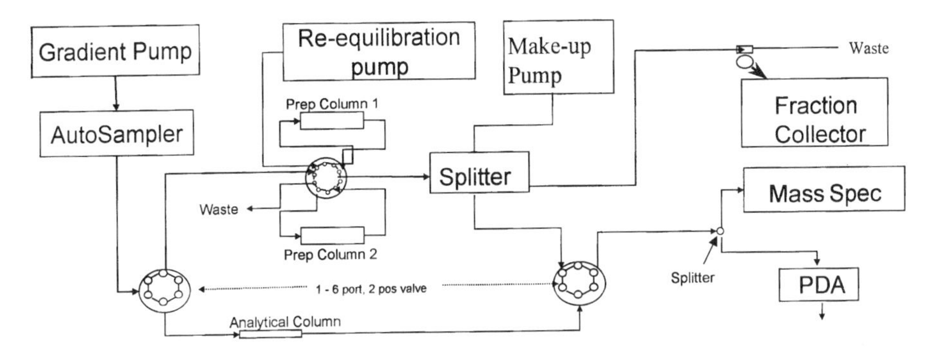

Figure 14 High-throughput mass-directed auto purification system schematic. System as shown is configured for preparative two-column regeneration and an analytical column for method development or to reanalyze fractions.

in the system decreases. Eventually the gradient back pressure can fall below the constant pressure of the makeup pump, resulting in splitter back streaming which prevents sample from getting to the MS probe and prevents late eluting peaks from being seen or collected. By using a lower viscosity organic solvent (e.g., 100% MeOH) as the makeup solvent, the back pressure is reduced to below that of the lowest point in the gradient, and back streaming is prevented. Subtle variations in split ratio delay times and band broadening are nominal across all gradient compositions.

Mass-directed autopurification systems are designed so that a peak is detected at the MS prior to reaching the fraction collector. Different flow rate–dependent delay times must be taken into account, and the software controls the synchronization between detection and collector trigger times. Figures 15 and 16 highlight the results of a mass-directed autopurification experiment (22). In this experiment, a three-component synthetic mixture was fractionated on the preparative scale and the fractions reanalyzed on the same system. Figure 15 shows the total ion chromatogram, UV signal, and fraction collection signal for the 20 mg/mL preparative run. As can be seen from the fraction collection signal, as the chromatography changes, the intelligent fraction collection compensates. The MS response directed the fraction collection based on the molecular weight of the three components in each instance. Figure 16 depicts the reanalysis of the second fraction on the analytical scale, on the same system. In spite of the two closely eluting peaks before and after, the data show a clean spectrum free of contamination.

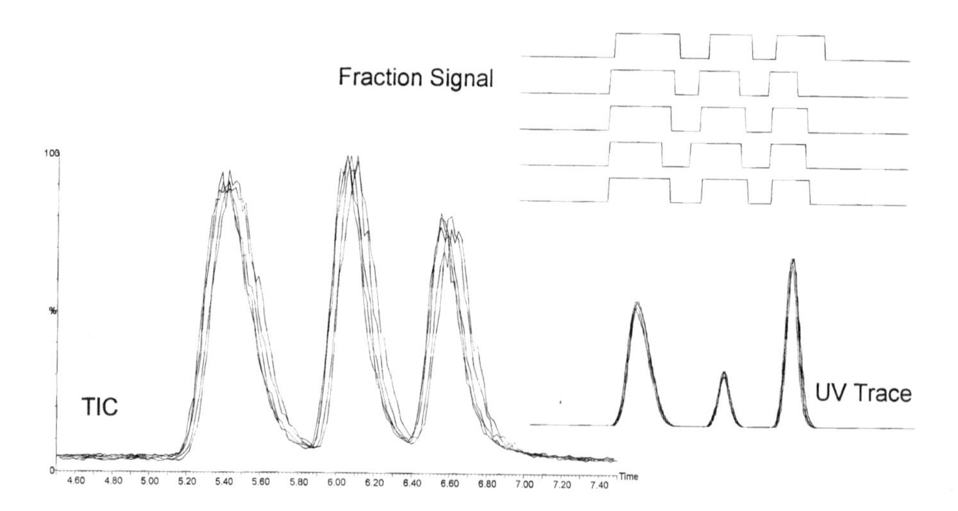

Figure 15 Mass directed fraction collection. Separation was performed on Waters Auto-purification System with FractionLynx™ software (Waters Corporation, Milford, MA) on a 19 by 50mm 7 micron particle size Symmetry® C18 column at ambient temperature. The mobile phase consisted of 0.1% phosphoric acid as the A solvent, and acetonitrile as the B solvent, run as a linear gradient from 0–100% B over 8 minutes at 20.0 mL/min. UV detection at 254nm, and a 10 mL injection was used. Peaks (20 mg/mL each in 50/50 methanol/water) in order of elution are terfenadine, diphenhydramine, and oxybutynin chloride. The two lower traces show the MS total ion chromatogram (TIC-lower left in figure-generated with an electrospray interface), and the UV trace (lower right, an extracted PDA channel) for five separate overlaid runs. A separate mass directed fraction collection signal for each of the five runs is shown in the upper right. Note the timing differences—as the preparative chromatography changes, the intelligent fraction collection compensates.

V. THE FUTURE OF LIQUID CHROMATOGRAPHY IN COMBINATORIAL CHEMISTRY

As stated earlier, LC has emerged as the technique of choice for assessing the quality of combinatorial libraries. While this trend will certainly continue, LC, either alone, or in combination with other techniques, will continue to evolve. The trend towards shorter, smaller internal diameter columns will continue, however hardware limitations such as pressure limitations, system volume, injection overhead times and data storage must be overcome. A balance be-

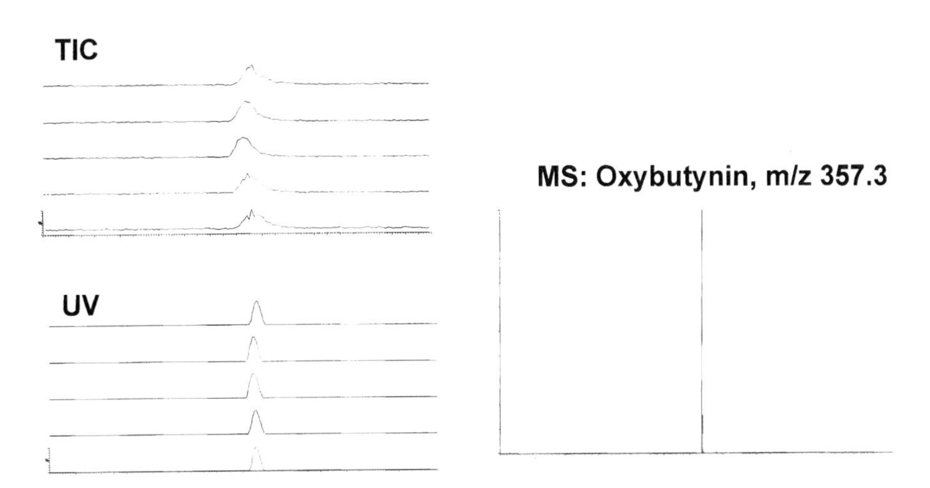

Figure 16 Analytical scale reanalysis of fraction two from Figure 14. The presence of a single compound (oxybutynin, m/z 357.3) on the TIC (top) is confirmed by the UV data (bottom), and the MS on the right lacks evidence of contamination from peaks one and three in the preparative run. Separation conditions that closely match those reported in Figure 2B were used.

tween cycle to cycle analysis times and peak capacity, whether obtained by an actual separation or by instrument specificity or deconvolution must be maintained. Recent reports of the use of capillary LC may also play a significant role in new developments in the field of combinatorial chemistry (23). Capillary LC is currently used mainly for sensitivity enhancements. Chromatographic theory predicts that reducing column diameter results in a mass sensitivity increase that is inversely proportional to the square of the column diameter ratios. However, as mentioned previously, but especially on the capillary scale, extra care should be taken to minimize extra column band broadening by using reduced diameter tubing, smaller volume detector cells, and proper fitting connections (9). In addition, capillary LC has a distinct advantage over analytical scale LC when interfaced to MS. Using analytical systems, sensitivity is sacrificed either by introducing the entire sample into the MS interface and losing much of the sample by high capacity pumping or by flow splitting just before the MS inlet. Low microliter per minute capillary flows are appropriate for direct interfacing, so virtually the entire sample is introduced into the MS ion source. Additional high throughput capillary LC techniques and methods currently in development may eventually extend the use of this technique to combinatorial applications (24). Miniaturization of different sorts is also under way that may eventually find LC applications in combinatorial chemistry. Microelectromechanical systems, or MEMS, or other ''chip-based'' technology may eventually be optimized for use in screening applications in the drug discovery process.

Detection schemes will also continue to evolve. While NMR is covered in another chapter in this volume, advances in NMR used as an LC detector have also been reported (25–28) that may eventually increase the utility of this tool for combinatorial chemistry. Used in both stopped flow and on line modes, LC/NMR can be extremely useful for structure elucidation provided the proper mobile phases or solvent suppression techniques are used (27).

Affinity chromatographic techniques are also finding applications. For example, peptide ligands can be immobilized on chromatographic media that will selectively bind and release target molecules under optimized conditions (29, 30). Stable, selective, high affinity ligands used in this manner can purify a target molecule from a complex mixture, even in the presence of closely related impurities. It is possible to increase or modulate the strength of binding and incorporate chromatographic selectivity against specific contaminants, which can accelerate process development and increase product recovery in drug discovery.

VI. CONCLUSION

LC characterization of combinatorial libraries is a vital tool in drug discovery, and, in one format or another, will remain so for a long time. High throughput methods, open access environments, mass directed auto-purification, and total system integration have lead to increased utilization of LC in ways that a few short years ago were unimaginable. This trend can only continue, as researchers strive for higher throughput, and wider applicability in solving challenges in drug discovery.

ACKNOWLEDGMENTS

The author would like to acknowledge the contributions of Chris Chumsae, Andrew Brailsford, Jeffrey Holyoke, and Beverly Kenney of Waters Corporation who collaborated in various ways in the data generated for use in this chapter.

REFERENCES

1. IM Mutton. Chromatogr 708(47):1–8, 1998.
2. JN Kyranos and JC Hogan, Jr. Anal Chem 70(11):389A–395A, 1998.
3. WK Goetzinger and JN Kyranos. Am Lab April:27–37, 1998.
4. R Cole, KA Laws, DL Hiller, JP Kiplinger, and RS Ware. Am Lab July:15–20, 1998.
5. ME Swartz, M Balogh, and B Kenney. Poster presented at the Eastern Analytical Symposium, Somerset NJ, November 1998, and Waters Corporation (Milford, MA) Literature Code M29.
6. JP Kiplinger, RO Cole, S Robinson, EJ Roskamp, RS Ware, HJ O'Connell, A Brailsford, and J Batt. Structure controlled automated purification of parallel synthesis products in drug discovery. Rapid Commun Mass Spectrom 12:658–664, 1998.
7. L Zeng, L Burton, K Yung, B Shushan, and DB Kassel. Automated analytical/preparative HPLC-MS system for the rapid characterization and purification of compound libraries. J Chromatogr A 794:3–13, 1998.
8. HN Weller, MG Young, SJ Michalczyk, GH Reitnauer, RS Cooley, PC Rahn, DJ Loyd, D Fiore, and SJ Fischman. Mol Div 3:1–24, 1997.
9. LR Snyder, JJ Kirkland and JL Glajch. Practical HPLC Method Development. Second Edition. New York: John Wiley and Sons, Inc., 1997.
10. JL Li and J Morawski. LC/GC Magazine 16(5):468–476, 1998, and Waters Corporation (Milford, MA) Literature Code Number T158.
11. C Chumsae and A Brailsford. Poster presented at ASMS, Houston, TX. June 1999.
12. Brochure No. BR25/DAM, Micromass Corporation, Manchester UK, April 1999.
13. A Stolyhwho, H Colin, and G Guichon. J Chromatogr 265:1–18, 1983.
14. A Stolyhwho, H Colin, M Martin, and G Guichon. J Chromatogr 288:253–275, 1984.
15. TH Mourey and LE Oppenheimer. Anal Chem 56:2427–2434, 1984.
16. M Lafosse, C Elfakar, L Morin-Allory, and M Dreux. J High Res Chromatogr 15:312–318, 1992.
17. CE Kibbey. Mol Div 1:247–258, 1995.
18. A Brailsford and C Chumsae. Poster presented at ASMS, Houston TX. June 1999.
19. EW Taylor, MG Qian, and GD Dollinger. Anal Chem 70:3339–3347, 1998.
20. R Bizanek, JD Manes, and EM Fujinari. Peptide Res 8:40–44, 1996.
21. WL Fitch, AK Szardenings, and EM Fujinari. Terahedron Lett 38:1689–1692, 1997.
22. ME Swartz and BF Kenney. Poster presented at the Eastern Analytical Symposium, Somerset, NJ, November 1998.
23. SA Cohen, BF Kenney, J Holyoke, TA Dourdeville, and D Della Rovere. LC/GC, 17(4S):S9–S16, 1999.

24. ME Swartz. Unpublished results, 1999.
25. PA Keifer. DDT, 2(11):468–478, 1997.
26. FS Pullen, AG Swanson, MJ Newman, and DS Richards. Rapid Comm Mass Spectr 9:1003–1006, 1995.
27. SH Smallcombe, SL Patt, and PA Keeifer. J Mag Res, Series A 117:295–303, 1995.
28. H Barjat, GA Morris, MJ Newman, and AG Swanson. J Mag Res, Series A 119: 115–119, 1996.

6
Capillary Electrophoresis in Combinatorial Library Analysis

Ira S. Krull
Northeastern University
Boston, Massachusetts

Christina A. Gendreau
Waters Corporation
Milford, Massachusetts

Hong Jian Dai
Shuster Laboratories
Quincy, Massachusetts

I. INTRODUCTION AND OVERVIEW: WHY HPCE FOR COMBINATORIAL MAPPING ANALYSIS?

Although capillary electrophoresis (CE),[1] also known as high-performance CE or HPCE, has been known and described in the literature for almost two decades, its use for combinatorial mapping is much more recent (1–14). There does not appear to be a previous review, other than that in *Analytical Chemistry* (an American Chemical Society journal), that describes the general applicability and applications of CE for these purposes (15). There are relatively few actual publications in the refereed literature that have utilized various CE modes to perform analysis of combinatorial maps. At the same time, there are

[1] All acronyms and abbreviations are defined in the glossary preceding the references.

many more CE papers that have used affinity recognition to identify an active antigen or antibody or receptor molecule, but not as part of a larger library of similarly structured compounds. Most papers in the literature deal with the synthesis and characterization of libraries in terms of chemical structures, not biological activity (15–17). Because sections of this book are devoted to analytical approaches to screen compounds in combinatorial libraries for activity, it represents the first major effort to describe how analytical chemistry can be applied to identify individual, active members of a chemical library. In this chapter, the focus is on describing how CE approaches (methods, instrumentation, protocols, techniques) can be applied to a combinatorial library and isolate, as well as characterize, only those active (lead) compounds in a given library. This is quite different from characterizing all of the members of a given library in terms of their individual structures or saying that a given structure is present in a particular library.

In general, CE is perhaps ideally situated for combinatorial map searching, as well as for providing structural information about lead/target compounds in that library. Any useful analytical approach, be it mass spectrometry (MS), nuclear magnetic resonance (NMR), Fourier transform infrared (IR), high-performance liquid chromatography (HPLC), or HPCE, must accomplish several ideal goals: (a) recognize those compounds that shall prove of biological interest and opportunity, using some type of molecular/biological recognition (antibody–antigen, receptor binding protein–antigen, drug–protein, and so forth); (b) separate the recognized members of the library from those showing little or no biological interactions with biopolymers or target drugs; (c) separate active, lead compounds from one another on some reasonable time scale; (d) indicate general level of molecular recognition (high vs. low); and (e) provide some type of structural information about the active compounds, using techniques such as NMR, MS, and FTIR. Quantitation of lead or active compounds is not really necessary for any library, since presumably one can make more of those interesting compounds at a later date, once they are shown to be recognizable and structurally determined. The successful analytical technique is more qualitative than quantitative, but ideally it will provide an indication of biological activity and structural information. One without the other is less than ideal for a useful, practical, and valuable analytical method of mapping a combinatorial library.

We use the word mapping here not so much to identity all of the structures present but rather to identity the structures of just those active (target, potential leads) compounds in that library. It has become difficult to determine the structures of individual members of a large, complex library; it is much easier to pinpoint which members of that library are indeed biologically active and then determine their individual structures. Although analytical techniques have been used mostly to identify potential pharmaceutical agents from a combinatorial library, it should be clear that these methods could be applied for virtually any purpose, so long as one has a recognition element or compound available. That is, libraries can be applied for the identification of agricultural chemicals, flavors, perfumes, fragrances, insecticides, insect pheromones, and so forth. What is crucial for the success of these methods really depends on the recognition element, and the process by which a compound or compounds will indicate biological or chemical activity for the stated purposes.

CE is potentially an ideal approach for mapping of combinatorial libraries because it is a purely liquid phase technique, as opposed to HPLC, which requires some sort of solid–liquid interaction, partitioning, diffusion, etc. There are arguments against using solid phase techniques, such as immunoaffinity chromatography (ICA) or high-performance affinity chromatography (HPAC) to isolate active, lead compounds, rather than a purely liquid solution approach. As Dunayevskiy has aptly described and discussed, solid phase binding assays, such as using immobilized antibodies to identify active antigens in HPAC or immunodetection (ID), can be influenced by the site or nature of the attachment of the ligand to the solid support (18).

This is a classic problem in all immunorecognition or immunodetection or affinity isolation techniques in HPLC or, in the future, in capillary electrochromatography (CEC). The support (bead) used to hold the ligand or receptor molecules in any affinity recognition framework can present conformational limitations that can at times prevent any or all binding to the ligand or receptor. Unfortunately, there can be no firm rule for the optimal offering of a compound to a receptor or vice versa. There are a large number of ways to immobilize a receptor protein or its ligand, but none of these will *guarantee* that the bound species can now be adequately recognized and captured by its partner (ligand receptor) (19–35). The area of immobilization of compounds to a solid support is immense, with a large number of linkage agents commercially available. It is never fully clear which of these will provide a bound species offering itself in the optimal way to its partner, now in a solution passing through that bed. It is always possible to use more than one method of immobilization of the same receptor protein in the hope that this will improve the chances of capturing any and all possible partners/ligands (19,21). However, these methods only *approximate* complete freedom from steric effects, and one is left with limits on 100%, true recognition interactions that occur in a purely solution phase recognition approach. For these very reasons, more and more applications of combinatorial library searching are being described using purely solution phase approaches.

Vouros and Dunayevskiy and others have utilized free solution or affinity CE methods combined (interfaced) with MS detection for searching libraries, as described below (18). However, CE or affinity CE is nothing but a separation technique, and it cannot and never will identify structures, especially when one is unsure of the contents of a given library. CE is a solution-based approach, though capillary gel electrophoresis (CGE) does use a polymeric gel in the capillary, which separates on the basis of differences in mass-to-charge or charge-to-mass ratios. This almost sounds like the basis of separations in MS. The reader must be referred to the available texts or review articles that describe how CE works, instrumental requirements, and the foundations or fundamentals of its separating abilities (36).

There are numerous variations on free solution CE (FSCE), such as micellar electrokinetic capillary chromatography (MECC or MEKC), where a moving, pseudostationary phase is added to the CE buffer, and secondary chemical equilibria or interactions ensue that effect separations of even neutral compounds, as well as ionic analytes. However, in general, CE utilizes truly homogeneous, solution phase separation approaches, *without* a stationary (permanent, fixed) phase, making it perhaps ideally suited for molecular recognition in searching combinatorial libraries.

In affinity CE (ACE), one adds to the running buffer or to the sample, the recognition element or ligand, and then looks for changes in mobility patterns of individual members of the sample (mobility shifts, as in flat-bed electrophoresis). This has been described in many publications, without searching combinatorial libraries, as a method to identify antibody–antigen partners, active antibodies, and active antigens. Thus, ACE is really a variation on CE, by virtue of adding something to the running buffer or, more often, to the sample that will then provide different mobility tendencies and migration patterns (mobility shifts). The two different electropherograms generated, with and without added ligand or receptor, provide differences that can be utilized to identify active partners, analogous to difference chromatography, used for many years in HPLC (19). A variation of affinity CE, if you will, permits the detection of low levels of enzymes by placing the enzyme substrate into the running CE buffer, something termed enzyme-modulated microanalysis (EMMA).

At the same time, there are some limitations in utilizing CE for library search routines. CE utilizes very small amounts of sample, nanoliter volumes, and thus very low levels of possible recognition elements. Because detection is usually through the capillary, except for MS, detection limits are very high (poor), making this a generally insensitive technique for trace analysis. There are some recent improvements in detector cell design, bubble cells, Z cells, rectangular capillaries, and so forth that have lowered detection limits for many analytes in CE (13,14). Also, since ultraviolet detection (UV)/fluorescence detection (FL)/electrochemical detection (EC) methods provide little, if any, actual structural information about a specific CE peak, applications of CE for combinatorial library searching really require interfacing with MS detection (37). MS will often provide much lower detection limits for library components, and (most importantly) it can often provide structural information and even absolute identification (38).

CE also has the ability of providing improved resolutions for similar structures, such as peptides, which may be difficult or impossible to separate by any and all HPLC methods (39). Because it moves analytes and the buffer by electroosmotic flow (EOF) rather than by pressurized flow delivery as in HPLC, individual peaks tend to be much narrower, sharper, symmetrical, and of higher plate counts (efficiency, N (number of theoretical plates or plate count for peak)). This also leads to improved resolutions of nearly identical compounds, assuming that they have some differences in mass-to-charge ratios or can interact with buffer components differently, as with their ligands or receptors.

It would therefore appear at the outset that CE may provide the best approaches for identifying lead, active compounds from large libraries of possible targets. There are, of course, some limitations in using current, solution phase methodologies. Primary among these possible or real limitations is the idea that activity of a given pool is dependent on the cumulative activity of all compounds present. It is not always guaranteed that one will locate the true, optimally active compound. One may have missed the best candidate of all, depending on how that library was synthesized and its final contents. It is also quite possible that in using a library of many compounds, recognition may be a result of synergistic or summary effects, rather than the true effect of just a single, individual compound. This is very common, and often the larger the library, the more likely synergism may play a role. It is also possible, at times, that finding synergistic compounds, more potent when administered together, is a desirable outcome. Though perhaps more difficult to sell to the Food and Drug Administration (FDA) for final, investigational new drug (IND) approval, it has been done many times.

This fault runs counter to the one just mentioned, which suggests that to find the true optimum compound one should be using larger and larger libraries, which only makes the possibility of synergism more and more likely. What a conundrum. These conflicting demands are not necessarily easily resolved, other than to utilize individual compounds for testing. However, that only defeats the goal of being able to screen larger and larger combination libraries in shorter and shorter times with less and less money and manpower. Thus, while using purely solution phase screening strategies possible in CE-MS seems ideal, clearly certain loopholes and pitfalls remain. Nothing in life or science is perfect, and we are surrounded by gray areas, where we would prefer black or white. There may be no simple way to resolve this lingering problem.

II. AFFINITY CAPILLARY ELECTROPHORESIS

This section is devoted to a description and summary of those literature reports wherein some form of affinity recognition has been described. None of these reports actually utilized a combinatorial library for such recognition; those applications follow. However, in order to appreciate how and why ACE plays such a large role in using CE-MS for combinatorial library searching, one must begin with the more general and simpler use of affinity recognition in CE. We describe here the use of some type of biorecognition element, be that antigen, antibody, receptor proteins, binding drug, or ligand, that interacts with a partner to provide altered electrophoretic migration tendencies and thus mobility differences in CE. As a result of temporary, noncovalent association or binding of the ligand to its receptor molecule, it experiences a real difference in its normal, free migration tendency and appearance in the final electropherogram. One can then imagine generating two difference electropherograms: that for the ligand alone in the absence of its receptor, and that in the presence of the receptor, where some degree of affinity or binding will occur. Depending on the migration tendency of the receptor molecule relative to that of the ligand, which is usually different, one should observe a mobility shift of the ligand. The degree of this mobility shift is directly related to the affinity constant of the ligand for its receptor; the dissociation constant of the complex; the nature of the buffer medium and its effect on complex mobility; differences in mass-to-charge ratios of the ligand, receptor, and ligand–receptor complex; and perhaps other CE conditions. In order to make use of affinity recognition in CE or ACE, one must have real, discernible differences in the mobilities of the free ligand vs. its ligand–receptor situation.

We know of no specific review articles that discuss only affinity CE, though there is a very recent review that describes the use of CE for antibody analysis, often with applications that utilize antibody–antigen recognition called immunoaffinity or affinity CE, as above (40). At times, recent texts also contain sections that discuss affinity recognition in CE, such as that by Righetti (10). Because most applications of CE for combinatorial library searching have involved some type of biological recognition, as opposed to molecular imprinting (recognition) methods, becoming more familiar with ACE and its variations can only help us to understand and appreciate why ACE-MS techniques may be quite successful for future library searching.

We will therefore start by describing some recent applications of affinity recognition in CE and what these teach us about utilizing their principles and practical approaches for eventual application to combinatorial libraries. There are far more papers that utilize some form of ACE, but far less that then apply those methods for combinatorial library searching. Although there are some reports on the use of CE for general, group separations of compounds in larger combinatorial libraries, it does not appear that CE alone will provide significant advantages in discerning the contents and structures of individual members of a larger library. Thus, we have entirely emphasized here only those reports that utilized CE for some type of affinity or immunoaffinity recognition of a specific antigenic species, whether that was a small or large molecule. While most such reports have utilized small antigens in ACE, it is entirely possible that larger protein libraries will also be successfully searched using ACE methods.

In recent years, ACE has been used to study ligand–receptor interactions and determine the binding constants of formed complexes (41–43). Some typical applications include enzyme–inhibitor, DNA–peptide, protein–sugar, peptide–drug, and antibody–antigen (Ab–Ag), etc. We will focus on Ab–Ag interactions.

The adsorption of proteins to the inner wall of the capillary is a major consideration in ACE assays (44,45). The interaction can lead to band broadening, irreproducible migration times, low resolution, and low recovery of the protein. Various approaches to control the problem include chemical modification to the inner surface of the capillary, choosing a proper buffer species and pH, and the use of buffer additives to reduce the protein interaction with the capillary wall. The buffer additive has to be cautiously chosen so as not to participate in the interaction between Ag and Ab.

Ab–Ag interactions and complexation kinetics vary for different systems (42). Some possible CE patterns are shown in Fig. 1 for a receptor–ligand system. For a fast-interaction rate system, each protein molecule spends the same time forming a complex(es) with the Ab, and although the migration time changes, the peak shape does not. Of course, it is also possible that several Ab–Ag complexes may form in a system-dependent way which can only complicate the final CE analysis. For a slowly interacting system, some protein molecules spend more time forming a complex with the Ab than the others, which leads to broad peaks, or a disappearance of the peak altogether, or even

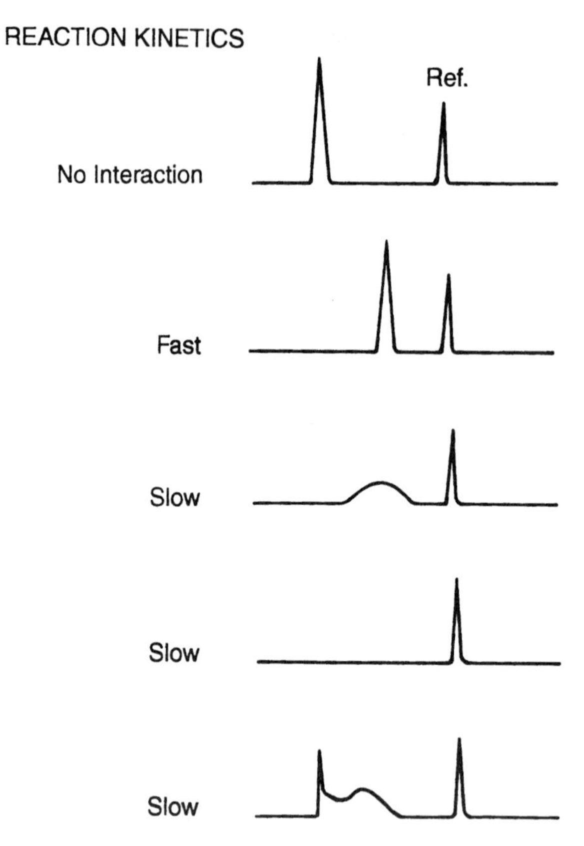

Figure 1 Schematic diagram of some of the possible interaction patterns in affinity CE with hypothetical, homogeneous receptor–ligand systems characterized by different reaction kinetics. Ref. is a noninteracting component, whereas the other peak represents a molecule interacting quickly (second panel from top) or more slowly (lower panels) with ligands of lesser electrophoretic mobility, which are present in the electrophoresis buffer (42). (Reproduced with permission from the copyright holder, Elsevier Science Publishers and *Journal of Chromatography*.)

separate peaks for complexed and uncomplexed proteins. As the assay protocols are different for slow and fast kinetic systems, some initial experimentation needs to be performed to determine the exact kinetics prevalent in a given situation. Experiments similar to those in Fig. 1 can be performed, and the kinetics can be determined based on the observed peak shape(s) (42).

In a system with fast kinetics, buffers containing different concentrations of Ag or Ab are prepared, and the sample having a fixed concentration of Ab or Ag is injected into the capillary. Scatchard analysis of the change in migration time of the Ab or Ag as a function of the concentrations of Ag or Ab in the buffer can then determine the binding constant.

In a system with slow kinetics, the Ag and Ab need to be preincubated before injection into the capillary (45,46). Fixed concentrations of Ag or Ab are then incubated with different concentrations of Ab or Ag. The quantity of free Ab or Ag can be determined by the peak area using a calibration plot. Scatchard analysis is made by plotting the total amount of Ab or Ag vs. bound Ab or Ag. For the intermediate kinetics system, separation conditions, such as applied voltage, length of the capillary, pH, and other factors, can be changed so that the system can be analyzed using one of the final experimental protocols.

III. SEPARATION AND DETECTION OF ANTIBODIES, ANTIGENS, AND ANTIBODY–ANTIGEN COMPLEXES BY HPCE METHODS

Reports on the separation of Abs by HPCE have mainly included the use of capillary zone electrophoresis (CZE) and capillary isoelectric focusing (CIEF), with much less being reported by CGE methods (40). The separation mechanism of CZE is based on electrophoretic mobilities of the sample components, which are affected by solvent characteristics, including pH, ionic strength, and viscosity. The separation in CIEF is based on differences in isoelectric points (p*I*). Most work on Ab separations by CE has been used to separate a monoclonal Ab because its homogeneity when compared with a polyclonal Ab often leads to a much higher specificity. This will also result in a higher sensitivity for the final immunoaffinity process as there will be a single peak or species detected, along with an improved possibility to develop simple and sensitive assay methods. Monoclonal Abs produced by hybridoma or genetic engineering techniques have often displayed a certain degree of microheterogeneity when analyzed by CE (47–56). This microheterogeneity may be caused by differences in glycosylation, variations in protein sequences, posttranslational modifications, improper folding, and other factors (57).

An Ab will interact with its Ag and form the usual Ab–Ag complexes. It is easy to see that if a polyclonal Ab were used for recognition with its Ag, the final electropherogram of the mixture could become quite complicated. For the above reasons, Nielsen et al. used the monoclonal Ab specific for human growth hormone (hGH) to study the CZE separation of Ag, Ab, and the various possible complexes (58). Theoretically, if the monoclonal Ab were used, only two types of complexes would be formed—those corresponding to reaction at the one or two Ag binding sites of each Ab molecule. Figure 2 illustrates a series of CZE electropherograms for immunoglobulin (IgG), hGH, and various mixtures containing an excess of IgG (Ab). This work demonstrated the ability of CZE to separate the Ag and Ab complexes from excess Ab and/or free Ag. In Fig. 2, it is apparent that incomplete resolution of all possible Ab–Ag complexes has occurred, for various reasons. If there are a large number of complexes formed (>2), then the resolution window between free Ab and Ag may be too small to resolve these completely. Alternatively, if the complexes are rapidly interconverting with one another, i.e., the kinetics of formation and dissociation are rapid, perhaps too fast for the resolution time scale, then these species would always produce broad, unresolved peaks that are never really separable.

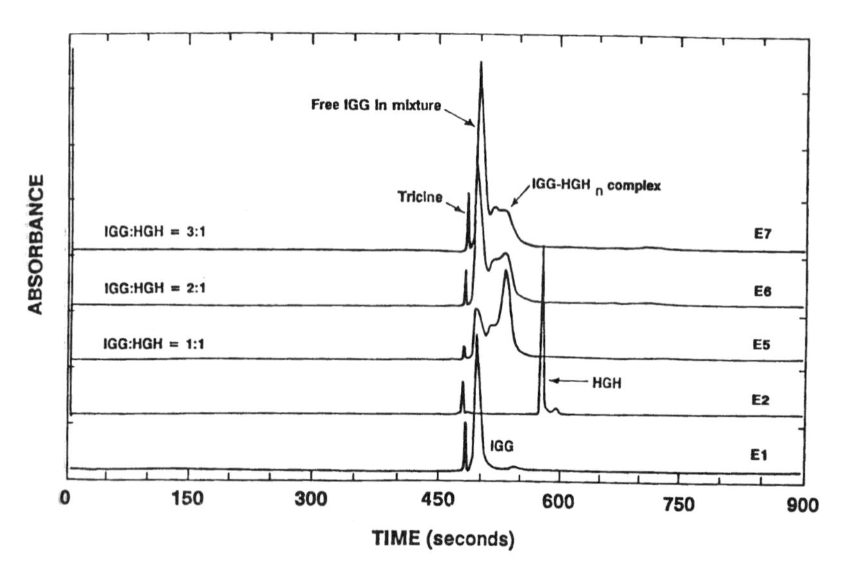

Figure 2 Electropherograms of IgG, hGH, and mixtures containing an excess of IgG. Experimental conditions used a fused silica capillary 100 cm long (80 cm to the detector) with 50 μm i.d. and 360 cm o.d. The buffer was 0.1 M tricine, pH 8.0, applied voltage was 30 kV, injection volumes were ~9 nl, detection was at 200 nm (52). (Reproduced with permission from the copyright holder, Elsevier Science Publishers and *Journal of Chromatography*.)

We have used bovine serum albumin (BSA) as an Ag and its monoclonal, anti-BSA, Ab to perform a separation study of the complexes by CZE(20,59). In this example (Fig. 3), an AccuPure Z1 reagent was used to suppress any unwanted protein and uncoated, silica wall interactions. In the four sequences illustrated, the first two (a, b) illustrate injection of purified (affinity chromatography) anti-BSA (a) and then BSA (b), whereas panel (c) illustrates a mixture of Ag and Ab with anti-BSA Ab in excess. Peaks 1 and 2, it is assumed, represent the two possible Ag–Ab complexes formed in solution. We are assuming that the earlier eluting complex is due to a 1:1 complex and peak 2 is perhaps due to a 1:2 complex of Ab–Ag. In Fig. 3, panel (d), an excess of BSA was present, leading to residual BSA peak, and now what appears to be only a single Ab–Ag complex, probably the 1:2 (peak 2), having two BSA molecules complexed with each Ab species because there is no residual Ab present. The BSA–anti-BSA system is perhaps an ideal example to study by CZE methods. These complexes appear to be quite stable under the migration conditions and times. There are fairly narrow, well-resolved, complex peaks, suggesting a lack of interconversion on this time scale.

However, in the absence of true identification of these complexes, it is not 100% possible to identify each of the complex peaks (20,59). Indeed, in none of the existing CE immunoaffinity studies reported have any Ab–Ag complexes been identified by light scattering or mass spectrometric methods (60a,b). This would require, for example, a size exclusion chromatographic (SEC) separation of a particular complex and then on-line characterization with isolation and reinjection under CZE conditions (60c).

In the case of BSA–anti-BSA, the two complexes were well separated. These separation conditions may not, however, differentiate any microheterogeneity, if present, of the Ab. Alternatively, the microheterogeneity of the Ab may be so little that it cannot be distinguished by these CZE conditions.

In Shimura and Karger's work (61) that dealt with the immunoassay (ACE or APCE) of hGH, they used CIEF to separate the complexes from Ag and Ab. To avoid any microheterogeneity of the Ab that might complicate their immunoassay, a Fab′ fragment from the Ab was used to interact with the Ag. The complexes thus formed and the excess tagged Fab′ fragments were all separated by CIEF. Figure 4 illustrates a set of typical CIEF electropherograms for this system, with conditions indicated. In panel (a), using a buffer without the Ag, only the two FL-tagged Fab′ species are present. In panel (b), now with met-rhGH first complexed with the TR-Fab′ species, two separate Fab′–Ag complexes appear, with perhaps some residual TR-Fab′ species/peaks still present. The presence of two TR-Fab′ species, panel (a), illustrates a generic problem in using immunoaffinity (Ab) recognition in all of CE:

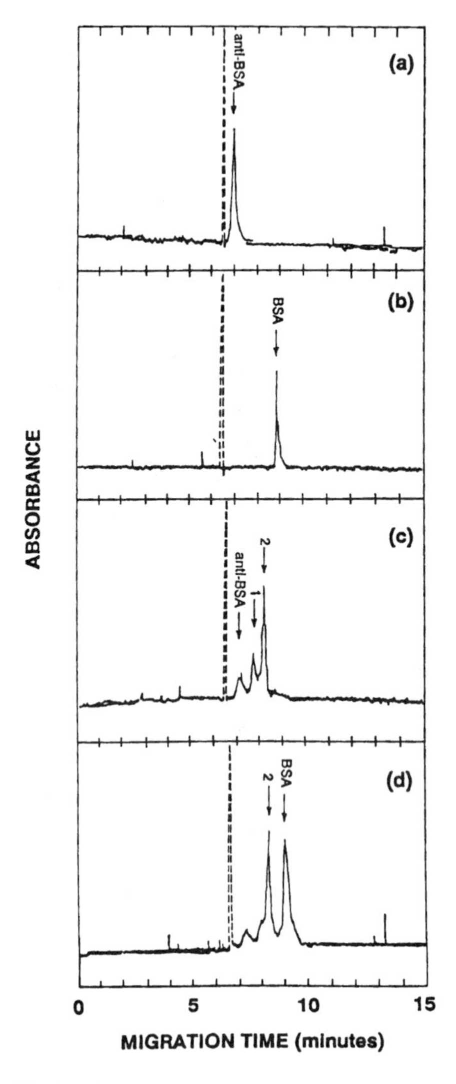

Figure 3 A series of CZE electropherograms for bovine serum albumin (BSA), anti-BSA, and complexes of BSA–anti-BSA. Operating conditions: uncoated capillary, 50 μm × 70 cm, 40 cm to detector, 15 kV applied voltage, UV detection at 214 nm, vacuum injection 30 kpa-s, buffer 60 mM phosphate, pH 7.8, 1 M AccuPure Z1-methyl reagent (Waters), sample dissolved in phosphate buffer, pH 7.0. (a) Injection of anti-BSA purified on protein G column first. (b) Injection of BSA, monomer purified by SEC first. (c) Mixture of BSA monomer and anti-BSA, with anti-BSA in excess. (d) Mixture of BSA monomer and anti-BSA, with BSA in excess (peak assignments: peak 1 = 1:1 complex; peak 2 = 1:2 complex, other peaks as indicated) (59).

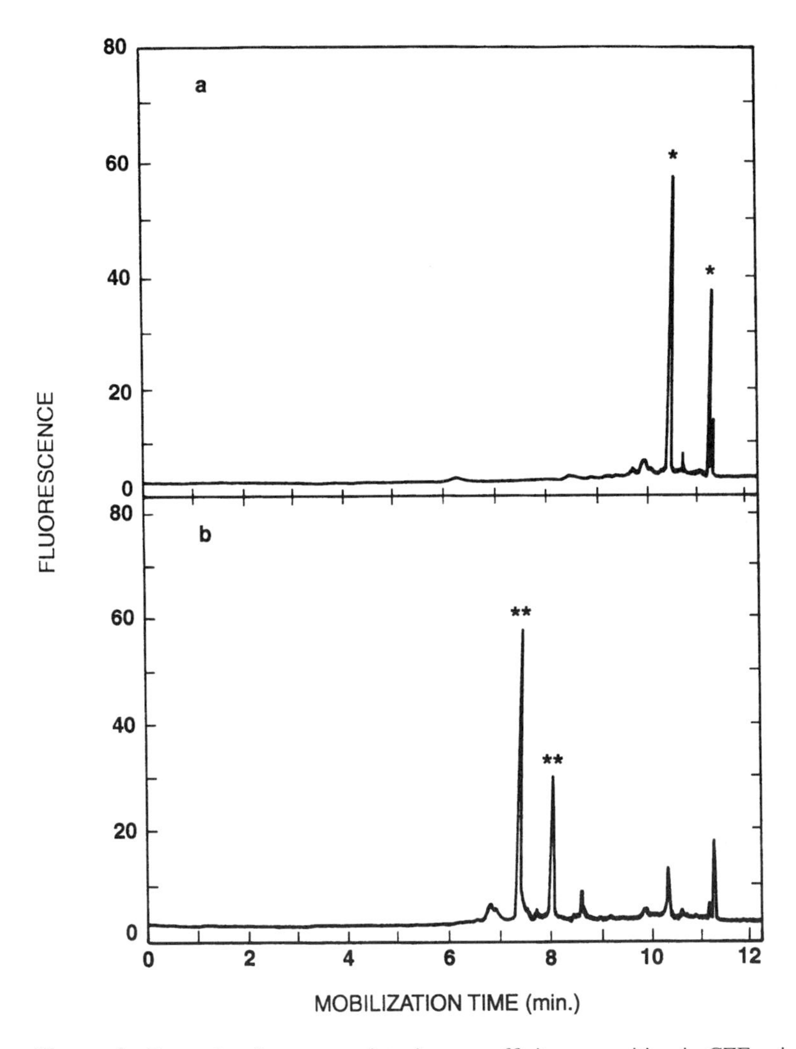

Figure 4 Example of an approach to immunoaffinity recognition in CZE using an FL probe Fab fragment of the Ab (61). CIEF analysis of crude TR-Fab′ and its complexes with met-rhGH. APCE was done with Pharmalyte 3–10 and using TR-Fab′ before purification as an affinity probe with (a) TTA-BSA buffer and (b) met-rhGH (1 μg/ml) as samples. The free TR-Fab′ and the complexes are marked by * and **, respectively. (Reproduced with permission from the copyright holder, American Chemical Society and *Analytical Chemistry*.)

the formation and presence of several complexes. Multiple complexes reduce sensitivity and specificity of the final assay, and only complicate the identification of a single, individual Ag, possibly present together with cross-reacting analogs that might also form Ag–Ab complexes with the same FL-tagged Ab (Fab′) species. Thus, though use of Fab′ species for the immunorecognition step does reduce microheterogeneity problems by removing the carbohydrate region of the Ab, solution tagging with common FL reagents, such as tetramethylrhodamine (TR), often leads to multiple products, which again complicate the final CE patterns (62,63).

There are several reports in the literature that have utilized CE-based techniques for recognition of a specific antigen and then performed an immunoassay on the now-isolated species. These have all involved a form of ACE, usually immuno-ACE, and thus could be applicable for combinatorial library searching in the future.

Immunoassays are based on the bioaffinity between an Ag and its corresponding Ab. Quantitation of Ag requires the ability to detect and discriminate between the Ag–Ab complex and either the free Ab or Ag. The high separating power of CE makes this a logical candidate for the development of immunoassays. In comparison with conventional immunoassays, a CE-based immunoassay has the following advantages: (a) It is fast and easy to automate. Fast complexation rates for a homogeneous solution system (preinjection) require less incubation time than for a heterogeneous, solution-solid phase system (conventional immunoassays on a plate/tube). (b) Multiple washing steps are eliminated. (c) Less sample is needed. (d) The high separation power of CE can discriminate between specific and nonspecific binding, protein variants, metabolites, and so forth. This last advantage is potentially significant when compared with current methods for performing immunoassays on a plate or in a tube or even by immunodetection in a flow injection mode (64). That is, in CE, there is the potential for baseline resolution and thus identification among the various possible antigens that could recognize a given Ab, such as protein variants, metabolites, decomposition products, and deamidation products. This is not possible in any of the current batch-type immunoassays.

For CE-based immunoassays, Fab fragments are usually preferred to the intact Abs, due to the multiple Ag binding sites on Abs and microheterogeneity of the Fc (crystalline or carbohydrate) portion of the Ab caused by the variations in glycosylation. That is why in almost all of the reported CZE electropherograms for intact Abs, such peaks are usually quite broad, no matter the specific separation conditions. Fab or Fab′ fragments are often, under very similar HPCE conditions, much sharper and better resolved, especially in CZE and CIEF. Papain and pepsin are the most commonly used reagents for Ab digestion. Even after digestion, it is possible that more than a single Fab or Fab′ species will be obtained. In the work of Schmalzing et al., three Fab fragments were observed, due to the fact that papain cleavage is not very specific, or derived from the microheterogeneity in the Fab fragments (65). Cation exchange separation was therefore performed to achieve a single Fab species.

Since Ag–Ab complex formation is a reversible reaction, a certain degree of complex dissociation will take place inside of the capillary. Once the complex dissociates, the Ag and Ab will rapidly move apart due to the differences in their electrophoretic mobilities. The level of the complex detected may actually be lower than that formed in the original mixture. The amount of decrease depends on the dissociation kinetics and the time required for separation. To obtain maximal complex signal, a short separation time is normally recommended (61,66).

At present, most CE-based immunoassays can be divided into two categories: competitive and direct immunoassays. For competitive immunoassays, the Ag molecules are first labeled with FL tags. Labeled Ag and a limited amount of Ab are then added to the sample to be analyzed. The labeled Ag will compete with the original/sample Ag in the sample for the limited binding sites on the Ab. After incubation, a small amount of the Ab-Ag mixture is injected into the capillary. Upon separation, using fluorescence detection, two peaks will appear in the electropherogram, one coming from the labeled Ag and the other from the Ab–Ag complex. The free Ag may elute together with the labeled Ag if the FL tag is a small molecule, or it may be separated from the labeled Ag, but under both conditions, no free Ag can be observed by FL detection. The amount of Ag in the sample is directly proportional to the free, FL-labeled Ag signal, and inversely proportional to the complex Ab-Ag (FL) signal. In principle, both signals can be used for quantitation. Most of the competitive immunoassays described are carried out on small molecules with an FL tag.

For example, Schultz et al. demonstrated a competitive immunoassay of insulin with a sensitivity around 10^{-9} M (66,67). Using this assay, the insulin content of single islets of Langerhans was determined. Fluorescein Isothiocyanate (FITC) tagging of insulin yielded at least three distinct products, all of which were separated by CZE, as illustrated in Fig. 5 (66). When these FL-tagged insulin species were then complexed with their Fab, at least two complexes were formed, appearing together with uncomplexed, FL-tagged in-

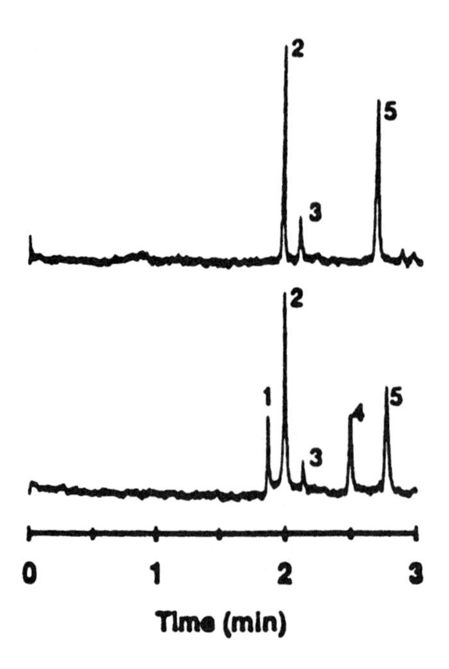

Figure 5 (Top) Electropherogram of 100 nM FITC-insulin under HPCE conditions that employed: uncoated capillaries, 25 μm i.d. and 150 μm o.d., total lengths of 25–30 cm, length to detector 12–15 cm, buffer of 0.05 M sodium phosphate with 0.025 M K_2SO_4 at pH 7.5, applied voltage 1000 V/cm, hydrostatic injection. (Bottom) Electropherogram of 100 nM FITC-insulin and 50 nM Fab. Peaks 2, 3, and 5 are FITC-insulin, while peaks 1 and 4 are due to the formation of the complex of Fab with FITC-insulin in peaks 2 and 5, respectively. An He-Cd laser was used as the excitation source (66). (Reproduced with permission from the copyright holder, American Chemical Society and *Analytical Chemistry*.)

sulin (FL detection) (Fig. 5). Perhaps these results highlight some of the lingering problems in using immunoassays and/or immunorecognition methods in HPCE, whereby contrary to batch-type immunoassays, the presence of several FL- or otherwise tagged Ag or Ab species in CE can lead to multiple, complex peaks, complicating the final results. These limitations, especially with larger Ags, also with tagged Abs, have not been resolved (62,63).

Schmalzing et al. developed a similar method to that above, but now for cortisol in serum (57). In this case, it was fairly easy to FL-tag the small cortisol molecule for this FL based CE-immunoassay, whereby a single final tagged antigen formed. This is an excellent example of how the combination of immunorecognition and immunoassay-based methods can be routinely applied with CZE separations for improved, perhaps absolute, analyte identification and full quantitation, using excellent calibration plots and validated methods.

Chen and Pentoney have reported a competitive immunoassay method for digoxin in serum with sensitivity in the low 10^{-11} M range (68). The high-resolution power of CE makes the simultaneous analysis of two, or more than two, species feasible, whereas this may not be the case in most other separation modes (e.g., HPLC). Chen and Evangelista also reported a simultaneous competitive immunoassay of morphine and phencyclidine in urine (69). The immunoassay could be performed routinely and reproducibly in less than 5 min with detection limits of 4 nM for phencyclidine and 40 nM for morphine.

The problem with labeling big molecules, such as proteins and antibodies, is that multiple products are typically obtained (as above) because most of the labeling reactions utilize the primary amino groups on the proteins and antibodies. The tags can bind in different amounts and at different locations to the proteins (62,70–73). In one illustrative example, this problem was avoided by using thiol groups at the hinge region of the Fab' fragment to react with the label (61), although several chemical reactions were involved to achieve the monothiol group on Fab'. Even here, two separate, tagged Fab' species were produced (Fig. 4), still complicating the overall CE assay and specificity.

Multiple labeled Ags (or Abs) lead to multiple peaks or, at times, a broad peak in the CE electropherograms. These species can then form more than one Ab–Ag complex, which complicates the analysis, due to perhaps insufficient resolution of the labeled Ags and the Ab–Ag complexes. Sometimes chemical modifications to the Ag (or Ab) are necessary to facilitate the separation by changing the final mass/charge ratio (74–76).

The principle of direct immunoassays is that Abs are first tagged with FL labels and then mixed with the Ag sample of interest. The Ab should be in excess. After incubation, the mixture is injected into the capillary, and the amount of Ag can be determined by the Ab–Ag complex signal. Competitive immunoassays are usually preferred for small Ags because the separation of the Ab–Ag complex from the Ab can be very difficult for small Ags. As above, Shimura and Karger reported a direct immunoassay of hGH (61). Due to the focusing effect of CIEF (Fig. 4) and the high sensitivity of CE–laser-induced fluorescence (LIF), a detection limit of 0.1 ng/ml was achieved. Similarly, Chen demonstrated the direct immunoassay of IgG with a detection limit of 6×10^{-10} (74).

A more universal immunoassay method in CE was developed by Reif et al. (76). In this approach, generic protein G was first tagged at several locations with solution FITC reagent. The heavily FITC-tagged protein G was then combined with the Ab, due to the affinity between protein G and the Fc portion of the antibody. This FITC, dual-protein reagent was then complexed with the Ag (analyte) in a CE-based assay. In this fashion, FITC-tagged protein G behaved as an FL label. The Ab did not have to be labeled for detection. This may have been the only sandwich immunoassay reported in CE. The term sandwich here refers to an approach that utilizes two antibodies surrounding the antigen, for improved specificity and identification of the correct analyte/antigen.

All of the work described above used an Ab-Ag reaction in solution. The same reaction can occur when Ab is immobilized onto the inner walls of the capillary. For example, Phillips and Chmielinska reported the analysis of cyclosporin in tears using this CE-based method (77). One third of the capillary was immobilized with the purified Ab. The sample to be analyzed was introduced into the capillary and incubated for 10 min. After rinsing the capillary with the buffer to remove the unbound materials, the buffer in the reservoirs was changed to an acidic solution. Under these conditions, the Ab–Ag immobilized complexes were broken. The active Ab remained immobilized on the CE capillary column, whereas the Ag was brought to the detector by EOF. Different from conventional, batch-type immunoassays, cyclosporin A (Cys A) and its metabolites could be differentiated due to the differences in their relative electrophoretic mobilities. As shown in Fig. 6, this is perhaps an ideal illustration of the power of combining Ab-based immunoassays with CE separations, and the inherent ability to resolve very similar species after a first immunorecognition step. It is apparent that this entire approach could be readily utilized for combinatorial library searching. In this particular instance, the assay did not use En or FL enhancement for detection, but rather represents a form of immunoaffinity extraction and preconcentration. This was then followed by the high resolving power of CE for whatever species (untagged) were first recognized by the immobilized Ab. It would, of course, be possible to first tag the Ag analytes in a sample and then introduce these into the immobilized Ab-CE system for improved isolation, separation, and detection. This approach, in general, is quite similar to the combination of immunoaffinity sample preparation followed by HPLC separation-detection methods (20,78–84).

Like conventional immunoassays, immunoassays in CE can be performed in different modes, such as sandwich, double sandwich, competitive, or direct on/off. However, the more species (analytes) that are possibly pres-

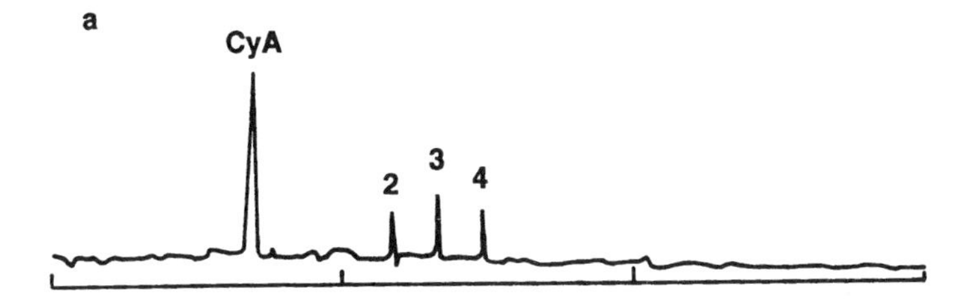

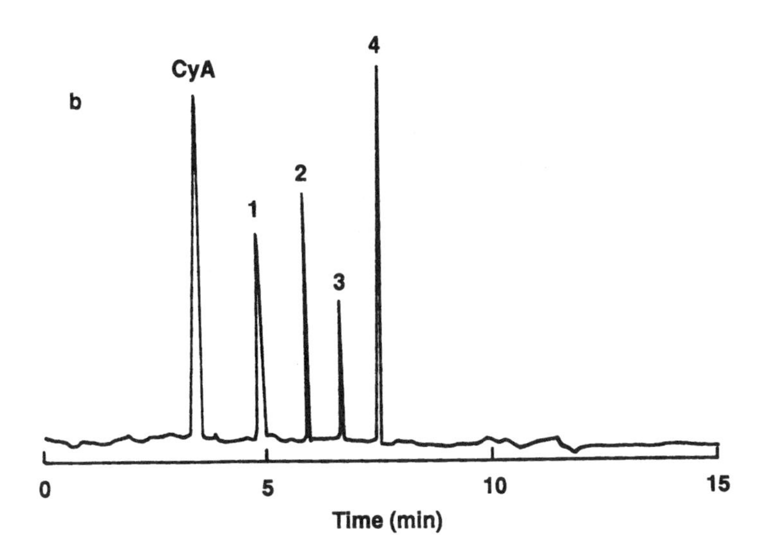

Figure 6 Immunoaffinity or immunoassay CE profiles of tear fluid obtained from (a) a patient with no clinical signs of CyA toxicity and (b) a patient during an episode of systemic toxicity. Relevant peaks: CyA, cyclosporin A; peaks 1–4 represent cyclosporin metabolites: 1, AM1, 2, AM9, 3, AM1c, 4, AM4N (77). (Reproduced with permission from the copyright holder, John Wiley & Sons, Ltd. and *Biomedical Chromatography*.)

ent, the more challenging and difficult the needed separation may become. Other than FL tags, CL, EC, and enzyme labels can also be employed. Much less has been described using such enhanced (signal-amplified) approaches in immunorecognition or immunoassays combined with CZE. Because of the amplification effect of enzyme–substrate reactions, enzyme labels should provide lower (better) detection limits (85).

IV. SPECIFIC APPLICATIONS OF HPCE TO COMBINATORIAL CHEMISTRY ANALYSIS

A search of the recent literature on the use of HPCE for the analysis of combinatorial libraries shows that as of today not a great deal of work has been done in this area. In fact, only seven published papers could be found, four of which are from the same group (86–88). However, as three of the seven were published in 1996 (88–90), it can be suggested that perhaps this research area is now picking up and will show more results in the near future.

The first published report on the use of HPCE to analyze a combinatorial library came in 1993 by Chu et al. (86). They evaluated using affinity capillary electrophoresis (ACE) for the identification, based on competitive binding, of tight-binding ligand(s) for a receptor in mixtures of equimolar ligand libraries. This has become the general approach in HPCE for determining binding constants, biologically recognized and presumably active, interacting partners, and as a general, potentially automatable, on-line screen of combinatorial libraries for active lead/target compounds. In this first example of using ACE to find active target compounds, vancomycin was used as the receptor and a small library was created consisting of 32 peptides to examine feasibility. This approach of library searching has now become the basis of a general method to perform drug screening and has since been commercialized (91). Of course, since this generalized approach appears to work for potential, lead drug candidates, it can be readily extended in the search for active agricultural chemicals, fragrances, perfumes, insect pheromones, affinity ligands, and so forth. The method is not limited to identifying only drug candidates. That is, it could just as readily be utilized for seeking agricultural compounds, perfumes, veterinary products, or any bioactive chemical compound.

Using this approach, vancomycin was first run in the electrophoresis buffer with a probe ligand, Fmoc-Gly-D-Ala-D-Ala (**L**), that showed high binding affinity for the target substrate, vancomycin. It was shown that the concentration of vancomycin affected the electrophoretic mobility of **L** (Fig.

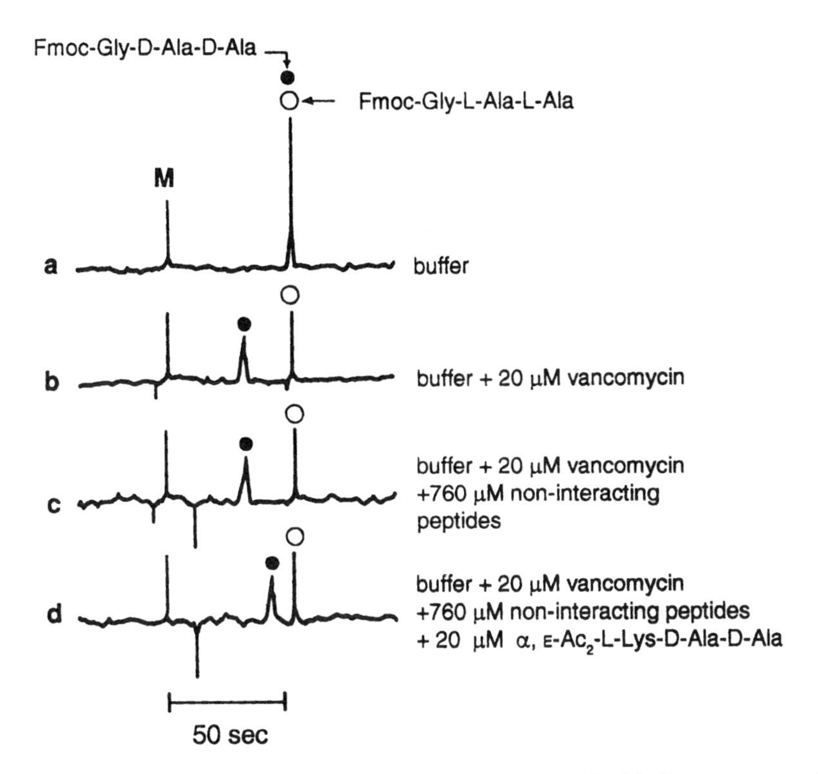

Figure 7 Illustration of CE-UV for identification of peptides binding to vancomycin by using affinity recognition in CE. How the concentration of vancomycin in the electrophoresis buffer (20 mM phosphate, pH 7.4) affects the electrophoretic mobility of Fmoc-Gly-D-Ala-D-Ala (L, black circles) but not Fmoc-Gly-L-Ala-L-Ala (open circles). Specific conditions indicated elsewhere (86a). (Reproduced with permission of the copyright holder, publisher and *Journal of Organic Chemistry*.)

7a, b), as can be seen in the shift of the **L** peak. When the peptide library was added, the electrophoretic mobility of **L** again was changed due to one or more of the peptides (**L′**) competing with **L** for vancomycin. Through the use of subsets of the library, it was possible to then specify (narrow possibilities) which of the peptide(s) was **L′** (Fig. 8). Figure 9 shows how the mixture of 32 peptides was split into 2 groups of 16 peptides each and analyzed by CE. One set of 16 could then be eliminated as no shift was seen. The remaining 16 were split into 2 groups of 8 peptides each and the above was repeated. Eleven experiments were necessary to determine the one peptide that was a

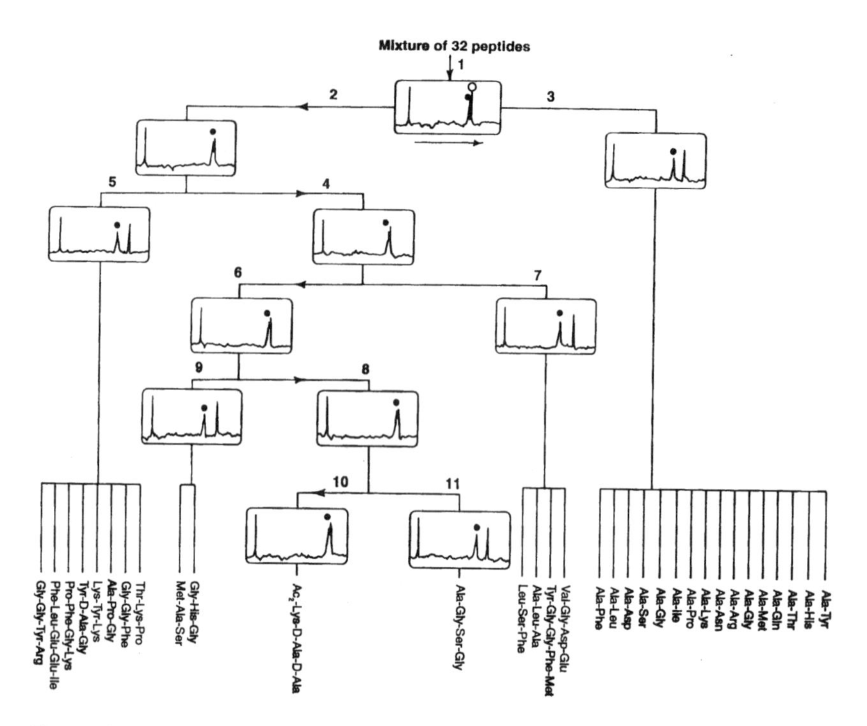

Figure 8 Stepwise elimination of noninteracting peptides from a mixture of 32 peptides and identification of one tight-binding ligand for vancomycin. Interpretation of each electropherogram is described in the text. Specific experimental conditions are described elsewhere (86a). (Reproduced with permission of the copyright holder, publisher and *Journal of Organic Chemistry*.)

tight binding receptor. In this way it was determined that **L′** was a,e-Ac2-L-Lys-D-Ala-D-Ala (Fig. 7c, d). Thus, it was shown that HPCE could be used to identify ligands from small libraries that bind most tightly to a receptor, such as vancomycin (86).

In 1995, a report using ACE-MS was published (87). Chu et al. reported on the development of a simple, one-step procedure for the on-line separation and identification of ligands that again bind most tightly to a receptor. Vancomycin was again chosen as the receptor, and it was used in the electrophoretic buffer to completely fill the capillary. In order to prevent vancomycin from flowing into the MS, the electrophoretic buffer pH was chosen to prevent the vancomycin from migrating. A neutral, hydrophilic, polymer-coated capillary was used to minimize EOF in the capillary and to therefore reduce the vanco-

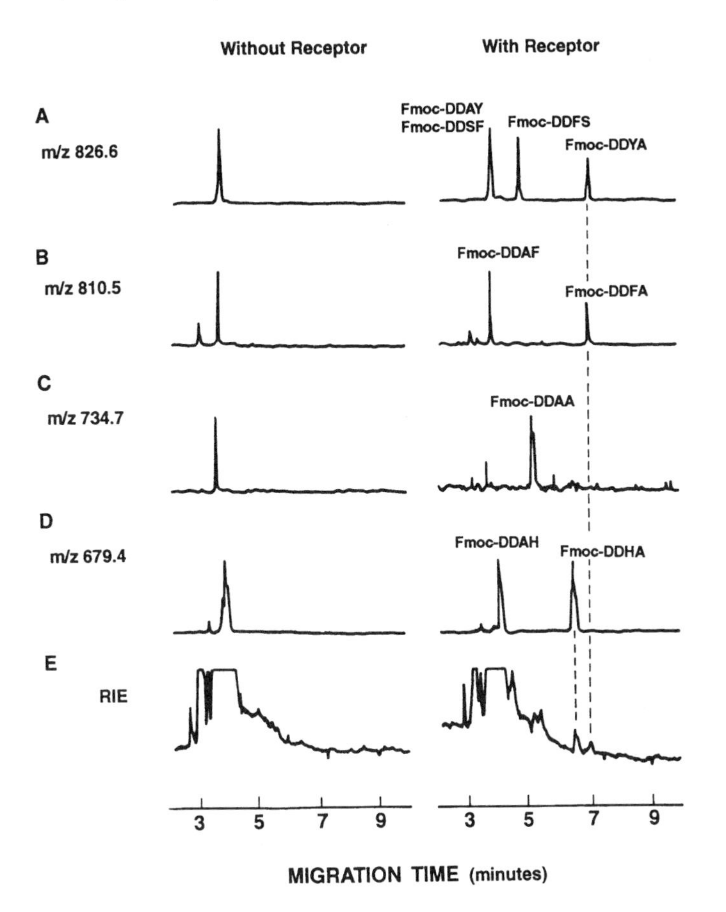

Figure 9 Affinity CE-MS (ACE) of a synthetic, all-D, Fmoc-DDXX library of 100 tetrapeptides using vancomycin as the receptor (A–D). Selected ion electropherograms for the masses are indicated; (E) reconstructed ion electropherogram for runs without (left) and with (right) vancomycin in the electrophoresis buffer. Specific ACE and MS conditions are indicated elsewhere (87). (Reproduced with permission of the copyright holder, the publisher and the *Journal of the American Chemical Society*.)

mycin flow to the MS. The idea of the research was that ligands that bound tightly to the receptor would be retained in the capillary for a longer period. The later eluting peptides could then be identified by MS. From a known 100-tetrapeptide all-D Fmoc-DDXX library, three peptides (Fmoc-DDYA, Fmoc-DDFA, and Fmoc-DDHA) were identified that bound tightly to the vancomycin. Figure 9a–d illustrates ion electropherograms for these three, both without (left) and with (right) receptor. Figure 9e shows the reconstructed ion electropherogram, again both with and without receptor. It was also shown that the binding was sequence-specific, as Fmoc-DDAY, Fmoc-DDAF, and Fmoc-DDAH did not bind to vancomycin in these experiments (87).

Chu et al. have also published more extensive results of using ACE-MS to successfully screen combinatorial libraries using larger numbers (500–1000 compounds) and a larger variety of amino acid residues (88). They felt that larger libraries would be difficult to analyze because the sensitivity of the MS instrument was not sufficient to handle many more compounds. They suggested that a more sensitive MS detection method, such as ion trap-MS, might increase the size of the library suitable for successful screening. Another way to increase the possible library size, they explained, was to remove the nonbinding peptides prior to introduction of the more interesting, strongly binding compounds, into ACE-MS. In order to remove many of the nonbinding ligands and thus possibly allow a larger initial library to be used, an affinity extraction step was developed in which the receptor was immobilized to a solid support. The peptides that did not bind were discarded, whereas the bound peptides were eluted and then analyzed with the ACE-MS method. This caused the binding peptides to be preselected and preconcentrated, which improved the capabilities of the ACE-MS method for screening combinatorial libraries (88).

Boutin et al. have also used HPCE [as well as NMR, MS, and tandem MS (MS/MS)] to analyze combinatorial libraries (89). Their goal was to provide a complete set of analytical data on a tetrapeptide library synthesized from 24 amino acids. The purpose of the work was to prove that all of the amino acids used in the synthesis were present in the resulting library, in the amounts theoretically calculated. Using HPCE, the library was split into different classes of compounds based on their net charges and the pH of the electrophoretic buffer. Integration of the areas under each curve, while taking into account the effect of the number of charges on the migration rate (amount $\approx$ area/time), allowed an estimation of the number of compounds in each class. Correlation coefficients of 0.98 and higher ($n = 10$) were found between the theoretically calculated number of compounds in each class and the values obtained by integration. HPCE was also used to show that the side chains of the amino acids were fully deprotected (89).

Dunayevskiy et al. showed the ability of HPCE-MS to determine the purity and composition of a library theoretically composed of 171 disubstituted xanthene derivatives, with the possibility of analyzing libraries of up to 1000 components (92). Previously, the ability of MS alone to analyze a library of up to 55 components was shown (93), but it was suggested that for more complex mixtures a second separation method needed to be added to better resolve the species. HPCE-MS was attempted. The investigators found that 124 of the possible 171 compounds overlapped in molecular weight, which would have made MS identification alone difficult if not impossible. When HPCE was used prior to MS, most of these 124 compounds were separated, with only eight MS/MS experiments needed to identify the 19 unresolved molecules with overlapping molecular weights. Figure 10 shows how the addition of an organic modifier, 40% methanol in this case, to the electrophoretic buffer improved the resolution. Figure 10a shows the comigrations without the organic modifier and 10b shows how the resolution was improved so that all six peaks could be distinguished when the modifier was added. This addition of an organic modifier caused only four MS/MS experiments to be needed for nine compounds (92).

Finally, Jung et al. published a communication showing that cyclohexapeptide libraries could be used as chiral selectors in HPCE (90). They explained that a lot of time and effort could be saved by using a library of possible chiral selector compounds, rather than attempting to test single compounds one by one. Furthermore, they suggested that some libraries might show cooperative effects between the components that could affect enantiose-

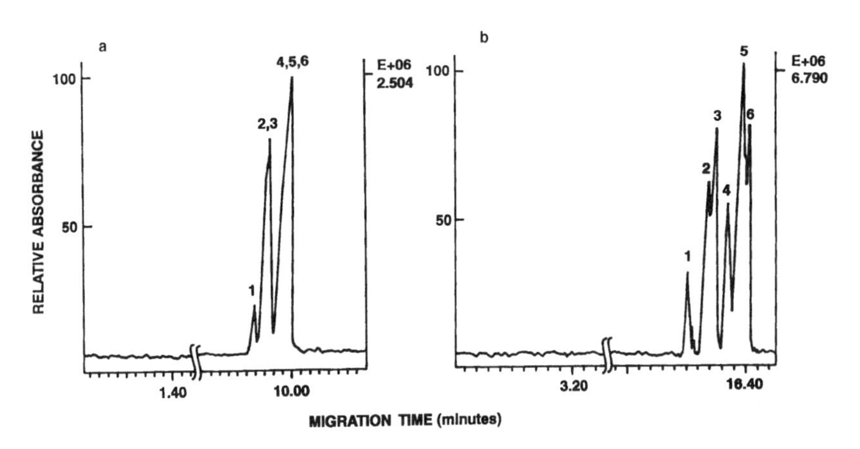

Figure 10 CE-MS electropherogram of different mixtures of xanthene derivatives in Tris-acetate buffer at pH 7.9. (a) Mixture 2, Ile/X/Ile(1), Ile/X/Pro (2), Ile/X/Ala (3), Pro/X/Pro (4), Pro/X/Ala (5), and Ala/X/Ala (6). (b) Mixture 2 dissolved in 20 mM Tris-acetate buffer at pH 7.9 containing 40% (vol/vol) MeOH. $E + 07 = 10^7$ (92). (Reproduced with permission of the copyright holder, publisher and the *Proceedings of the National Academy of Science USA*).

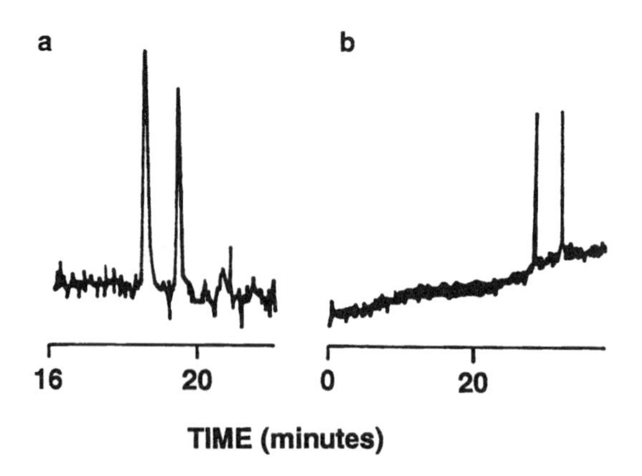

Figure 11 Enantiomeric resolution of (a) DNP-D, L-glutamic acid with c(DFXXXa) and (b) DNP-D, L-glutamic acid with c(RKXXXa). Specific conditions indicated elsewhere (90). (Reproduced with permission of the copyright holder, publisher and *Angew. Chem. Int. Ed. Engl.*).

lectivity. They successfully separated 2,4-dinitrophenyl-D,L-glutamic acid, when they ran a cyclopeptide library composed of c(DFXXXa) in the running buffer at 20 kV (Fig. 11a). They improved the resolution and selectivity of this separation when they changed the library to c(RMXXXa) and used inverse polarity at −10 kV (Fig. 11b). Work is now in progress on the use of their method to screen for the most effective chiral selector and to then identify such component(s) (90).

From these examples, it can be seen that the use of HPCE to separate, identify, and analyze compounds in combinatorial libraries is on the increase. The questions are how to further improve HPCE approaches for combinatorial library searching, how to further improve the crude preseparation of uninteresting compounds in the library from more interesting and potentially binding ligands, and then how to improve the overall HPCE separations prior to use of MS or MS/MS routines to identify each individual binding library component.

V. FUTURE ROLE OF HPCE IN COMBINATORIAL LIBRARY ANALYSIS AND INTERPRETATION

CE appears to offer some very exciting and scientifically significant advantages insofar as being able to select (recognize) bioactive drug candidates and resolve these from less interesting subsets of a combinatorial library population. There are some lingering problems in using CE-based techniques, such as requiring a major instrumental commitment of CE and MS components. These ACE-MS techniques also require sophisticated operators, i.e., trained personnel familiar with both CE and MS operations and sample requirements. There is the need to develop good separation conditions for the library in the presence of the receptor molecule(s), so that just those compounds actually interacting with the receptor will be retarded or retained and separated from less interesting library components. There is of course also the need for purified receptor molecules, but that is true in any library search method, be that LC-, MS-, or CE-based. There are alternative techniques that might work, at times as well or even better than ACE-MS, such as affinity filtration or affinity membrane separation methods, prior to HPLC-MS or CE-MS, where the affinity step is performed apart from the separation instrument (precolumn). Because there are numerous separation conditions already available for many library mixtures, these can be readily utilized for ACE-MS library search methods/conditions, perhaps with little further method optimization. There is little question but that ACE-MS methods do work, though the maximum size of the library that can be successfully searched remains a question. There are also very few research groups that are routinely utilizing ACE-MS for combinatorial library searching, although this appears from the available literature to be a completely viable and successful technique, perhaps for a wide variety of compound (drug) types. There appears to be a growing role for ACE and ACE-MS in combinatorial library searching, and one would expect that its application will only continue to grow in the future.

VI. CONCLUSIONS

This chapter reviews some of the basic principles of CE operations, and how CE can be used to separate individual library members on the basis of their interaction and recognition by a receptor molecule. Also reviewed are some of the basic needs or requirements for a successful library searching approach in CE, such as having the receptor molecule in the sample before injection or in the CE buffer during separation. Both are totally viable approaches and have been described in the literature. The literature involving affinity approaches in CE, antigen–antibody recognition, antibody–drug interactions, resolution of ligand–receptor complexes from other components of the sample, and then the use of ACE and ACE-MS to resolve active, receptor-binding species from nonactive components of the original library mixture, has also been reviewed.

In addition, this chapter describes how resolved library components can then be structurally identified by various MS approaches, so that active, lead compounds cannot only be shown active against a particular target receptor, but their actual structures can then be determined with almost 100% success using modern MS approaches and structure software. These techniques are clearly very new, but they draw on older, more established CE and MS methods, and therefore the hyphenated methods of ACE-MS also appear to be perfectly usable and reliable. These analyses, now possible by ACE-MS, are nothing more than separation/detection hyphenated methods, which are now combining an affinity recognition step in the ACE portion with structural determinations via the MS portion of the ACE-MS system. However, the affinity recognition step in the ACE part simplifies or isolates interesting, perhaps target, compounds from all other library components, making the job of the MS much simpler and more straightforward. The affinity recognition step and the less demanding MS analysis are perhaps the real attributes of using CE and ACE-MS for combinatorial library searching to isolate active drug candidates against specific receptors (or ligands).

GLOSSARY

Ab = antibody
Abs = antibodies
ACE = affinity capillary electrophoresis
Ag = antigen
Ab–Ag = antibody-antigen complex
Ab–En = antibody–enzyme conjugate
anti-BSA = antibody to BSA
anti-Ab = antibody of Ab
APCE = affinity probe capillary electrophoresis
BSA = bovine serum albumin
CE = HPCE = capillary electrophoresis
CEC = capillary electrochromatography
CGE = capillary gel electrophoresis
CIEF = capillary isoelectric focusing
CL = chemiluminescence detection
CZE = capillary zone electrophoresis
Cys A = cyclosporin A
EC = electrochemical detection
ELISA = enzyme-linked immunosorbent assay

EMMA = enzyme-modulated microanalysis
En = enzyme
EOF = electroosmotic flow
Fab = fragment of intact antibody containing recognition region (epitope)
Fab' = tow Fab fragments held together by at least one disulfide bridge
Fc = crystalline fragment of intact antibody containing carbohydrate regions
FL = fluorescence detection
FITC = fluorescein isothiocyanate
FMOC-Cl = 9-fluorenyl methyl chloroformate
FMOC = 9-fluorenyl methyl formyl (grouping)
FSCE = free solution capillary electrophoresis
FTIR = Fourier transform infrared
hGH = human growth hormone
HPAC = high-performance affinity chromatography
HPIAC = higher-performance immunoaffinity chromatography
HPLC = high-performance liquid chromatography
HRP = horseradish peroxidase (enzyme)
IACE = immunoaffinity capillary electrophoresis
ICE = immunoassay CE
ICA = immunochromatographic analysis
ID = immuno-detection
IgG = immunoglobulin
immuno-ACE = immunoaffinity capillary electrophoresis
LIF = laser-induced fluorescence (detection)
MECC = micellar electrokinetic capillary chromatography
MEKC = micellar electrokinetic chromatography
MS = mass spectrometry or spectrometer
MS/MS = tandem mass spectrometry
NMR = nuclear magnetic resonance
RIA = radioimmunoassay detection
SEC = size exclusion chromatography
TR = tetramethylrhodamine FL tag (probe)
UV = ultraviolet detection

ACKNOWLEDGMENTS

Our knowledge and appreciation of affinity recognition in HPLC and HPCE has been attributable to several firms and individuals. For example, we must indicate our appreciation to numerous individuals within PerSeptive Biosystems, Inc., especially M. Meyes, R. Mhatre, T. Naylor, M. Vanderlaan, S. Martin, F. Regnier, and others, who over the years have provided us with guidance, suggestions, encouragement, and materials. Professor H. Zou is acknowledged for his early collaborations in the areas of HPIAC and ID. D. Fisher also collaborated on some of the early developments and optimization in utilizing ID in a postcolumn, HPLC format. Professor G. Li also collaborated on some of the early developments in utilizing HPIAC and ID, prior to interfacing with HPLC and then to our own development of affinity and immunoaffinity recognition studies in HPCE areas. Several graduate students and postdoctoral fellows or visiting scientists worked with us on the development of affinity CE applications, such as R.-L. Qian, R. Strong, B.-Y. Cho, H. Zou, and X. Liu. Certain early drafts of sections of this review were prepared by R.-L. Qian and X. Liu, for which we are grateful.

Finally, financial support and technical collaborations have been provided NU in antibody areas by Pharmacia and Upjohn Pharmaceutical Company, through the Animal Health and Drug Metabolism Division (J. Nappier and G. Fate). Additional antibody analysis collaborations have been possible through SmithKline Beecham Pharmaceuticals (D. Nesta and J. Baldoni). These contracts have allowed us to become involved in affinity and immunoaffinity CE areas. Isco Corporation, Thermo Separation Products (Thermo Quest), and Waters Corporation have all donated major instrumentation, materials, and supplies to our efforts in the areas of affinity CE. Colleagues at Supelco, Phase Separations, Ltd., J&W Scientific, and Unimicro Technologies have all donated coated or packed capillaries for studies in CE and CEC. We are very appreciative of all these collaborations and technical/financial assistance in developing CE, CIEF, and, most recently, ACE approaches for proteins and antibodies.

REFERENCES

1. JW Jorgenson, M Phillips, eds. New Directions in Electrophoretic Methods. ACS Symposium Series, Volume 335. Washington, D.C.: American Chemical Society, 1987.
2. CS Horvath, JG Nikelly, eds. Analytical Biotechnology: Capillary Electrophoresis and Chromatography. ACS Symposium Series, Volume 434. Washington, D.C.: American Chemical Society, 1990.
3. SFY Li. Capillary Electrophoresis: Principles, Practice and Applications, Amsterdam: Elsevier, 1992.

4. PD Grossman, JC Colburn, Eds. Capillary Electrophoresis: Theory and Practice. San Diego, CA: Academic Press, 1992.
5. N Guzman, ed. Capillary Electrophoresis: Technology. New York: Marcel Dekker, 1993.
6. R Weinberger. Practical Capillary Electrophoresis. San Diego, CA: Academic Press, 1993.
7. JP Landers, ed. CRC Handbook of Capillary Electrophoresis: Principles, Methods, and Applications. Boca Raton, FL: CRC Press, 1994.
8. DN Heiger. High Performance Capillary Electrophoresis: An Introduction, A Primer. 2nd ed. Waldbronn, Germany: Hewlett-Packard, 1992.
9. DR Baker. Capillary Electrophoresis. Techniques in Analytical Chemistry Series. New York: John Wiley and Sons, Inc., 1995.
10. PG Righetti, ed. Capillary Electrophoresis in Analytical Biotechnology. CRC Series in Analytical Biotechnology. Boca Raton, FL: CRC Press, 1996.
11. P Camillari, ed. Capillary Electrophoresis: Theory and Practice. Boca Raton, FL: CRC Press, 1993.
12. RA Mosher, W Thormann. The Dynamics of Electrophoresis. Weinheim, FRG: VCH, 1992.
13. KD Altria, MM Rogan. Introduction to Quantitative Applications of Capillary Electrophoresis in Pharmaceutical Analysis: A Primer. Fullerton, CA: Beckman Instruments, 1995.
14. KD Altria, Ed. Capillary Electrophoresis Guidebook: Principles, Operation, and Applications. Methods in Molecular Biology Vol. 52. Totowa, NJ: Humana Press, 1996.
15. LA Thompson, JA Ellman. Synthesis and applications of small molecule libraries. Chem Rev 96:555–600, 1996.
16. JC Hogan, Jr. Combinatorial chemistry in drug discovery. Nature Biotech 15: 328–330 1997.
17. S Borman. Combinatorial chemistry, Special Report. C&EN, 43-62 (February 24, 1997).
18. YM Dunayevskiy. Applications of mass spectrometry in combinatorial chemistry research. Northeastern University, Ph.D. dissertation, December 1996.
19. IS Krull, BY Cho, R Strong, M Vanderlaan. Principles and practice of immunodetection in FIA and HPLC. LC/GC Int Magazine 10:278–292, 1997.
20. H Zou, Y Zhang, P Lu, IS Krull. Characterization of immunochemical reaction for human growth hormone with its monoclonal antibody by perfusion protein G affinity chromatography and capillary zone electrophoresis. Biomed Chromatogr 10:78–82, 1996.
21. BY Cho, R Strong, H Zou, DH Fisher, J Nappier, IS Krull. Immunochromatographic analysis of bovine growth hormone releasing factor involving RP-HPLC-immunodetection. J Chromatography: 743:181–194, 1996.
22. H Zou, Y Zhang, P Lu, and IS Krull. Perfusion immunoaffinity chromatography and its application in analysis and purification of biomolecules. Biomed Chromatography 10:122–126, 1996.

23. RE Strong, BY Cho, D Fisher, JL Nappier, IS Krull. Immunodetection approaches and high performance immunoaffinity chromatography for an analog of bovine growth hormone releasing factor at trace levels. Biomed Chromatography, 10:337–345, 1996.
24. RL Lundblad. Chemical Reagents for Protein Modifications. 2nd ed. Boca Raton, FL: CRC Press, 1991.
25. Pierce Chemical Company, a. Pierce Catalog and Handbook of Life Science and Analytical Research Products. (Pierce Chemical Company, Rockford, IL, 1994–96); b. Pierce Molecular Biology Catalog of Molecular Biology Research Products. (Rockford, IL, 1994-97).
26. MD Savage, G Mattson, S Desai, GW Nielander, S Morgensen, EJ Conklin. Avidin-Biotin Chemistry: A Handbook. Rockford, IL: Pierce Chemical Company, 1992, pp. 175–190.
27. PF Dimond. ImmunoDetection I: Perfusion immunoassays for biomolecular detection. BioConcepts Technical Newsletter. 1(2), Cambridge, MA: PerSeptive Biosystems, 1993.
28. ID, Immunodetection, Real-Time Immunoassay. Cambridge, MA: PerSeptive Biosystems Technical Literature, 1994.
29. Integral, Micro-Analytical Workstation Applications Guide. Cambridge, MA: PerSeptive Biosystems, Inc., 1994.
30. PerSeptive Biosystems Technical Operating Instructions, ImmunoDetection Sensor Cartridge. Cambridge, MA: PerSeptive Biosystems, 1992.
31. CB Quern. Immunodetection technology II: Enzyme-labeled perfusion immunoassays. BioConcepts Technical Newsletter. 2:1–4, Cambridge, MA: PerSeptive Biosystems, 1994.
32. UB Sleytr, P Messner, D Pum, M Sara, eds. Immobilized Macromolecules: Application Potentials. Heidelberg: Springer-Verlag, 1993.
33. RF Taylor, Ed. Protein Immobilization: Fundamentals and Applications. New York: Marcel Dekker, 1991.
34. M Husain, C Bieniarz. FC site-specific labeling of immunoglobulins with calf intestinal alkaline phosphatase. Bioconj Chem 5:482–490, 1994.
35. GT Hermanson, AK Mallia, PK Smith. Immobilized Affinity Ligand Techniques. 2nd ed. San Diego, CA: Academic Press, 1997.
36. IS Krull, J Mazzeo. Capillary electrophoresis: the promise and the practice. Nature 357:92–94, 1992.
37. JR Chapman, ed. Protein and Peptide Analysis by Mass Spectrometry. Methods in Molecular Biology. Volume 61. Totowa, NJ: Humana Press, 1996.
38. JT Watson. Introduction to Mass Spectrometry. 3rd ed. Hagerstown, MD: Lippincott-Raven Publishers, 1997.
39. BY Cho, R Strong, IS Krull. HPCE of a fermentation derived, cyclic peptide analog, animal growth promotor. J Chromatography. B, 697:163–174, 1997.
40. IS Krull, X Liu, J Dai, C Gendreau, G. Li. HPCE methods for the identification and quantitation of antibodies. J Pharm Biomed Anal. 16:377–393, 1997.

41. Y Chu, WJ Lees, A Stassinopoulos, CT Walsh. Using affinity capillary electrophoresis to determine binding stoichiometries of protein-ligand interactions. Biochemistry 33:10616–10621, 1994.
42. NHH Heegaard. Determination of antigen–antibody affinity by immuno-capillary electrophoresis. J Chromatography 680:405–412, 1994.
43. J Liu, KJ Volk, MS Lee, EH Kers, IE Rosenberg. Affinity capillary electrophoresis applied to the studies of interactions of a member of heat shock protein family with an immunosuppressant. J Chromatography 680:395–403, 1994.
44. AM Arentoft, H FroKiaer, S Michaelsen, H Sorensen, S Sorensen. High-performance capillary electrophoresis for the determination of trypsin and chymotrypsin inhibitors and their association with trypsin, chymotrypsin and monoclonal antibodies. J Chromatography 652:189–198, 1993.
45. NHH Heegaard, FA Robey. Use of capillary zone electrophoresis to evaluate the binding of anionic carbohydrates to synthetic peptides derived from human serum amyloid P component. Anal Chem 64:2479–2482, 1992.
46. NHH Heegaard, FA Robey. Use of capillary zone electrophoresis for the analysis of DNA-binding to a peptide derived from amyloid P component. J Liq Chromatogr 16:1923–1939, 1993.
47. NA Guzman, L Hernandez, BG Hoebel. Capillary electrophoresis, a new era in microseparations. BioPharm 2:22–37, 1989.
48. NA Guzman, MA Trebilcock, JP Advis. Capillary electrophoresis for the analytical separation and semi-preparative collection of monoclonal antibodies. Anal Chim Acta 249:247–255, 1991.
49. SK Hjerten, F Kilar, JL Liao, A Chen, C Siebert, M Zhu. Carrier-free zone electrophoresis, displacement electrophoresis and isoelectric focusing in a high-performance electrophoresis apparatus. J Chromatography 403:47–61, 1987.
50. T Wehr, M Zhu, R Rodriguez, D Burke, K Duncan. High performance isoelectric focusing using capillary electrophoresis instrumentation. Am Biotech Lab. 8:22–29 1990.
51. M Zhu, R Rodriguez, T Wehr. Optimizing separation parameters in capillary isoelectric focusing. J Chromatography 559:479–488, 1991.
52. BJ Compton. Electrophoretic mobility modeling of proteins in free zone capillary electrophoresis and its application to monoclonal antibody microheterogeneity analysis. J Chromatography 559:357–366, 1991.
53. SJ Harrington, R Varro, TM Li. High-performance capillary electrophoresis as a fast in-process control method for enzyme-labelled monoclonal antibody conjugates. J Chromatography 559:385–390, 1991.
54. MA Costello, C Woititz, J DeFeo, D Stremlo, LFL Wen, DJ Palling, K Iqbal, NA Guzman. Characterization of humanized anti-TAC monoclonal antibody by traditional separation techniques and capillary electrophoresis. J Liq Chromatography 15:1081–1097, 1992.

55. NA Guzman, J Moschera, K Iqbal, AW Malick. Effect of buffer constituents on the determination of therapeutic proteins by capillary electrophoresis. J Chromatography 608:197–204, 1992.
56. (a) TL Huang, PCH Shieh, N Cooke. Isoelectric focusing of proteins in capillary electrophoresis with pressure-driven mobilization. Chromatographia 39:543–548 1994; (b) TL Huang, PCH Shieh, N Cooke. The separation of hemoglobin variants by capillary zone electrophoresis. JHRCCC 17:676–678, 1994.
57. D Schmalzing, W Nashabeh, XW Yao, R Mhatre, FE Regnier, NB Afeyan, M Fuchs. Capillary electrophoresis-based immunoassay for cortisol in serum. Anal Chem 67:606–612, 1995.
58. RG Nielsen, EC Rickard, PF Santa, DA Sharknas, GS Sittampalam. Separation of antibody-antigen complexes by capillary zone electrophoresis, isoelectric focusing and high-performance size-exclusion chromatography. J Chromatography 539:177–185, 1991.
59. R Qian, I Krull. Unpublished results, 1994–1996.
60. (a) IS Krull, R Mhatre, J Cunniff. LC/GC Magazine. 12:914, 1994; (b) IS Krull, R Mhatre, J Cunniff. LC/GC Magazine. 13:30, 1995; (c) RL Qian, R Mhatre, IS Krull. Characterization of antigen–antibody complexes by size exclusion chromatography (SEC) coupled with low angle light scattering photometry (LALLS) and viscometry. J Chromatogr. A, 787:101–119, 1997.
61. K Shimura, B Karger. Affinity probe capillary electrophoresis: analysis of recombinant human growth hormone with a fluorescent labeled antibody fragment. Anal Chem 66:9–15, 1994.
62. (a) G Li, J Yu, IS Krull, S Cohen. An improvement for the synthesis of a styrene-divinylbenzene, copolymer based, 6-aminoquinoline carbamate reagent. Application for derivatization of amino acids, peptides, and proteins. J Liquid Chromatogr 18:3889–3918, 1995; (b) IS Krull, R Strong, Z Sosic, BY Cho, S Beale, S Cohen. Labeling reactions applicable to chromatography and electrophoresis of minute amounts of proteins. J Chromatography B, Biomed Applics. B, 699: 173–208, 1997.
63. PR Banks, DM Paquette. Monitoring of a conjugation reaction between fluorescein isothiocyanate and myoglobin by capillary zone electrophoresis. J Chromatography 693:145–154, 1995.
64. IS Krull, BY Cho, R Strong, M Vanderlaan. Principles and practice of immunodetection in FIA and HPLC. LC/GC Int Magazine 10:278–292, 1997.
65. (a) E McCarthy, G Vella, R Mhatre, YP Lim. Rapid purification and monitoring of immunoglobulin from ascites by perfusion ion-exchange chromatography. J Chromatography 743:163–170, 1996; (b) R Mhatre, W Nashabeh, D Schmalzing, X Yao, M Fuchs, D Whitney, F Regnier. Purification of antibody Fab fragments by cation-exchange chromatography and pH gradient elution. J Chromatography 707:225–231, 1995.
66. NM Schultz, RT Kennedy. Rapid immunoassays using capillary electrophoresis with fluorescence detection. Anal Chem 65:3161–3165, 1993.
67. NM Schultz, L Huang, RT Kennedy. Capillary electrophoresis-based immunoas-

say to determine insulin content and insulin secretion from single Islets of Langerhans. Anal Chem 67:924–929, 1995.

68. FA Chen, SL Pentoney, Jr. Characterization of digoxigenin-labeled beta-phycoerythrin by capillary electrophoresis with laser-induced fluorescence application to homogeneous digoxin immunoassay. J Chromatography 680:425–430, 1994.
69. FA Chen, RA Evangelista. Feasibility studies for simultaneous immunochemical multianalyte drug assay by capillary electrophoresis with laser-induced fluorescence. Clin Chem 40:1819–1822, 1994.
70. UB Sleytr, P Messner, D Pum, M Sara, eds. Immobilised Macromolecules: Application Potentials. Heidelberg: Springer-Verlag, 1993.
71. RF Taylor, ed. Protein Immobilization: Fundamentals and Applications. New York: Marcel Dekker, 1991.
72. J Turkova, I Vins, MJ Benes, Z Kucerova. Review: Oriented immobilization of enzymes and other biologically active proteins through their antibodies. Int J Biol Chem 1:1–15, 1994.
73. SS Wong. Chemistry of Protein Conjugation and Cross-Linking. Boca Raton, FL:CRC Press, 1993.
74. FA Chen. Characterization of charge-modified and fluorescein- labeled antibody by capillary electrophoresis using laser-induced fluorescence. Application to immunoassay of low level immunoglobulin A. J Chromatography 680:419–423, 1994.
75. FA Chen, JC Sternberg. Characterization of proteins by capillary electrophoresis in fused-silica columns: review on serum protein analysis and application to immunoassays. Electrophoresis 15:13–21, 1994.
76. O Reif, R Lausch, T Scheper, R Freitag. Fluorescein isothiocyanate-labeled Protein G as an affinity ligand in affinity/immunocapillary electrophoresis with fluorescence detection. Anal Chem 66:4027–4033, 1994.
77. TM Phillips, JJ Chmielinska. Immunoaffinity capillary electrophoretic analysis of cyclosporin in tears. Biomed Chromatography 8:242–246, 1994.
78. BY Cho, R Strong, H Zou, DH Fisher, J Nappier, IS Krull. Immunochromatographic analysis of bovine growth hormone releasing factor involving RP-HPLC-immunodetection. J Chromatography 743:181–194, 1996.
79. (a) LJ Janis, FE Regnier. Immunological-chromatographic analysis. J Chromatography 444:1–11, 1988; (b) LJ Janis, FE Regnier. Dual-column immunoassays using protein G affinity chromatography. Anal Chem 61:1901–1906, 1989.
80. CL Flurer, M Novotny. Dual microcolumn immunoaffinity liquid chromatography: an analytical application to human plasma proteins. Anal Chem 65:817–821, 1993.
81. M de Frutos, FE Regnier. Tandem chromatographic-immunological analyses. Anal Chem 65:17A–25A, 1993.
82. M de Frutos. Chromatography-immunology coupling, a powerful tool for environmental analysis. Trends Anal Chem 14:133–140, 1995.

83. A Farjam, GJ De Jong, RW Frei, UA Brinkman, W Haasnoot, ARM Hamers, R Schilt, FA Huf. Immunoaffinity pre-column for selective on-line sample pretreatment in high-performance liquid chromatography determination of 9-nortestosterone. J Chromatography 452:419–433, 1988.
84. A Farjam, AE Brugman, A Soldaat, P Timmerman, H Lingeman, GJ De Jong, RW Frei, UAT Brinkman. Immunoaffinity precolumn for selective sample pretreatment in column liquid chromatography: immunoselective desorption. Chromatographia 31:469–477, 1991.
85. RE Strong, BY Cho, D Fisher, JL Nappier, IS Krull. Immunodetection approaches and high performance immunoaffinity chromatography for an analog of bovine growth hormone releasing factor at trace levels. Biomed Chromatography 10:337–345, 1996.
86. (a) YH Chu, LZ Avila, HA Biebuyck, GM Whitesides. Using affinity capillary electrophoresis to identify the peptide in a peptide library that binds most tightly to vancomycin. J Org Chem 58:648–652, 1993; (b) LZ Avila, YH Chu, EC Blossey, GM Whitesides. Use of affinity capillary electrophoresis to determine kinetic and equilibrium constants for binding of arylsulfonamides to bovine carbonic anhydrase. J Med Chem 36:126–133, 1993.
87. YH Chu, DP Kirby, BL Karger. Free solution identification of candidate peptides from combinatorial libraries by affinity capillary electrophoresis/mass spectrometry. J Am Chem Soc 117:5419–5420, 1995.
88. YH Chu, YM Dunayevskiy, DP Kirby, P Vouros, BL Karger. Affinity capillary electrophoresis-mass spectrometry for screening combinatorial libraries. J Am Chem Soc 118:7827–7835, 1996.
89. JA Boutin, P Hennig, PH Lambert, S Bertin, L Petit, JP Mahieu, B Serkiz, JP Volland, JL Fauchere. Combinatorial peptide libraries: robotic synthesis and analysis by nuclear magnetic resonance, mass spectrometry, tandem mass spectrometry, and high-performance capillary electrophoresis techniques. Anal Biochem 234:126–141, 1996.
90. G Jung, H Hofstetter, S Feiertag, D Stoll, O Hofstetter, KH Wiesmuller, V Schurig. Cyclopeptide libraries as new chiral selectors in capillary electrophoresis. Angew Chem Int Ed Eng, 35:2148–2150, 1996.
91. HK Associates, Inc., Cambridge, MA, 1996–1997.
92. YM Dunayevskiy, P Vouros, EA Winter, GW Shipps, T Carell, J Rebek Jr. Application of capillary electrophoresis-electrospray ionization mass spectrometry in the determination of molecular diversity. Proc Natl Acad Sci USA 93:6152–6157, 1996.
93. YM Dunayevskiy, P Vouros, T Carell, EA Wintner, J Rebek Jr. Characterization of the complexity of small-molecule libraries by electrospray ionization mass spectrometry. Anal Chem 67:1100–1104, 1995.

7

Finding a Needle in a Haystack: Information Management for High-Throughput Synthesis of Small Organic Molecules

David Nickell
Parke-Davis Pharmaceutical Research
Ann Arbor, Michigan

Unless it produces action, information is overhead
from an interview with F. Dressler in the *Wall Street Journal*, May 9, 1997

I. INTRODUCTION: WHAT IS A HAYSTACK?

The process of drug discovery is like searching for a needle in a haystack. Many compounds must be tested before a single marketable drug is identified. Sorting through the information generated by the discovery process is also analogous to finding a needle in a haystack. With the appropriate tools, we can improve our chance of finding the needle. It is the purpose of this paper to enhance the reader's knowledge regarding the issues surrounding information management for high-throughput organic synthesis and to describe a hypothetical information management system.

Information and the generation of knowledge based on it are integral to the research and development process in the pharmaceutical industry. Without effective management of our information resources, these assets will have little more value than the background noise of everyday life. Because of the introduction of new technology (e.g., high-throughput screening and organic synthesis) to the drug discovery process, we must adapt our current information management systems.

New drug development in the pharmaceutical industry is undergoing rapid evolution. Recent data indicate that the average time from discovery to marketing approval has increased to 13–15 years (1). Because of the dropout rate for new drug candidates, many thousands of compounds must be evaluated before approval is given to market one new chemical entity. New technologies are being introduced in the pharmaceutical industry to bring new drugs to market in a more cost-effective and rapid manner.

The introduction of high-throughput screening (HTS) technology in the early 1990s (2) forced the pharmaceutical industry to reevaluate much of its standard practices. Researchers could now evaluate thousands or even hundreds of thousands of compounds per month. Because of the large amounts of data produced by HTS, in-place information management systems were not adequate for storage or retrieval of the data for analysis. A cottage industry grew from the variety of systems developed to manipulate and store the data from HTS.

However, the high capacity of HTS uncovered a second limiting process. Medicinal chemists were not able to generate compounds at a pace that could utilize the capacity of HTS. The development of combinatorial chemistry technology and its introduction into the modern medicinal chemistry laboratory gave the practicing medicinal chemist the opportunity to meet this challenge.

Combinatorial chemistry is a term that has come to mean the utilization of novel chemistry, reaction equipment, automation, and advanced information systems to increase productivity and efficiency of synthetic chemistry (3). It is also generically referred to as high-throughput organic synthesis (HTOS). Both terms have been used to describe the disciplines of ''parallel synthesis'' whereby libraries of individual compounds are synthesized (e.g., see Refs. 4 and 5) and that of ''combinatorial synthesis'' whereby libraries consisting of mixtures or pools of compounds are produced by the combinatorial mixing of chemical building blocks (e.g., see Refs. 6 and 7). A number of reviews on combinatorial chemistry are available in the chemical literature (8–19). Edwards has reviewed the combinatorial chemistry alliances in the 1990s (20). A compendium of solid phase chemistry publications has been assembled by James (21).

High-throughput organic synthesis is a marriage of science and technology. It requires the collaboration of many disciplines including chemistry, robotics, and information management technology. Automation of HTOS must be approached from a high-level view. Like retrosynthetic analysis in which synthetic pathways to a target molecule are identified by a series of steps ''in which a target structure can be transformed in a succession of steps through simpler structures into a starting point for a synthesis'' (22), analysis of an information system for HTOS can be deconvoluted to component parts. In the current environment, many pharmaceutical companies are approaching the development of an information system from the reverse direction (e.g., buying automated synthesizers, then designing the information system). This approach is of the ''penny wise, pound foolish'' school of thought because it sacrifices long term system integration for short-term compound generation. Companies who develop automated synthesis systems from the bottom up often find themselves awash in data after running the automated synthesizers for a few months.

Often the starting point in these analyses is how to automate existing processes. This may not be practical and may involve a great deal of time and resources. It is often better to rethink a traditional process from an automation point of view. A classic example of this is the effort that many companies have invested in the automation of a traditional aqueous workup. Mixing of biphasic mixtures has led to many problems. A better solution, in this case developed by rethinking the problem, is to use solid phase extraction (SPE) technology.

As described by Peccoud, ''automation of molecular biology spans at least three levels—automation of instruments, automation of experiments and automation of the laboratory'' (23). The same statement can be made about HTOS. Many of the systems in development today address only the automation of the instruments. Only a few companies are taking the next step to automate chemical experiments. The integration of all instruments used in the experiments with centralized data and control systems will form the basis for future automated HTOS laboratories.

HTOS experiments are unique in that they are automation-limited. Because the reactions cannot proceed more quickly than the robotics can deliver reagents or transfer reagents to reaction stations, the limiting factor becomes the automation. This is an unexpected attribute of the automated process as common wisdom dictates that automation will facilitate a process rather than be a rate-limiting step. It should be noted that automation will not necessarily increase the speed of the synthetic process. A well-trained chemist can often exceed the rate at which a liquid-handling robot can prepare reagents or reactions. The primary advantages of the use of robotics in organic synthesis are improved accuracy and liberating a chemist from repetitive procedures.

The pharmaceutical industry is a knowledge-based endeavor. In fact, it can be said the product of pharmaceutical research and development is information and the knowledge gained from it. Information derived from different types of bioassays and chemical entities is used to make decisions about which compounds will be advanced along the drug development pathway. However, information is only as valuable as the increase in the knowledge it produces. Knowledge, being defined as information and the understanding inferred from the relationships between pieces of information, is gained from the use of information (e.g., interpretation). The amount of knowledge gained can be described as a function of the available hardware, the software used to store and interpret the data, and data accessibility (in other words, how available is the information when it is needed?). Because the value of information is time-dependent (i.e., it has its maximum value in a finite window of opportunity) it must be readily accessible. In recent years, a significant amount of resources have been directed in the industry to ''manage'' this information as it has been recognized that the knowledge generated by individual pharmaceutical companies is an important corporate resource.

Like many aspects of the drug discovery process, HTOS is information-intensive. The focus of HTOS is the conversion of a set of reagents (building blocks) to a set of products. A traditional chemical synthesis, in which only one compound is synthesized at a time in a serial manner, also produces and utilizes information. A chemist would record the list of reagents to be used and the amounts required, how the reagents are to be combined, and a list of the desired products. Along with information about performing the synthesis, the chemist will record data about the chemical analysis and structure of the products. For several hundred years the traditional method of recording this information was to painstakingly describe each experiment in a laboratory notebook. While this process required time and energy on the part of the chemist, it was not an impossible burden. With the introduction of HTOS, the manual documentation of a chemical synthesis is a very daunting if not impossible task. For example, let us consider a five-step synthesis performed using traditional techniques. Each step of the synthetic process would typically be recorded on an individual page in a laboratory notebook and would require four to six manual calculations. If we now perform an HTOS experiment in which 96 compounds are to be synthesized the volume of information to be recorded balloons rapidly. A total of 480 pages would be required to record this information along with 384–576 calculations! As one can see from this simple example, a chemist performing HTOS experiments would soon be overwhelmed by the amount of information necessary for each HTOS experiment. To develop an automated system to relieve the chemist of this information management burden, we must first examine the synthetic process in general as any HTOS information management system must model the work flow for the synthetic process.

II. BUILDING A HAYSTACK: THE SYNTHETIC PROCESS

Any chemical synthesis can be resolved into three basic processes as shown in Table 1. The information associated with each process must be captured and stored in an automated procedure.

The *design process* is a melding of target selection with synthetic methodology. For a traditional synthesis, the potential targets are only limited by the chemist's skill. Because the glassware used for traditional syntheses is of modular design and can be built into a large number of configurations, the reactor design generally does not impose any restrictions on the reactions that can be run. Thus the chemist can select any method for synthesis of the desired targets.

HTOS, however, introduces a new set of restrictions on synthetic methodology. The commercial availability of chemical building blocks becomes a major factor in library design. Due to the long lead times for custom-synthesized building blocks, the synthesis of compound libraries for the drug discovery are often restricted to commercially available starting materials. Availability of building blocks also impact the selection of the chemistry used to synthesize the final products. The synthetic methodologies are restricted to those chemical reactions that use the available building blocks. The synthetic transformations must also be selected with regard to compatibility of the intermediate and final products with the conditions used in the reaction sequence. In general, regardless of a manual or automated operation, chemical transformations should be selected that provide the best chance of success (e.g., high

Table 1 Basic Process of a Chemical Synthesis

Process	Activities
Design	Selecting compounds to make and how to synthesize them
Creation	Synthesizing and purifying the products
Validation	Confirming that you actually made what you intended

yield, limited reaction steps, solution or solid phase techniques, easy to handle reagents, common solvents, etc.).

Library design is also influenced by the purpose of the library to be synthesized. For lead generation studies a diverse set of compounds is desired, whereas for lead optimization studies, the library should represent compounds that will provide a representational structure–activity relationship (SAR) for the structural scaffold under investigation. For compounds to be used in HTS, members of the library should be chosen that will provide the maximum information content when evaluated in the screening system.

During the HTOS design process, it is important to capture the information used to make the choices of compounds to be included in the library. These decision points can often be used to influence the selection of the compounds to be included in the next-generation library. Along with the information about the decision making process, a representation of the compounds to be included in the library must be generated. This may be a fully enumerated structure set (useful for small libraries) or generic representations of libraries such as Markush structures. When properly stored, these virtual libraries can be used for compound searching and registration.

The *creation* of the target compounds is the next process in a chemical synthesis. In a traditional synthesis, a chemist would assemble the synthesis apparatus, charge a round-bottomed flask with the necessary reagents, adjust the temperature of the reaction then allow it to proceed for an appropriate period of time. When the reaction is complete, the chemist would work up the reaction mixture and purify the crude reaction product if necessary. The procedure followed would be recorded in a laboratory notebook along with any purification strategy.

While an HTOS experiment and a traditional synthesis have much in common in the creation phase, there are some features that are unique to HTOS. For example, because HTOS libraries are spatially oriented (i.e., each compound or discrete mixture of compounds has a specific location in the reaction block), locations for each member of the library must be determined and tracked throughout the process. Pirrung and Chen (24) and Deprez et al. (25) have used ''indexed'' combinatorial libraries, in which product mixtures are tested and their activities used as indices to the rows or columns of a two-dimensional matrix reflecting the activities of individual compounds. Unless special tagging techniques are utilized, e.g., radiofrequency encoding (26,27), chemical tagging (28), etc. (29), the locations of the final products in the library array must be accurately tracked by an information management system. Tracking reagents to be used as well as the products to be made in each step of an HTOS experiment is often the primary objective of the software used to drive automated synthesizers.

The physical format of the compound library must also be determined before starting an HTOS experiment. The decision as to whether to synthesize mixtures of compounds or discrete chemical entities is related to the purpose for synthesizing the library. Libraries of mixtures allow for the screening of more compounds than discrete libraries, but large numbers of compounds per pool can adversely affect the detection of active compounds. The testing of mixtures also introduces the possibility of signal-to-noise ratio deterioration or the introduction of false positives (30). Within the pharmaceutical industry today, the trend is to produce smaller libraries of well-defined (discrete) compounds using parallel synthesis techniques (31).

Validation of the products of a synthetic sequence has historically been the crown jewel in the synthetic process. Traditionally, a chemist would rigorously attempt to produce the purest compounds in the highest yield before submitting them for biological testing. Validation technologies have advanced from the use of melting points as an indicator of purity to sophisticated spectrographic techniques in use today. All of these techniques had one aspect in common, i.e., they all relied on examining the compounds as neat compounds or solutions of free compounds. Advances in analytical technology not only allowed analyses of smaller amounts of compounds but also produced greater amounts of information. Even for the medicinal chemist using traditional chemical techniques, information overload is becoming a problem. It is not uncommon to find in the offices of today's chemist stacks of notebooks of analytical data corresponding to the compounds he or she has synthesized.

The adoption of the solid phase synthesis technology by practicing medicinal chemists has forced a reevaluation of reaction and compound validation. While traditional analytical techniques can be used for solution phase chemistry (32), there is a lack of analytical support for solid phase chemistry (33). Techniques for analyzing small molecules produced on solid supports while still attached to the resin include microanalysis (34), infrared (IR) spectroscopy (35–38), nuclear magnetic resonance (NMR) spectroscopy (39–66), and color detection reagents for reactive functional groups (67,68). Solid phase syntheses have also been analyzed by cleavage of a small sample from the support followed by high-performance liquid chromatography (HPLC), gas chromatography (GC), thin-layer chromatography (TLC), or quantitative NMR (69) analysis or monitoring the reaction solution for decrease in reagents or generation of byproducts (38). Minimal amounts of sample cleaved from the resin can be analyzed by matrix-assisted laser description ionization

(MALDI)–time-of-flight (TOF) MS (70), although the volume of data generated for a library by this technique can require gigabytes of disk storage space.

Validation of the compounds in HTOS libraries has become an area that is receiving more attention. Because of the movement away from pools of compounds (mixtures) to discrete compounds for SAR development, quality control and quality assurance of the samples has become an issue. The issue of quality vs. quantity in HTOS has been discussed in a recent paper by MacDonald et al. (71). Bauer has described an information management system that incorporates several quality checks on the data generated by HTOS systems (72).

III. FINDING THE NEEDLE: A MODEL HTOS INFORMATION MANAGEMENT SYSTEM

The work flow of the HTOS process must be the guiding principle when designing a system to manage HTOS information (73). The system must be designed for maximum flexibility to allow for integration of new functionality. Maximum flexibility is provided by incorporating a system architecture that is both open and modular. An open system provides ''hooks'' to the outside world. An HTOS information management system must be able to communicate to programs outside its scope using standard protocols. Providing connectivity to the outside world also provides a foundation for the second cornerstone of an information management system: modularity.

An open system with modular architecture is the model followed by desktop computer hardware and software designers. It is now possible to assemble a PC entirely from components manufactured by different vendors. Object-oriented programming techniques also allow for the use of modularized ''components,'' thus saving the programmer many hours of labor. When modularity is designed into an HTOS information management system, it can be easily expanded to accommodate new processes and equipment. It is a safe assumption that chemists will want to assemble an HTOS system based on the best available technology. This means that the design module may be purchased from vendor A while the robot is purchased from vendor B. If these two components cannot be tightly integrated, one of the suppliers will lose a sale. Open and modular systems will also allow the system to grow as needed. Much like the traditional laboratory glassware commonly used today, the chemists will assemble the components that are needed, replacing those that are obsolete. As with desktop computers, proprietary closed HTOS systems are only an intermediary design phase on the road to open modular systems.

An HTOS information management system must incorporate modules that can be used to design products and synthetic routes, manage the data necessary to perform the actual synthesis, and interface with analytical instruments to provide bidirectional information transfer. Although not part of the synthetic process, the information management system must also provide an automatic mechanism to register the new synthetic targets into compound registration systems. Manual registration will not be possible due the increase in the number of compounds per chemist that will result from the widespread adoption of HTOS techniques.

Figure 1 represents a schematic diagram for a model HTOS information management system. It comprises a series of replaceable modules that reflect

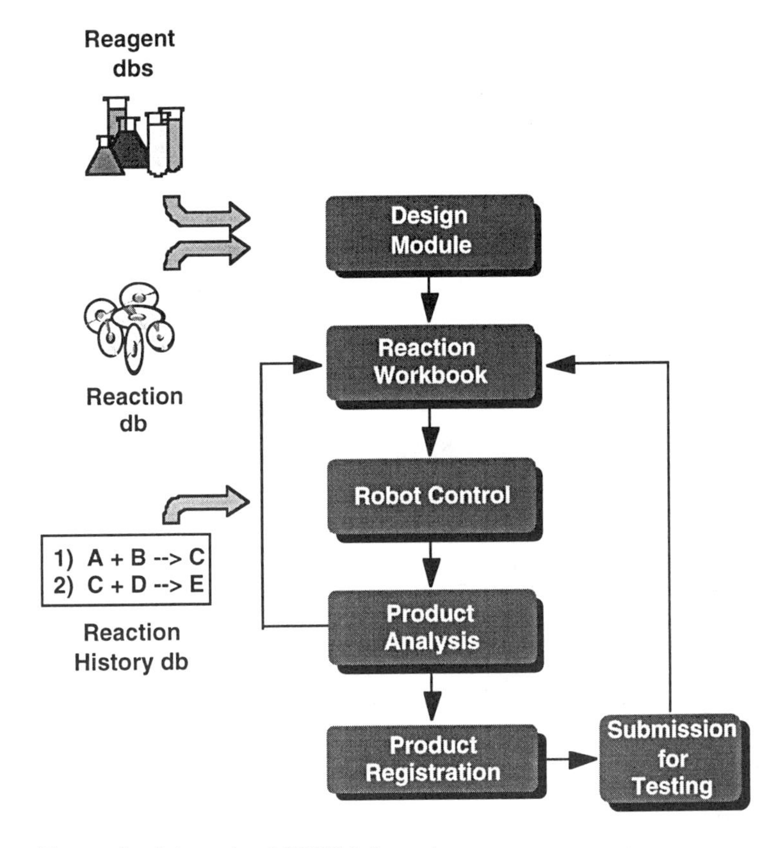

Figure 1 Schematic of HTOS information management system.

the work flow of the synthetic process. The reaction workbook is the central ''hub'' in which data about the experiments are collected. It is also the primary interface to the system for the chemist. Each module will be described in the following sections.

A *design module* will provide the capability to select the compounds and the method by which they will be synthesized. In order to perform this task, the design module must be linked to external databases that can provide data about the availability of chemical reagents and information about synthetic methods. Once the products are selected and the synthetic method selected, this information will be collated in the reaction workbook. The function of the design module is complementary to diversity assessment tools.

The best analogy for a *reaction workbook* is the typical laboratory notebook but modified for HTOS so that it can easily manipulate the data for multiple reactions. A reaction workbook should provide, at a minimum, the functions shown in Table 2.

A reaction workbook module should function in either a standalone desktop environment or connected to an enterprise-wide information system. It must be an ''intelligent'' system to prevent the chemists from making mistakes that because of the repetitive use of data can cause cascading errors in reagent delivery and process timing. Using the information input into the system, it will automatically perform the necessary calculations such as the volume of reagents to be transferred and the total amounts of neat reagents needed to prepare these solutions.

Product and reagent location tracking is one of the most important functions of a reaction workbook due to the spatial orientation of the product arrays generated by HTOS techniques. The locations of reagents together with the volumes to be transferred and the sequence in which the reagents are to be delivered will be used to automatically create synthetic procedures for a liquid-

Table 2 Reaction Workbook Functionality

Perform necessary synthetic calculations
Track the locations of the products and the reagents
Associate the reagents with the products
Enumerate the products based on the available reagents
Store the synthetic procedures
Provide operational instructions for the associated automation
Collect analytical data on the products
Correlate screening data with products

handling robot. An ''outline'' for the synthetic procedure must be entered into the system by the chemist or downloaded from a database of synthetic procedures. Typically, the outline of the procedure will be entered by the chemist in the form of commands like ''add the generic building block A.'' The reaction workbook module will first calculate the volumes of reagents to be delivered and determine to which locations they should be dispensed. It will then create the necessary liquid-handling robot commands based on the locations of each of the building blocks. When complete, the list of action steps in each individual procedure will correspond to the physical steps a chemist would perform if he carried out the synthesis manually. These ''recipes'' can be stored for future reference and use. The reaction workbook module will use the recipes to generate robot-specific commands for the automation associated with the information management system. These synthetic procedures for the automated synthesizers may also become the ''documentation'' for the HTOS libraries. These procedures document what a chemist was attempting to make and how he or she tried to make it. For larger libraries, this method may be the only practical way to document what was made as enumeration of every chemical structure in a large library would require the dedication of significant computer resources.

A reaction workbook module will also become the primary vehicle for compound registration because of its links to corporate compound registration systems. To facilitate the registration of chemical structures, a reaction workbook must be ''chemically aware.'' In other words, it should recognize chemical structures and be able to derive information (e.g., molecular weight, R-group tables, etc.) from them. Ideally the reaction workbook should be ''reaction''-based. The chemist should be able to specify the desired chemical transformation using chemical reaction notation, specify a list of reagents to be used, and then automatically enumerate the individual products. A reaction workbook performs a yeoman's service for the chemist performing an HTOS experiment. It dutifully records and tracks the information used in the synthesis. While performing many behind-the-scenes calculations and data manipulations, it functions on the very simple assumption that the chemist has previously selected the reagents to be used in the design module and thus all possible compounds will be prepared from the list of selected building blocks. No facility need be incorporated to allow for selection of individual molecules or sublibraries if the libraries are small. If, on the other hand, the libraries are large, a design module should be used to create sublibraries that are of a size that can be produced on an automated synthesizer.

The control of the system automation is the function of the *robot control* module. This module will not only control the liquid-handling robots but will also schedule operations for multiple workstations (e.g., weighing, multiple reaction stations, etc.). It will also interface with reaction history databases. These databases consist of two components: (a) robot commands to perform the desired synthesis; (b) a transaction log. The log database will be useful in determining if the process was successfully completed or for diagnosing the cause of any system failures. Because there is not a standard communication protocol that can be used to issue commands and receive feedback from robotic peripherals, it is difficult to use a standard control system to control the automation. Some companies that are developing automation systems are investigating the integration of LabVIEW (74) as robot control software. However, within the manufacturing industries there are a group of real-time machine and process control vendors (75) whose software might be adaptable for HTOS automation control.

After the library has been synthesized, the *product analysis* must be completed to validate the library members. During the development of HTOS technologies, two philosophies have developed regarding the characterization of products. One view is that the components of a library need not be analyzed. The compounds are synthesized, then screened for activity in a bioassay. If a sample shows activity, it can be resynthesized and fully characterized using traditional techniques. This methodology provides the most effective way to screen large numbers of compounds.

With the movement toward smaller libraries of discrete compounds, library validation has become a more important issue. To maximize the information content of every HTOS experiment, each compound in a library must be analyzed for chemical structure and purity. Many of the analytical instruments available today are designed to function in a serial mode. Thus the instrument must acquire information about a single compound and then move on to the next, one at a time. The analysis of HTOS libraries is perhaps the weak link in the high-throughput discovery process. While screening and chemical synthesis have adopted parallel paradigms to enhance the speed of the processes, purification and analysis techniques are still serially oriented. This is an area where future parallel automated technology will have a significant impact.

To move data effectively, bidirectional interface between the analytical instruments and the reaction workbook module is needed to provide a mechanism to submit the libraries for analysis and to integrate the resulting data with the synthesis data. The data could be transferred and collected in the reaction workbook module where a chemist can review the data for integrity and forward it to corporate data repositories. A fully automated system that can review the analytical data and identify the library members that pass selection criteria would greatly enhance the throughput of the validation system.

Product registration has also become an issue in HTOS. For those companies who wish to register each sample produced in an HTOS experiment, a chemical structure along with a unique sample identifier must be generated for each sample. The samples must be associated with a particular plate or synthesizer as well as the sample's location within the array. The entire HTOS experiment must also be given a unique identifier. The compounds must also be identified as having been generated from an HTOS experiment. This is important because the samples are registered with proposed structures as the chemist truly has no idea if the synthesis produced the desired compounds. All of these ''tags'' are necessary so that an individual sample can be uniquely identified. For large libraries, this approach may not be practical.

Many groups register only the compounds from an HTOS experiment that are bioactive. After the samples are fully characterized, they are registered in a traditional manner. Because the ''hit'' rate from HTS is typically less than 1%, this method does not seriously overload in-place registration systems. The disadvantage to this method is that the information that a compound has been synthesized is lost. Not having access to that information can result in the repetition of experiments.

Whichever method is selected for registering HTOS samples, the process should be as automated as possible. Because the reaction workbook module will contain all the information necessary to register a library, a compound registration mechanism should be incorporated within it.

IV. BETTER METHODS FOR FINDING THE NEEDLE: THE FUTURE

For the near future, fully automated systems are not practical. Although work is proceeding on fully integrated high-throughput drug discovery systems (76), none has yet reached the marketplace because the decision-making algorithms and software are not yet in place. Effective integration between information systems and automation also needs to be developed. For the time being, the medicinal chemist will provide the decision making ''brains'' to analyze the data generated by HTOS and HTS systems and the integration between the components. This human–machine interface will provide the ultimate in modularity as different chemists can interact with the system in different ways depending on their scientific background and knowledge of the drug discovery process.

Although the volume of data generated by HTOS experiments can easily overwhelm a poorly designed information management system, it is not the only problem one needs to resolve. As mentioned earlier in this chapter, the value of information is derived from the knowledge gained from it. To improve the knowledge gained from the HTOS experiments, appropriate tools must be available to analyze the experimental data. To date, these tools are not available. For this reason, HTOS technology is not being used to its fullest potential.

Well-designed data analysis tools must incorporate three principle components: (a) the tools must have the ability to handle large data-sets; (b) the tools must be capable of representing the data using a visual paradigm; (c) the tools must incorporate algorithms that can derive relationships between different types of data. Typically, scientists are still using the tools that were developed 20 years ago. These tools (e.g., spread sheets) were adequate for the interpreting of small numbers of data but are not up to the job when analyzing billions of data points. Such tools must be able to visualize (77) and compare the MS data or the NMR data for all of the components of a library in an easy-to-comprehend manner. Analysis tools must also be able to visually represent different types of data in a way that will facilitate its interpretation. For example, how can the calculated log P's of a series of compounds produced by parallel synthetic techniques be represented? How can the HTS results be represented and be compared to the substitution pattern at a particular variable site of a synthetic scaffold? These are the sort of problems with which medicinal chemists are currently struggling by manual methods. Although commercial software developers are beginning to address this issue (78), there is no solution available that is fully integrated with an HTOS information management system.

HTOS technologies have forced the practicing chemist to reevaluate how he or she does business. While a number of interesting puzzle pieces have been identified, it is not clear as to how they fit together. This is one of the most exciting challenges facing us today and in the near future.

REFERENCES

1. JA DiMasi. Trends in drug development costs, times, and risks. Drug Info 29: 375–384, 1995.
2. M Reichman, AL Harris. Practical high-throughput screening. In: W Moos, M Pavia, B Kay, A Ellington, eds. Annual Reports in Combinatorial Chemistry and Molecular Diversity, Vol. 1. Leiden, The Netherlands: ESCOM, 1997, pp 273–286.

3. SH DeWitt, AW Czarnik. Automated synthesis and combinatorial chemistry. Curr Op Biotech 6:640–644, 1995.
4. SH DeWitt, JS Kiely, CJ Stankovic, MC Schroeder, DMR Cody, MR Pavia. DIVERSOMERS: an approach to non-peptide non-oligomeric chemical diversity. Proc Natl Acad Sci USA 90:6909–6913, 1993.
5. BA Bunin, JA Ellman. A general and expedient method for the solid phase synthesis of 1,4-benzodiazepine deviates. J Am Chem Soc 114:10997–10998, 1992.
6. M Pirrung, J Chau, J Chen. Discovery of a novel tetrahydroacridine acetylcholnesterase inhibitor through an indexed combinatorial library. Chem Bio 621–626, 1995.
7. T Carell, E Wintner, A Sutherland, J Rebek, Y Dunayevskiy, P Vouros. New promise in combinatorial chemistry: synthesis, characterization, and screening of small-molecule libraries in solution. Cur Bio 5:171–183, 1995.
8. A number of relevant articles can be found in AW Czarnik, SH DeWitt, eds. A Practical Guide to Combinatorial Chemistry. Washington, DC: ACS Books, 1997.
9. WA Warr. Combinatorial chemistry and molecular diversity. An overview. J Chem Inf Comp Sci 37:134–140, 1997.
10. F Balkenhol, C von dem Dussche-Hunnefeld, A Lansky, C Zechel. Combinatorial synthesis of small organic molecules. Angew Chem Int Ed Eng 35:2288–2337, 1996.
11. A number of excellent references are located in the March 1996 issue of Acc Chem Res, which is dedicated to combinatorial chemistry.
12. A tabular collection of reactions used in solid phase organic synthesis can be found in P Hermkens, H Ollenheijm, D Rees. Solid-phase organic reactions: a review of the current literature. Tetrahedron 52:4257–4554, 1996.
13. Thompson, JA Ellman. Synthesis and applications of small molecule libraries. Chem Rev 96(1):555–600, 1996.
14. G Lowe. Combinatorial chemistry. Chem Soc Rev 245:309–382, 1995.
15. NK Terrett, M Gardner, DW Gordon, RJ Kobylecki, J Steele. Combinatorial synthesis—the design of compound libraries and their application to drug discovery. Tetrahedron 51(30):8135–8173, 1995.
16. MA Gallop, RW Barrett, WJ Dower, SPA Fodor, AM Gordon. Applications of combinatorial technologies to drug discovery 1. J Med Chem 37:1233–1251, 1994.
17. EM Gordon, RW Barrett, WJ Dower, SPA Fodor, MA Gallop. Applications of combinatorial technologies to drug discovery 2. J Med Chem 37:1385–1401, 1994.
18. JS Fruchtel, G Jung. Organic chemistry on solid supports. Angew Chem Int Ed Eng 35:17–42, 1996.
19. X Will, I, Pop, L Bourel, D Horvath, R Baudelle, P Melnyk, B Deprez, A Tartar. Combinatorial chemistry: a rational approach to chemical diversity. Eur J Med Chem 31:87–98, 1996.

20. MG Edwards. Combinatorial chemistry alliances in the 1990s: review of deal structures. In W Moos, M Pavia, B Kay, A Ellington, eds. Annual Reports in Combinatorial Chemistry and Molecular Diversity, Vol. 1. Leiden, The Netherlands: ESCOM, 1997, pp 314–320.
21. IW James. A Compendium of solid phase chemistry publications. In: W Moos, M Pavia, B Kay, A Ellington, eds. Annual Reports in Combinatorial Chemistry and Molecular Diversity, Vol. 1. Leiden, The Netherlands: ESCOM, 1997, pp 326–344.
22. EJ Corey, AK Long, SD Rubenstein. Computer-assisted analysis in organic synthesis. Science 228:408–418, 1985.
23. J Peccoud. Automating molecular biology: a question of communication. Biotechnology 13:741–745, 1995.
24. MC Pirrung, J Chen. Preparation and screening against acetylcholinesterase of a non-peptide "indexed" combinatorial library. J Am Chem Soc 117:1240–1245, 1995.
25. B Deprez, WL Bourel, H Coste, F Hyafil, A Tartar. Orthogonal combinatorial chemistry libraries. J Am Chem Soc 116:5405–5406, 1995.
26. KC Nicolaou, XY Xiao, Z Parandoosh, A Senyei, M Nova. Radiofrequency encoded combinatorial chemistry. Angew Chem Int Ed Eng 34:2289–2291, 1995.
27. EJ Moran, S Sarshar, JF Cargill, MM Shahbaz, A Lio, AMM Mjalli, RW Armstrong. Radio frequency tag encoded combinatorial library method for discovery of tripeptide-substituted cinnamic acid inhibitors of the protein tyrosine phosphatase PTP1B. J Am Chem Soc 117:10787–10788, 1995.
28. MHJ Ohlmeyer, RN Swanson, LW Dillard, JC Reader, G Asouline, R Kobayashi, M Wigler, W C Still. Complex synthetic chemical libraries indexed with molecular tags. Proc Natl Acad Sci USA 90:10922–10926, 1993.
29. For a review of current encoding technologies in use for combinatorial chemistry see AW Czarnik. New combinatorial methods—encoding. Curr Op Chem Biol 1:60–66, 1997.
30. NK Terrett, D Bojanic, D. Brown, PJ Bungay, M Gardner, DW Gordon, CJ Mayers, J Steele. Bioorgan Med Chem Lett 5:917–922, 1995.
31. P Meyers. Combinatorial chemistry: the road to hitsville? Pharmaceut News 3: 16–18, 1996.
32. CE Kibbey, Analytical tools for solution-phase synthesis. In: AW Czarnik, SH DeWitt, eds. A Practical Guide to Combinatorial Chemistry. Washington, DC: ACS Books, 1997, pp 199–248.
33. For a recent review of analytical methods used for the quality control of combinatorial libraries, see WL Fitch. Analytical methods for the quality control of combinatorial libraries. In: W Moos, M Pavia, B Kay, A Ellington, eds. Annual Reports in Combinatorial Chemistry and Molecular Diversity, Vol. 1. Leiden, The Netherlands: ESCOM, 1997, pp 59–68.
34. Although a number of groups have investigated microanalysis of resin-bound small molecules, the technique has not provided reliable results. SH DeWitt, personal communication, 1997.

35. JR Hauske, P Dorff. A solid phase CBZ chloride equivalent—a new matrix specific linker. Tetrahedron Lett 36:1589–1592, 1995.
36. B Yan, G Kumaravel, H Anjaria, A We, RC Petter, CR Jewell Jr, JR Wareing. Infrared spectrum of a single resin bead for real-time monitoring of solid phase reactions. J Org Chem 60:5736–5738, 1995.
37. B Yan, G Kumaravel. Probing solid phase reactions by monitoring the IR bands of compounds on a single ''flattened'' resin bead. Tetrahedron 55: 843–848, 1996.
38. K Russell, DC Cole, FM McLaren, DE Pivonka. Tools for combinatorial chemistry. In: IM Chaiken, KD Janda, eds. Molecular Diversity and Combinatorial Chemistry: Libraries and Drug Discovery, Washington, DC: ACS Books, 1996, pp 246–254.
39. EC Blossey, RG Cannon. Synthesis, reactions, and 13C FT NMR spectroscopy of polymer-bound steroids. J Org Chem 55: 4664–4668, 1990.
40. C Boojamra, K Burow, L Thompson, J. Ellman. Solid-phase synthesis of 1,4-benzodiazepine-2,5-diones. Library preparation and demonstration of synthesis generality. J Org Chem 62:1240–1256, 1997.
41. JA Boutin, P Hennig, P-H Lambert, S Bertin, L Petit, J-P Mahieu, B Serkiz, J-P Volland, J-L Fauchere. Combinatorial peptide libraries: robotic synthesis and analysis by nuclear magnetic resonance, mass spectrometry, and high-performance capillary electrophoresis techniques. Anal Biochem 234:126–141, 1996.
42. FGW Butwell, R Epton, EJ Mole, N Mazaffar, S Phillips. Deprotection studies in ultra-high load solid gel phase peptide synthesis. 1. Carbon-13 NMR investigation of the efficacy of boron trifluoride based side-chain deprotection cocktails. Innovation Perspect Solid Phase Synth., 1st Int Symp, SPCC UK, Birmingham, U.K., 1990.
43. JK Chen, SL Schreiber. Combinatorial synthesis and multidimensional NMR spectroscopy: an approach to understanding protein-ligand interactions. Angew Chem Int Ed Eng 34:953–969, 1995.
44. L Dapremont, AM Valerio, AM Bray, KM Stewart, TJ Mason, AW Wang, NJ Maeji. Multiple synthesis using the multipin method. Physiol Chem and Phys Med NMR 339–343, 1995.
45. J Ellman, D Mendel, P Schultz, D Wemmer, B Volkman. Site-specific isotopic labeling of proteins for NMR studies. J Am Chem Soc 114:7959–7961, 1992.
46. R Epton, P Goddard, KJ Ivin. Gel phase carbon-13 NMR spectroscopy as an analytical method in solid (gel) phase peptide synthesis. Polymer 21:1367–1371, 1980.
47. W Ford, S Yacoub. A 13C NMR method to determine the origin of cross-linked chloromethyl polystrenes used in polymer-supported synthesis. J Org Chem 46: 819–821, 1981.
48. W Ford, S Mohanraj, EC Blossey. Carbon-13 NMR spectra of polymer-support reagents–methods and procedures. Macromol Synth 10:91–96, 1990.
49. E Giralt, J Rizo, E Pedroso. Application of gel-phase 13C-NMR to monitor solid phase peptide synthesis. Tetrahedron 40:4141–4152, 1984.

50. MF Gordeev, DV Patel, J Wu, EM Gordon. Approaches to combinatorial synthesis of heterocycles: solid phase synthesis of pyridines and pyrido [2,3-d] pyrimidines. Tetrahedron Lett 37:4643–4646, 1996.
51. C Johnson, B Zhang. Solid phase synthesis of alkenes using the Horner-Wadsworth-Emmons reaction and monitoring by gel phase P NMR. Tetrahedron Lett 36:9253–9256, 1995.
52. AJ Jones, CC Leznoff, PI Svirskaya. Characterization of organic substrates bound to cross-linked polystyrenes by 13C NMR spectroscopy. Org Magn Reson 18:236–240, 1982.
53. WM Kazmierski, RD Ferguson, RJ Knapp, GK Lui, HI Yamamura, VJ Hruby. Reduced peptide bond cyclic somatosatin based opioid octapeptides. Synthesis, conformational properties and pharmacological characterization. Int J Pept Protein Res 39:401–414, 1992.
54. T Keating, R Armstrong. A remarkable two-step synthesis of diverse 1,4-benzodiazepine-2,5-diones using the Ugi four-component condensation. J Org Chem 61:8935–8939, 1996.
55. PA Keifer. Influence of resin structure, tether length and solvent upon the high-resolution 1H NMR spectra of solid phase synthesis resin. J Org Chem 61:1558–1559, 1996.
56. GC Look, CP Holmes, JP Chin, MA Gallop. Methods for combinatorial organic synthesis: the use of fast 13C NMR analysis for gel phase reaction monitoring. J Org Chem 59:7588–7590, 1994.
57. C Mapelli, MD Swerdloff. Monitoring of conformational and reaction events in resin-bound peptides by 13C NMR spectroscopy in various solvents. In: Peptides. Leiden, The Netherlands: ESCOM, 1995, pp 316–319.
58. S Garigipati, B Adams, L Adams, SK Sarkar. Use of spin echo magic angle spinning 1H NMR in reaction monitoring in combinatorial organic synthesis. J Org Chem 61:2911–2914, 1996.
59. A Meissner, P Bloch, E Humpfer, M Spraul, OW Sorensen. Reduction of inhomogeneous line broadening in two-dimensional high-resolution MAS NMR spectra of molecules attached to swelled resins in solid phase synthesis. J Am Chem Soc 119:1787–1788, 1997.
60. MM Murphy, JR Schullek, EM Gordon, MA Gallop. Combinatorial organic synthesis of highly functionalized pyrrolidines: identification of a potent angiotensin converting enzyme inhibitor from a mercaptoacyl proline library. J Am Chem Soc 117:7029–7030, 1995.
61. GC Look, MM Murphy, DA Campbell, MA Gallop. Trimethylorthoformate: a mild and effective dehydrating reagent for solution and solid phase imine formation. Tetredron Lett 36:2937–2940, 1995.
62. RC Anderson, JP Stokes, MJ Shapiro. Structure determination in combinatorial chemistry: utilization of magic angle spinning HMQC and TOCSY NMR spectra in the structure determination of Wang-bound lysine. Tetredron Lett 36: 5311–5314, 1995.
63. RC Anderson, MA Jarema, MJ Shapiro, JP Stokes, M Ziliox. Analytical tech-

niques in combinatorial chemistry: MAS CH correlation in solvent-swollen resin. J Org Chem 60:2650–2651, 1995.
64. SK Sarkar, RS Garigipati, JL Adams, PA Keiffer. An NMR method to identify nondestructively chemical compounds bound to a single solid phase synthesis bead for combinatorial chemistry applications. J Am Chem Soc 118:2305–2306, 1996.
65. D Sherrington, P Dennison, C Snape, R Law. Quantitative solid-state 13C NMR analysis of the functional groups present in divinylbenzene networks. Polymer Prepr 34:558–559, 1993.
66. C Dhalluin, I Pop, B Depreze, P Melnyk, A Tartar, G Lippens. Magic-angle spinning NMR spectroscopy of polystyrene-bound organic molecules. In: M Chaiken, KD Janda, eds. Molecular Diversity and Combinatorial Chemistry: Libraries and Drug Discovery. Washington, DC: ACS Books, 1996, pp 255–272.
67. AA Virgilio, JA Ellman. Simultaneous solid phase synthesis of β-turn mimetic incorporating side-chain functionality. J Am Chem Soc 116:11580–11581, 1994.
68. SS Chu, SH Reich. NPIT: a new reagent for quantitatively monitoring reactions of amines in combinatorial synthesis. Bioorg Med Chem Lett 5:1053–1058, 1995.
69. SH DeWitt, MC Schroeder, CJ Stankovic, JE Strode, AW Czarnik. DIVERSOMER technology: solid phase synthesis, automation, and integration for the generation of chemical diversity. Drug Dev Res 33:116–124, 1994.
70. BJ Egner, GD Langley, M Bradley. Solid phase chemistry: direct monitoring by monitoring by matrix-assisted laser desorption/ionization time of flight mass spectrometry. A tool for combinatorial chemistry. J Org Chem 60:2652–2653, 1995.
71. AA MacDonald, DG Nickell, SH DeWitt. Combinatorial chemistry: quality vs. quantity. Pharmaceut News 3:19–21, 1996.
72. BE Barr. Automation issues at the interface between combinatorial chemistry and high-throughput screening. In: IM Chaiken, K D Janda, eds. Molecular Diversity and Combinatorial Chemistry: Libraries and Drug Discovery. Washington, DC: ACS Books, 1996, pp 233–243.
73. For an overview of the information management issues associated with combinatorial chemistry, see SM Muskal. Information management. In: AW Czarnik, SH DeWitt, eds. A Practical Guide to Combinatorial Chemistry. Washington, DC: ACS Books, 1997, pp 357–397.
74. National Instruments (6504 Bridge Point Parkway, Austin, TX 78730, http://www.natinst.com).
75. For examples, see the software packages from Think and Do Software, Inc. (4750 Venture, Ann Arbor, MI 48108, http://www.ThinkandDo.com) or Nematron (5840 Interface Dr., Ann Arbor, MI 48103, http://www.nematron.com.).
76. DK Agrafiotis, RF Bone, FR Salemme, RM Soll, US Patent 5,463,564, 1995.
77. New paradigms for visualization of large amounts of data are being developed by several companies. For examples, see InXight's series of VizControls (InXight

Software, Inc., 3400 Hillview Ave., Palo Alto, CA 94304, http://www.InXight.com) or Perspecta's Smart Content System (Perspecta, Inc., 600 Townsend St., Suite 170E, San Francisco, CA 94103, http://www.perspecta.com).

78. Oxford Molecular Group Oxford, U.K. recently introduced DIVA (a visualization and analysis tool for chemical and biological data) at the 210th ACS national meeting in San Francisco.

8

Bioanalytical Screening Methodologies for Accelerated Lead Generation and Optimization in Drug Discovery

James N. Kyranos and Stewart D. Chipman
ArQule, Inc.
Medford, Massachusetts

I. HISTORICAL PERSPECTIVE

The testing of small, synthetic organic molecules for their ability to modify the biological activity of enzymes, receptor, etc., for use as human therapeutics is the process typically referred to as drug discovery. This process has evolved significantly over the last several decades with the advent of increasingly sophisticated biological and chemical methodologies, as well as laboratory scale automation. In the first half of the twentieth century, drug discovery was primarily performed by chemists and pharmacologists. The pharmacologist's tool of choice was in vivo bioassays to test mixtures of compounds isolated from natural product sources by natural product chemists or small organic compounds that had been synthesized individually by organic chemists. The realization that protein–protein and protein–small molecule interactions transmitted information in cellular biochemical pathways was the seminal observation that led to the "lock-and-key" hypothesis of biomolecular interactions. This concept that the interaction of specific shapes, charges, etc., on biological molecules can serve to control cellular and organism metabolism is now well established. The discovery of small organic molecules that antagonize or agonize biochemical interactions relevant to disease processes is the goal of scientists engaged in modern drug discovery. Further advances in protein purification and structural analysis has led to an even greater understanding of these molecular recognition elements which, along with advances in computer-aided design software/hardware, has lead to the development of rational design of small organic molecules for synthesis and testing.

The first biochemical assays were developed using isolated cells, cellular extracts or purified proteins and were conducted in glass test tubes. The continuous incremental improvement of the test tube format for bioassay has lead to the 96-well-plate standard currently favored by most scientists. In this chapter we will review the current status of bioassays used in modern drug discovery and how these assays have been automated to accelerate the pace of drug discovery. We will also discuss how high-throughput screening (HTS) has driven the need for varying types of high-throughput organic synthesis (HTOS) and how HTS and HTOS are being integrated into integrated industrial processes to accelerate the discovery of human therapeutics. Finally, we will discuss new methodologies that address the question of how scientists will test the large quantity of chemical compounds being generated by combinatorial chemistry approaches against the enormous volume of therapeutically relevant targets being discovered by the discipline of genomics and proteomics.

II. INTRODUCTION

Drugs are chemical compounds that modulate the activity of proteins and other targets associated with a disease state to achieve a desired therapeutic response. The discovery and development of drugs has traditionally been an inefficient and expensive process. Recent studies have shown that time frames of 10–15 years from the discovery of a validated molecular target to the market introduction of a Food and Drug Administration (FDA)–approved drug are typical, with the average cost estimated to be in excess of $300 million (1).

The traditional drug discovery process is set in motion by the realization of an unmet therapeutic need (Fig. 1). The first major step in the process is the identification of one or more molecular targets whose role in the pathophysiology of the disease process has been established. The discovery of novel genes in the human genome has been greatly accelerated by gene sequencing and sequence analysis. However, the role of many of these genes remains ambiguous, limiting their utility as drug discovery targets. Differential expression analysis of diseases versus normal tissues has provided more insight into

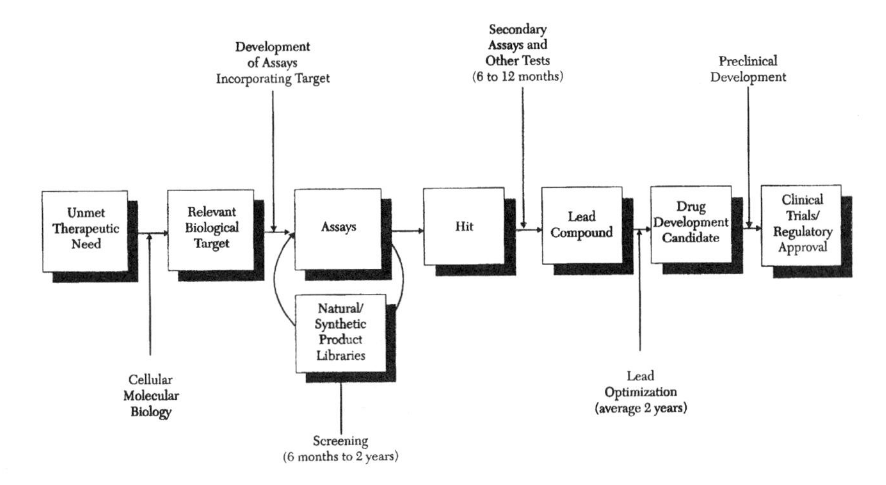

Figure 1 Traditional drug discovery process.

the role of novel molecular targets in human disease (2), but finding *validated* targets will remain one of the rate-limiting steps in the process for some time.

Once a suitable drug discovery target (i.e., enzyme, receptor, etc.) has been identified, the second step is to establish an efficient biological test or ''screen'' against which natural product extracts or small molecules can be assayed for agonism or antagonism. Typically these bioassays can take many forms, mirroring the diversity of the function of the biological targets themselves. In Sec. III we will describe several representative types of assays that have become popular in recent years. Once a suitable bioassay has been developed natural product extracts and small organic molecules are screened for activity against the target in order to generate biologically active structures or ''hits.''

Many automated robotic devices have been developed to accelerate the screening process, frequently referred to as high-throughput screening, or HTS. A discussion of automated robotic assay systems available will be found in Sec. IV. The advent of HTS systems with capacities approaching 10^5 compounds per day has driven the need for high-throughput organic synthesis (HTOS) and analysis. Further, the development of combinatorial chemistry and automated synthesis/analysis has necessitated strategies for integration of HTOS and HTS to maximize efficiency. The integration of these discovery technologies, and others, for accelerated lead generation and optimization will be discussed in Sec. V, with equal emphasis on technical developments and resource management.

The next major step in the process is lead optimization. Hits are optimized for potency against the primary therapeutic target, specificity against related and unrelated biological targets, and cellular toxicity/permeability. Those compounds with suitable characteristics for a potential drug are referred to as lead compounds. Typically during optimization a wider range of assay types are utilized to evaluate a smaller set of chemical compounds; this is an iterative design, synthesis, and testing cycle involving the biologist and the computational and medicinal chemists. Leads with appropriate properties are advanced to surrogate and functional animal models of the human disease to evaluate toxicity and bioavailability (including adsorption, distribution, metabolism, excretion), as well as disease modification or efficacy (3). Compounds that successfully meet the criteria of disease modification (efficacy) and safety are ready for Investigational New Drug (IND) application and further human clinical testing prior to market approval by the FDA.

In the early 1980s, high throughput biological screening using automated workstations and robotic systems increased the productivity of the primary screening biological component to the point where synthesis of new chemical entities became the rate-limiting step to the whole process. Moreover, as more biological assays were developed, optimized, and routinely performed on the robotic platforms, the limitation of the current synthetic approach became more obvious. In an effort to increase the productivity of synthetic chemistry, alternative modes of compound synthesis were investigated.

By the early 1990s split and pool combinatorial techniques had been developed for synthesis of a large collection of mixtures of peptides and small molecular weight compounds that could be used to identify initial leads against biological targets (4,5). Although combinatorial chemistry successfully provided large numbers of initial hits from the mixtures that were screened, the follow-up necessary to identify the active component from a complicated mixture was and still is quite challenging. Moreover, the screening of large mixtures is prone to a higher probability of false positive results due to incremental, additive effects of a large number of minor interactions. False negative results can be observed due to antagonist and agonist effects in the same mixture or due to a single compound that potently binds being offset by a larger number of compounds that bind weakly (6). Because of the above challenges, the pharmaceutical industry continues to prefer acquisition of chemical libraries of relatively large quantities (>10 mg) that contain single, well-characterized compounds having relatively high purity of the component of interest.

The requirements of single-component-per-well, highly characterized, pure compounds led to the development of high-throughput, parallel, combinatorial, organic synthesis borrowing many of the concepts developed for HTS (7). The ability to synthesize several thousand analogs simultaneously also increased the need to develop high-throughput analytical characterization as well as information management techniques and software to deal with the copious volumes of data. The development of HTOS, analytical chemistry, and chemical information management as drug discovery tools continues to narrow the numbers gap between chemical synthesis and biological testing.

Within an organization, the appropriate sizing of each of the four main disciplines (genomics, organic synthesis, biological evaluation, and preclinical testing) is critical for maximally efficient high-throughput drug discovery (Fig. 1). Appropriate management of a team of scientists who possess diverse skills, along with the numerous discovery tools they utilize, becomes a critical determinant of success in any drug discovery organization (8).

Finally, recent advances in genomics and synthetic chemistry that have yielded hundreds of new biological targets and thousands of new chemical entities focus attention back to the biological screening component as the starting point for the next major advancement in the drug discovery field. The conventional methods utilized for HTS need to be modified and/or replaced with economical, faster, high-resolution techniques that increase the productivity and efficiency of the drug discovery process. Many of the rapid, high-resolution methods presently under development use highly sensitive detection techniques such as mass spectrometry or fluorescence. These in turn allow an increase in the physical density of discrete samples per unit space (i.e., 1564 wells per plate) and limit the consumption of precious biological target and generation of chemical, biological, and radiological waste. We will review the past, present, and future of biological testing methodologies.

III. ASSAY DEVELOPMENT

The binding of a protein to its ligand is driven by a number of highly specific interactions, i.e., ionic interactions, hydrogen bonding, etc. These highly specific interactions serve to transmit biological ''information.'' This signal transduction occurs by several differing mechanisms including protein conformational change, substrate cleavage, or substrate modification. Therefore, it is of little surprise that the development of biological assays to discovery inhibitors of bimolecular interactions is a process almost as heterogeneous as the function of the proteins themselves. Two types of biological targets for which drug discovery assays are commonly developed are receptors and enzymes. We will present several examples of typical assays that have been developed for these classes of biological targets. However, the list is by no means exhaustive and each biological target will overlay unique considerations.

The simplest form of bioassay would detect any binding event between a receptor/enzyme and a ligand/substrate. For certain applications such a ‘‘nonfunctional’’ assay will provide useful information, and later in this chapter we will discuss extensively this approach to screening. However, only binding events that directly or indirectly interfere with the functionally relevant, ligand-binding surface (e.g., active site) of the protein will interfere with biological activity and therefore be of potential therapeutic interest. Whether the highest affinity binding ligands detected in these nonfunctional assay are those that interfere with binding at the protein's active site (since this is the region of the protein surface with the most binding interactions) is a matter of unresolved speculation. However, molecular bioassays that directly measure binding of a substrate to its enzyme's active site have been employed as the primary lead discovery assay in drug discovery. Molecular bioassays are based on the measurement of direct, selective binding of a labeled ligand to the protein's active site or the competitive displacement of the direct binding of a labeled ligand.

Enzymes are proteins that catalyze the formation or cleavage of chemical bonds. A wide range of enzymatic assays are utilized in modern drug discovery; for proteases the most simple measure the release of a quenched chromophore, fluorophore, or radioisotope from a peptide substrate. These styles of homogeneous assays are simple, with excellent signal-to-noise profiles. Fluorescence-quenched substrates (e.g., aminomethylcoumarin-AA_1-AA_2-AA_3-dinitrophenol-AA_4-AA_5) have gained immense popularity because the emission spectra of the aminomethylcoumarin in the intact substrate is entirely quenched by the dinitrophenol moiety (9). Substrate cleavage of amino acid 1–3 will result in the release of the fluorescence quench with a greater than 1000 times increase in the emission signal of the fluorophore. Such a homogeneous assay requires no separation or washing steps, yields a high signal-to-noise ratio, and thus is very amenable to robotic HTS automation of protease inhibition assays.

Another significant class of enzymes—the phosphotransferases (10)—catalyze the addition or cleavage of phosphate from amino acids and nucleotides. There has been an explosion in the identification of protein kinases, i.e., enzymes that frequently perform an intercellular signaling role to transmit information generated by ligands binding to the cell membrane to the cytoplasm or nucleus. Many investigators have described heterogeneous assays where they monitored the inhibition of activity of therapeutically interesting kinases by the incorporation of radioactively tagged ^{32}P from the substrate, requiring a subsequent separation of the free and bound ^{32}P by centrifugation or filtration (11). More recently, homogeneous-style kinase assays have been developed that require no separation step; enzyme activity is measured by fluorescence energy transfer (12). In this assay format a fluorescent molecule is attached to a substrate that can be phosphorylated. Upon addition of phosphate by the kinase, the substrate is recognized by a antibody with a second fluorescent molecule, which when in close proximity to the first fluorescent molecule causes a shift in its emission wavelength.

Receptors are proteins that bind ligands and transmit information via changes on other portions of the protein surface. The specific binding of a ligand is favored by a surface-to-surface fit that maximizes the ionic interactions, hydrogen binding, and other molecular interactions. Once a ligand has been identified that exhibits reasonably specific binding characteristics, then a competitive assay can be assembled (13). Homogeneous-style assay formats (i.e., scintillation proximity assay, or SPA) have been described for receptor binding that offer the advantages of the homogeneous format and the clean signal-to-noise ratio of radiochemical assays (14). For example, in the SPA assay, a receptor is bound to a polyvinyltoluene-based scintillation bead via one of several specific, high-affinity interactions such as protein A-IgG or avidin-biotin. These SPA bead–receptor complexes are then incubated with a control ligand that is radiolabeled and of known affinity. Once the radioligand is bound to the receptor and thus is in close proximity to the SPA bead, the energy of the decaying radionuclide is transferred to the SPA bead and the energy is converted to a light signal. This light signal is quantitated in a scintillation counter. A decrease or increase in the assay signal associated with the addition of a chemical entity is correlated with antagonism or agonism of the bead, respectively. Other similar formats are available; for instance, the scintillant can be embedded into the floor of the 96-well-plate (15) or, alternatively, the scintillant can be replaced with a fluorescence molecule to enable a fluorescence resonance energy transfer (FRET) system (12). These assay formats offer both advantages and disadvantages depending on the exact assay conditions.

A. The Test Tube?

The desire to test increasingly larger numbers of chemical entities, combined with the attendant high cost of reagents needed, has resulted in the evolution in the size of the container in which bioassays are conducted. A migration toward smaller and smaller tube sizes has also been facilitated by technological advances in liquid-handling devices and detection devices. As multiple test tube (8-ml working volume) assays became more common, vendors began to develop 6-well (2-ml working volume), 24-well (0.5-ml working volume), and finally 96-well (250-μl working volume) well plates. Scientists have adopted the standard 96-well plate because at this point liquid handling in the 1- to 100-μl range can be easily performed with manual and robotic liquid-handling devices (16). In addition, the advent of microprocessor-controlled fiberoptic detection devices with precise positional control has enabled the manufacture of detection devices capable of obtaining excellent visible, ultraviolet, or fluorescent signals from assays run in the 96-well-plate format (17).

It now appears that with current conventional liquid-handling and detection devices it is possible to utilize the 384-well (70-μl working volume) format (18). Further miniaturization of the plate format has yielded the 1536-well plate (15-μl working volume) that is currently being utilized by several investigators (19). However, utilization of this format requires nonconventional ''ink-jet''-style liquid handlers and imaging-type detection systems (20,21). These types of systems are currently available to lab scientists but at premium prices compared to 96-well technology. Several fundamental issues also remain as roadblocks for this technology, including sample evaporation, solvent compatibility with the ''biojet'' dispenser, and positional accuracy of the dispenser for each well. The decision to migrate to denser well formats may be driven in large degree by one's assay format, reagent costs, and research budget.

The next generation of assay technology will likely be a move away from plate-based screening altogether. Although massively parallel plate technology has afforded extraordinary throughputs, the inherent limitations of performing assays in wells will push many researchers to examine methodologies that do not require the use of tubes. We will discuss some of these advanced screening methodologies below.

IV. AUTOMATION OF BIOASSAYS

A. Manual Assays

As manual, multichannel pipeting devices and plate-based detection systems became staple tools in the modern discovery laboratory, bioassays for moderate throughput have been developed in a wide variety of formats. The flexibility of these detection devices allowed assays that utilize visible-, fluorescence-, luminescence-, and radioactive-based readouts. Manual pipeting allows for fine adjustment in the pipeting parameters for detergent, organic solvents, and other additives to the aqueous solution. These pipeting parameters include aspirate speed, delay time, ejection speed, and blowout speed and volume. Adjusting a mechanical pipeting device to accommodate the large variety of variables that it will encounter demonstrates the utility and flexibility of the human interface.

B. Liquid-Handling Robotics

Repetitive pipeting by humans is an error-prone activity. In order to eliminate the errors inherent in human high-throughput liquid handling, many manufactures have developed a wide variety of robotic devices. The most simplistic of these devices is the plate-based liquid-handling robot. Several examples of the more popular version of these devices include Beckman BioMek, Gilson Model 215, Packard Multiprobe, and Tecan Genesis (22). These liquid-handling workstations offer single-probe, 8 fixed probes, 4 or 8 variable span probes, and 96 fixed probes pipeting capability. Other options include fixed stainless steel probes, disposable pipet tips (20, 200, or 1000 μl size), a wide variety of liquid reservoirs with temperature control and mixing, system fluidics versus mechanical piston pipeting control, etc. All of these liquid handlers have computer-controlled functionality, with some of the better software offering GUI interfaces to facilitate programming of the instrument. The size of the pipeting deck varies greatly for these instruments. Some of the instruments are designed with the flexibility to adapt to many different microtiter plates, test tubes, and reservoirs, whereas others are dedicated to only the 96-well-plate format. Capacity will vary greatly, some liquid handlers being designed for only a few plates before a change of the deck is required and others allowing for longer periods of unattended operation. Which particular instrument will best fit your appropriate needs is probaly based on your specific needs.

C. Integrated HTS Robotics

Further bioassay automation requires the assembly of integrated robotic systems, such as those supplied by Zymark, Beckman/Sagian, or Robocon (23). These custom or semicustom robotic systems increase the bioassay throughput capability of the standalone liquid-handling robots by incorporating pipeting stations, light/temperature/atmosphere–controlled incubators, wash stations, detection devices, plate storage carousels, and plate transfer systems onto one system. Computerized control of the various subsystem routines allows for optimum scheduling and thus maximum throughput of the assay. These HTS robots are capable of operating for up to a 24-hour unattended duty cycle. Throughputs for a typical assay on these systems can range from several thousand per day for a ELISA-style assay to 15,000 per day (and up) for a homogeneous assay. These systems range from the highly customized component ones whereby a team of hardware engineers and software programmers are required for appropriate support, to the more ''off-the-shelf'' systems that can be maintained and operated by biological scientists with minimal mechanical skills.

V. INTEGRATION OF HTOS AND HTS FOR ACCELERATED LEAD GENERATION AND OPTIMIZATION

A number of research teams, including scientists at ArQule (7,24), have developed and reduced to practice the automated, parallel, solution phase synthesis of large (10^3–10^4) compound sets, or arrays, of small organic molecules. In this chemical synthesis scheme, each array of molecules is defined by a unique chemical scaffold or ''molecular core'' to which are attached diverse pharmacophoric side groups. These arrays are created by combinatorially reacting a series of functionally identical but structurally diverse building blocks to produce a single compound. The compound is then chemically analyzed by high-performance liquid chromatography (HPLC) and mass spectrometry (MS), and can be purified if necessary. The compounds are logically arranged in spatially addressable 96-well microtiter plates with a single compound per well (Fig. 2). This spatially addressable array format yields an ''information-rich'' array whereby chemical structure, mass, and other properties are stored in a chemical database that is searchable via a unique identifier. Many of these chemical arrays, each based on a unique molecular core, have been assembled into large (typically more than 250,000 compounds) chemical compound sets. This large compound set has been utilized to discover novel lead structures for pharmaceutical drug discovery that can be further optimized using traditional medicinal chemistry approaches.

Spatially addressable chemical arrays are inherently information-rich; this property results from utilizing reagents, arranged in the x, y, and z axis of the combinatorial cross, that have a high degree of structural relatedness (i.e., methyl, ethyl, isopropyl, etc.). As a result, the neighbors of a given product in this combinatorial array will also have a high degree of structural relatedness. This property of spatially addressable arrays provides an enormous potential for gaining structure–activity relationship (SAR) data from the primary bioassay. A very simple example of this principle can be demonstrated where a combinatorial array is created by crossing 8 related reagents on the

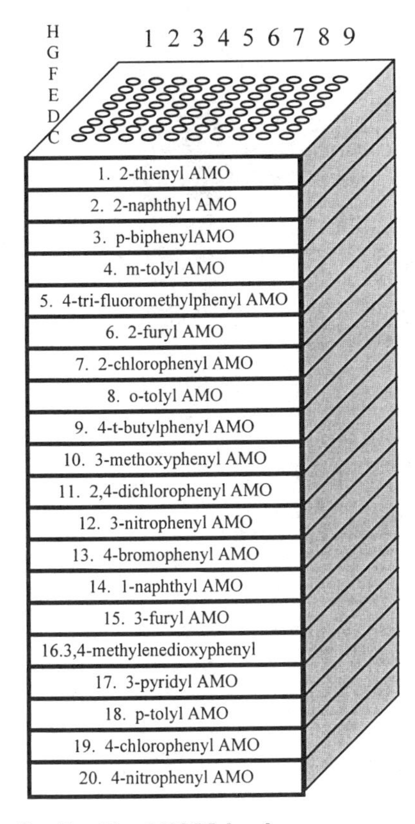

8 x 10 x 20 = 1600 Molecules

α-Keto esters

1 methyl 3-tri fluoromethylbenzoylformate
2 methyl benzoylformate
3 methyl 4-bromobenzoylformate
4 methyl 2,4-difluorobenzoylformate
5 methyl 4-nitrobenzoylformate
6 methyl 4-tert-butylbenzoylformate
7 methyl 3-methylbenzoylformate
8 methyl 3-methoxybenzoylformate
9 methyl 3-fluorobenzoylformate
10 methyl 4-methoxybenzoylformate

Diamines

A N,N'-dimethyl-1,2-ethylenediamine
B N,N'-dimethyl-1,3-propanediamine
C N,N'-dimethyl-1,6-hexanediamine
D piperazine
E homopiperazine
F 1,4-diaminobutane
G 1,3-cyclohexanebis(methylamine)
H 1,3-diaminopropane

AMO Precursor

EtO
N
R
O
O

Figure 2 Layout of a spatially addressable chemical compound array.

y axis and 10 related reagents on the x axis to yield 80 compounds. These chemical compounds were assayed versus a cysteine protease and the resultant data can be observed in Fig. 3. The SAR data obtained from this primary screen were used to rapidly accelerate the subsequent lead optimization efforts that resulted in a further 10-fold or higher gain in potency against the target protease.

A. Design and Synthesis

Several considerations drive the design and synthesis of spatially addressable, combinatorial arrays (25). The selection of building blocks and backbone

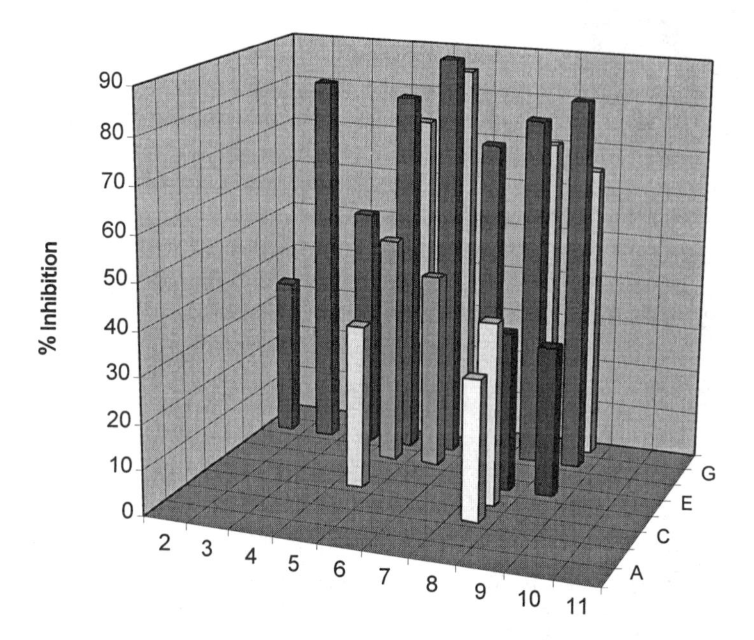

Figure 3 Percentage inhibition of cysteine protease activity by array X at 10 μM.

chemistries is driven by a combination of factors. These include years of accumulated medicinal chemistry knowledge, structure-guided design approaches based on high-resolution data of the target protein (when available), and biological assay data for inhibition of activity of the target protein (or related proteins) with closely related compounds (Table 1). In addition, there are practical criteria for synthesis that also drive the selection of a particular chemistry

Table 1 Factors Guiding the Design and Automated Synthesis of a Spatially Addressable Chemical Array

Chemical Array Design
• Medicinal chemistry experience
• Structure-guided design
• Biological data
Chemical Array Synthesis
• Synthetic feasibility
• Compatibility of chemistry with automated hardware
• Availability of building block reagents

and the inclusion of particular building blocks. These criteria include the general synthetic feasibility of the chemistry as regards the current portfolio of automated chemical synthesis hardware and the range of chemical transformations feasible on that platform. Finally, one must consider the availability of novel building block reagents and their influence on the chemical diversity of the array.

The work flow in synthetic organic chemistry can be broken down to a series of typical unit operations. Chemical synthesis, purification, and analysis can be leveraged by the application of laboratory scale automation to these unit operations to enhance their efficiency. Finally, one can apply the basic principle of computerized information management to enhance the level of process control. The integration of these technological approaches results in what can be referred to as industrialized HTOS. This approach to the chemical synthesis process has enabled the automated parallel synthesis, purification, and analysis of more than 2×10^5 compounds per year. Similar approaches in the electronics industry and biological testing disciplines have resulted in enormous productivity gains.

B. Automated High-Throughput Organic Synthesis

The discipline of combinatorial chemistry has automated a number of the unit operations in synthetic organic chemistry using a workstation approach (Table 2). Heavy emphasis has been placed on the adaptation of ''off-the-shelf,'' laboratory scale automation for the specific needs of synthetic chemistry to further simplify this approach. However, where necessary extensive customization or ground-up engineering of modules has been undertaken by research teams, as well as various automation vendors (26). Further addition of the

Table 2 Automated Molecular Assembly Process Workstations

Chemical inventory management
Weighing and dissolution
Open-well chemical synthesis
Thermal control and agitation
Centrifugal solvent evaporation
Liquid-liquid extraction
High-throughput preparative chromatography
Analytical chemistry
Array replication, shipping, and tracking

chemical design and synthesis principles outlined above and computerized process control has resulted in the integration of these workstations into a HTOS process. One possible iteration of the elements of this process and their interrelationships is represented in Fig. 4. In the figure they are represented as a linear process, but in actuality the order of the units operations in the process is infinitely flexible to enable the real-time adapting of the chemistry process to observations made by operators and analysts. Thus we are able to adopt commonly accepted industrial production practices to gain efficiency while essentially performing chemical research. Clearly, other chemical unit operations remain to be automated and integrated into the process. Automation of some of these unit operations will be a highly complex task and will involve both hardware and instrument control software engineering of the robotic unit operation. However, the integration of these modules into an HTOS process will be facilitated by the flexible architecture of the computerized information management software. It has been developed for maximal process control, along with standardized reaction block hardware designed to allow the physical reaction module to be quickly and easily moved from one station to another.

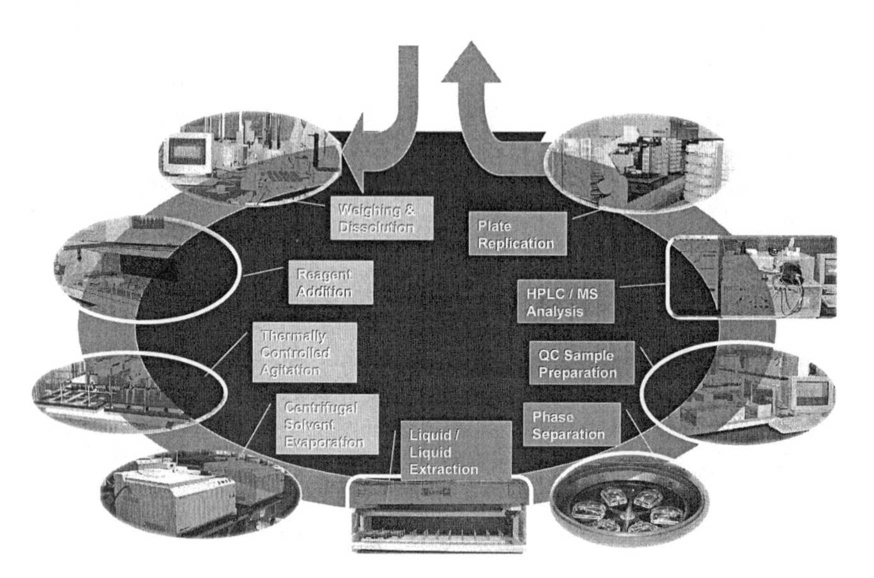

Figure 4 Industrialized high-throughput organic synthesis.

C. Information Management

In order to control the industrialized HTOS process scientists have developed software applications that facilitate the design and capture of the intellectual process that a chemist goes through when designing combinatorial chemical arrays (these have been termed array information management software, or AIMS). These software applications operate on conventional computer operating systems and computer/LAN hardware and offer a degree of process control with all of the attendant advantages. This software allows the chemist to develop an array production module (APM) or ''recipe'' that captures the set of mechanical instructions that the robotic instrument control software will require in order to execute the chemical synthesis on the automated platform. The AIMS facilitates and institutionalizes the APM knowledge base in a highly automated HTOS environment.

Once APMs have been created in the AIMS, a process control and monitoring software (PCMS) module can be constructed that provides for two-way communication between a database that contains the APMs and the instrument control software. This software allows for the retrieval of APM data from the AIMS by a production scientist and transfer to the instrument control software. Electronic transfer of APMs between unit operations in this manner allows for (a) charting of APM progress on the production floor and (b) collection of APM audit data. The PCMS will facilitate maximal process control for the automated chemical synthesis platform. Functionalities for the array information management and process control management software are summarized in Table 3.

The guidance system for the design, synthesis, and analysis of chemis-

Table 3 Summary of AIMS/PCMS Functionality

Array Information Management System
- Retrieval of chemical reagent data from chemical information database.
- Creation of array production models (APMs) by chemists to capture array layout and mechanical instructions.
- Ability to monitor the status of APMs on the production floor.
- Retrieval and reporting of array production history.

Process Control and Monitoring System
- Retrieval of APM details by instrument control applications from array information management software.
- Charting of array production progress through production and QC.
- Persistence of audit data from instrument control applications tracking APM natural history.

tries via HTOS is a series of interrelated information management systems. In addition to the AIMS/PCMS described above, organizations employing HTOS laboratories as one component of an integrated drug discovery platform have established fairly traditional chemical, analytical, biological, and corporate business information databases. Once chemicals have been synthesized and analyzed they are automatically registered in a chemical information management database, such as an ISIS-searchable Oracle relational database. An essential feature of such a database is the assignment of a unique identifier for each compound in the chemical compound set. This unique identifier provides links to information about the chemistry backbone, size of the array, and other chemical, physical and biological properties, as well as business data contained in the chemical and inter-linked databases. In drug discovery-based organizations, SAR data that are stored in an easily searchable (i.e., via a web browser) database can be utilized by medicinal chemists, biologists, and computational chemists to enhance the design of future arrays. Readily shared SAR data also facilitate the acceleration of lead optimization and preclinical studies by a large and diverse group of scientists.

D. Integration of HTOS and MHTS

The advent of ever increasing HTS methodologies has driven the need for high-throughput organic chemistry via combinatorial and parallel synthesis approaches. However, even with HTS technologies, the testing of 10^5–10^6 chemical compounds against multiple biological targets poses several significant hurdles, primarily the cost and availability of key reagents. Several strategies have been typically employed to manage the biological testing of large chemical compound sets against multiple biological targets. Single compound per bioassay per well is the most straightforward. The advantages are that no deconvolution is required and the potential for ''masking'' of bioactivity is minimized. Single compound per bioassay fits particularly well with the information-rich nature of spatially addressable chemical arrays. Essentially the entire primary bioassay provides extensive SAR data as was shown in Fig. 2, with the negative bioassay data also adding value for the subsequent lead optimization activities. However, the cost of this approach is significantly higher than that of the alternative strategy of compound pooling. Pooling of between 3 and 10 compounds per bioassay has been utilized to quickly and efficiently assay large compound sets (27,28). The major disadvantages are the need for subsequent deconvolution of positive readouts, potential for masking of one compound's activity by others and, specifically for spatially addressable chemical arrays, the information content of the compound set is partially lost.

In order to efficiently bioassay large compound arrays while preserving the integrity of the SAR information content of the spatial array, we have found that multiple, parallel, high-throughput bioassays are an attractive alternative. We will review several case examples of multiplexed bioassays with spatially-addressable chemical arrays and discuss issues related to data management, prevention of readout crosstalk and multiplexing of primary and secondary assays.

E. Multiplexed High-Throughput Screening (MHTS)

The concept of multiplexed bioassays to accelerate assays has been utilized previously. One of the best examples is DNA sequencing whereby an assay is performed using the method developed by Sanger with four different fluorescence-labeled reporters for each A, T, C, or G nucleotide (29). The reactions are loaded onto a 6% denaturing polyacrylamide gel and electrophoresed. As the fluorescence-labeled DNA moves past the laser in an Applied Biosystems DNA Sequencer, each dye is excited and emits a characteristic wavelength enabling nucleotide base assignment. The analogous use of multiplexed bioassays for HTS of spatially addressable chemical arrays preserves the information-rich nature of the array while allowing pooling of the assays to save costs.

Although MHTS bioassays can be multiplexed (2–4 per well) to accelerate testing of spatially addressable combinatorial libraries ($>10^5$ compounds) versus multiple biological targets, there are some practical limitations. For instance, the approach is more amenable to technically straightforward bioassays (i.e., SPA, colorimetric, or time resolved fluorescence (TRF)) since the assay conditions are more likely to be similar. Multiplexing is best employed when combining primary bioassays that utilize the same or different reporter systems (see below). In theory, multiplexing of first- and second-degree assays is possible, but again assay conditions must be similar, which is less often the case. Some of the inherent advantages of MHTS assay format include reduced consumable and personnel costs versus single compound/single bioassay, and an alternative to compound pooling that maximally preserves the information-rich nature of spatially addressable style combinatorial arrays. Single-compound-per-well testing prevents masking of activity and eliminates the need for deconvolution.

In the first example we have multiplexed two primary bioassays that utilize different reporter systems, i.e., visible and/or fluorescence readout. In vitro serine and metalloprotease assays were run in 96-well plates either separately and read in the appropriate detector, or multiplexed and read first in the spectrophotometer and then in a spectrofluorometer. Common assay conditions were developed and optimized for both enzymes. The serine protease assay substrate (e.g., AA_1-AA_2-AA_3 ~ pNA) had a visible readout at A_{405}; the metalloprotease assay substrate (i.e., AMC-AA_1-AA_2-AA_3-DNP-AA_4-AA_5) had a fluorescence readout (Ex_{320}, Em_{405}). ArQule compounds were dissolved in dimethylsulfoxide (DMSO), and tested at a final concentration of 10 μM (final solvent: 2% DMSO in aqueous pH 7.5 buffer). Serine- and metalloprotease-positive control inhibitors were spiked into a plate containing DMSO at a range of concentrations that yielded from ~20 to ~90% inhibition. The result from the separate and multiplexed assays is shown in Fig. 5; clearly, the observed inhibition in the multiplexed assay is quantitatively similar to the individual assays. In this particular example, the signal for the multiplexed fluorescence assay is reduced somewhat due to absorption by the chromogenic substrate; however the signal-to-noise ratio (*S*/*N*) remained acceptable.

In the second example, two primary, in vitro serine protease assays with identical *p*-nitroaniline-based chromogenic (A_{405}) readouts were either assayed separately or multiplexed. Positive control inhibitors specific to each enzyme were added to the assay at an appropriate concentration to determine assay

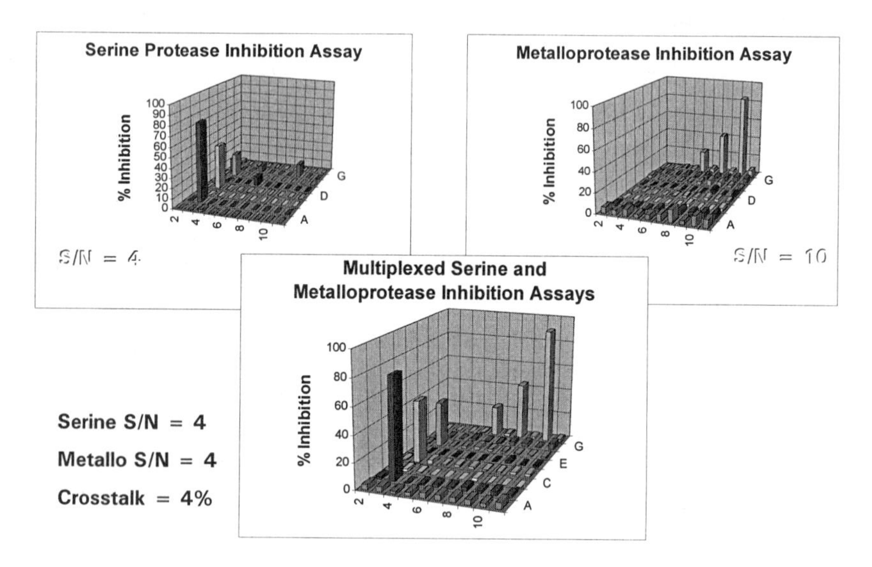

Figure 5 MHTS of first-degree bioassays: different reporter systems.

Table 4 Multiplexed High-Throughput Screening of First-Degree Bioassays—Same Reporter Systems

	Assay 1	Assay 2	Multiplexed assays (1 and 2)[a]
1 μM Inhibitor A	79 ± 2	0	43 ± 2
10 μM Inhibitor B	0	48 ± 3	20 ± 1
	(*S*/*N* = 11)	(*S*/*N* = 11)	(*S*/*N* = 9)

[a] Cross-talk between both enzymes vs. both substrates = 4%.

cross-talk and readout specificity. This represents a more difficult data interpretation scenario because any positive inhibition data will require separate follow-up testing for confirmation. In this case of identical readouts, the percentage inhibition determined for a given compound is mathematically halved. Experimental data for this assay (Table 4) demonstrated that when assayed separately the positive control inhibitor yielded the expected percent inhibition with a very high *S*/*N* value of 11. When the two assays were multiplexed inhibition by either positive control inhibitor was detectable (i.e., signal was well over the noise) *S*/*N* was only slightly compromised. Cross-talk was only 4% in this assay format.

F. Considerations for Establishing Multiplexed Assays

Several considerations for multiplexing assays can be determined from the above data. Clearly, multiplexing different assay readouts is optimal (e.g., chromogenic with fluorogenic or radiolabeled), since the possibility for cross-talk is minimized and the readout is unambiguous. Usually, to meet this criteria one would choose different assay classes (e.g., metalloprotease with serine protease), since the possibility for cross-talk would be minimized. Several conditions are necessary to determine if two or more assays are appropriate for multiplexing. These include the following: (a) cross-talk between enzymes and substrates must be minimal (<5%); (b) enzymes must not cleave or modify other enzymes or receptors in the assay during storage on the robot; (c) cross-talk between readouts (e.g., excitation/emission) must be minimal; (d) multiplexed ELISAs must confirm antibody specificity between antigens; and (e) multiplexed assays must utilize similar assay conditions (i.e., pH, time, buffers, etc.).

VI. ADVANCED TECHNOLOGIES FOR THE FUTURE OF BIOANALYTICAL TESTING

Many of the high-resolution screening methods presently under development employ high-sensitivity techniques, such as mass spectrometry or fluorescence detection. These sensitive techniques limit the use of precious biological target and/or increase the physical density of discrete samples per unit space. The latter consideration limits the use of solvents and generation of waste. Because of its inherent high sensitivity and direct mass measurement capabilities, mass spectrometry has been found at the forefront of new screening methodology development. The following examples represent the different approaches that have been taken with mass spectrometry.

A. Direct Protein-Ligand Binding

Many traditional bioassays rely on a competitive binding process whereby a ligand with known affinity for the target protein is incubated with a single chemical compound or a mixture. If the compound(s) of interest are interacting with the target protein with greater affinity, then the ligand will be excluded from binding. By measuring the difference in signal between the control sample (containing only the ligand of known affinity) and the sample with ligand and novel compound(s), any reduction in the interaction of the ligand is attributed to interaction by the chemical entity of interest. Although this indirect approach works fairly well with single-compound systems, it is more difficult with mixtures because of the necessary deconvolution of the mixtures. Moreover, this competitive binding process is not an option when a known binder for a target is not known.

However, by infusing a preincubated mixture of the receptor and ligand(s) into the mass spectrometer via electrospray ionization, the receptor–ligand complex can be seen directly (30,31). Typically, electrospray ionization of a large macromolecular protein will yield a mass-to-charge envelope of the molecular weight of the protein divided by the number of possible charges (between 2 and 30) that can be stabilized on that molecule. When the receptor–ligand complex is formed and survives the electrospray process, the molecular weight of the complex will be greater than the free receptor by the weight of the interacting ligand. A new peak, representative of the complex, will be seen in the receptor–ligand mass spectrum if the effective mass increase of the ligand divided by the number of charges on the complex is greater than the mass difference between two adjacent charge states of the free receptor. Using this approach, one can directly observe the receptor–ligand complex and determine which chemical entity is interacting with the receptor. Since the mass and the charge states for a given receptor are known and the masses of the combinatorial chemical entities are also known, then all possible combinations of receptor–ligand masses can be determined and monitored.

This direct receptor–ligand binding approach has been used to identify compounds with affinity for the receptor; however, by using infusion techniques that are very labor-intensive the overall throughput is rather limited. Recently, Li et. al. (32) automated the procedure by incorporating flow injection analysis using a conventional autosampler to increase the throughput. By injecting a small aliquot of the receptor–ligand solution in the flowing stream to the mass spectrometer, the analysis of each sample was conducted in under 3 min, thus gaining a significant increase in throughput.

B. Immunoaffinity Extraction LC/MS

The immunoaffinity extraction LC/MS technique (33) is a variation of affinity chromatography. The receptor of interest is loaded on the column, followed by injection of the combinatorial library of small molecular weight compounds. The compounds that interact with the receptor are captured on the affinity column while the rest of the library is removed from the column. Following the loading step the pH of the mobile phase is changed, which strips the affinity column of the receptor–ligand complexes and loads the complexes onto a second column containing a restricted access media (RAM) stationary phase (34). The RAM column separates the receptor from the ligands and the ligands are then flushed onto a reverse phase column for separation and identification by MS. Using this approach a large number of potential ligands can be rapidly screened against a variety of receptors. Moreover, by incorporating a number of switching valves and computerized control the whole process can be automated to provide a relatively high-throughput screen.

C. Multidimensional Chromatography/Mass Spectrometry

This technique uses size exclusion and reverse phase chromatography in tandem to separate the receptor bound ligand followed by mass spectrometry to identify the particular ligand from a combinatorial library (35). The protein target is incubated with a single compound or mixtures under conditions that optimize protein–ligand binding. An aliquot of the mixture is then loaded onto a size exclusion column that separates the protein and protein–ligand complex from the smaller molecular weight unbound ligand. The protein and the protein–ligand complex eluding from the size exclusion column in the void volume is directed onto a reverse phase column, whereas the retained low molecular weight compounds are directed to waste. Once the protein–ligand complex is directed to the reverse phase column with a high organic mobile phase, the complex is disrupted and the ligands that were originally bound to the protein are released. The reverse phase column then separates the different ligands from each other as well as from the protein and the column flow is directed into the mass spectrometer for identification of the ligands.

D. Bioaffinity Characterization Mass Spectrometry

In bioaffinity characterization mass spectrometry (BACMS) the combinatorial library is incubated with the protein under suitable conditions and the mixture is then injected into the mass spectrometer under electrospray mode (36,37). The protein–ligand complex can be isolated and then energized to dissociate the complex. Because of the relatively low concentrations of complexes that are formed, a Fourier transform ion cyclotron resonance (FTICR) instrument is used to accumulate the protein–ligand complex of interest in the FTICR cell. Once enough signal is accumulated in the cell, the complex is dissociated and the ligand is identified by high-resolution MS and MS/MS experiments. The benefit of this methodology is that, like all other solution-based characterizations, there is no stearic constraint on complexation due to a solid support tethering the protein.

E. Affinity Capillary Electrophoresis/MS

Once again, the strength of this technique is that the protein–ligand complex is allowed to form in solution, thereby removing any steric interference that may by possible in solid phase affinity chromatography (38,39). In this technique, an electrophoresis capillary is filled with an aliquot of the receptor solution, which has been pH-adjusted so that the receptor is neutral. The combinatorial library is then injected as a small plug in the capillary and the electrophoresis is started. As the library components move through the electrophoresis capillary, the ones that interact with the receptor are slowed down and come off the column later than those that do not interact with the receptor. By using a mass spectrometer as the final detector, the identity of the ligands that interacted can be determined for subsequent synthesis and additional study. Moreover, by appropriately adjusting the mobilities of both the ligands and the receptor, a variation known as the ''plug-plug'' approach can be performed. In this mode, both the ligands and the receptor are introduced in the capillary and allowed to interact at certain points in the capillary. By using this approach, weaker interacting ligands can be identified.

F. Immunoaffinity Ultrafiltration LC/MS

In this procedure the receptor is incubated with the combinatorial library under the appropriate conditions that will encourage receptor–ligand binding. Following incubation, the solution is transferred to centrifugal filters and centrifuged at 10000*g* for several minutes. The filter allows small molecular weight components—including the unbound ligands—to pass through the filter and retains compounds greater than the molecular weight cutoff of the filter, including free receptor and receptor–ligand complexes (40). Because of the inherent dead volume of the filter several aliquots of appropriate buffer are passed though the filter to remove any residual free ligand(s) solution. The affinity complex is then disrupted with an appropriate solvent and the solution is again centrifuged. This time the ligands that were originally bound to the receptor are allowed to pass in to a clean receptor tube and the receptor is still maintained on the filter. The solution containing the ligands extracted from the receptor–ligand complexes is then analyzed by HPLC and identified by MS using electrospray ionization. This procedure tends to select library components with the greatest affinity for the target receptor.

G. Pulsed Ultrafiltration Mass Spectrometry

The pulsed ultrafiltration MS technique is similar to the immunoaffinity ultrafiltration LC/MS process in that both techniques use a membrane with molecular weight cutoffs to separate free small molecular weight ligands from receptors or receptor–ligand complexes. The pulsed ultrafiltration process utilizes a typical HPLC/MS system, except that the chromatographic column is replaced by a pulsed ultrafiltration chamber. This chamber consists of either a 1.0-in.-diameter in-line solvent filtration unit that uses a high molecular weight cutoff filter membrane, or a home-built chamber that incorporates mechanical agitation (41). The mixture of receptor and a small molecular weight combinatorial library are first incubated and the solution is loaded into the pulsed ultrafiltration chamber. After the chamber is washed with an appropriate solvent for several minutes to remove any unbound ligands, the mobile phase solution is changed to one that will disrupt the receptor–ligand complex and release the free ligands. The free ligands are then transported to the mass spectrometer and identified by MS or MS/MS analysis. Alternatively, the receptor can be loaded into the pulsed ultrafiltration chamber and an aliquot of a combinatorial library can be passed through the chamber. The ligands that have a high affinity for the receptor will bind, whereas the others will pass through the chamber. Then by changing the mobile phase and releasing the bound ligands to be identified by MS, the receptor is now available for an additional aliquot of the same or a different combinatorial library to be screened. This is valid as long as the conditions used to disrupt the receptor–ligand complex does not permanently denature the receptor.

ACKNOWLEDGMENTS

The authors acknowledge their colleagues at ArQule, Inc., many of whom have made significant contributions to this body of scientific work.

REFERENCES

1. P Banerjee, M Rokofsky. Re-inventing drug discovery: the quest for innovation and productivity. (1997) *www.ac.com/services/pharma/phar_dd.html*
2. *www.lifespan.com*
3. *www.navicyte.com*
4. RA Houghten, C Pinilla, SE Blondelle, JR Appel, CT Dooley, JH Cuervo. Generation and use of synthetic peptide combinatorial libraries for basic research and drug discovery. Nature 7;354(6348):84–86, 1991.
5. *http://www.pcop.com/chemistry.html*, *http://www.combichem.com/home.html*, *http://www.trega.com/technology.htm*
6. S Kaur, L McGuire, D Tang, G Dollinger, V Huebner. Affinity selection and mass spectrometry-based strategies to identify lead compounds in combinatorial libraries. J Protein Chem 16(5):505–511, 1997.
7. JC Hogan. Combinatorial chemistry in drug discovery. Nat Biotechnol. 15(4): 328–330, 1997; JC Hogan. Directed combinatorial chemistry. Nature 384(6604 Suppl):17–19, 1996.
8. P Banerjee. Re-Inventing Drug Discovery: The Quest for Innovation and Productivity Presentation at Discovery '98: Emerging Technologies for Drug Discovery. NMHCC Biotechnology Conferences. May 18, San Diego.
9. CG Knight. Fluorimetric assays of proteolytic enzymes. Meth Enzymol 248:18–34, 1995.
10. T Hunter. The Croonian Lecture 1997. The phosphorylation of proteins on tyrosine: its role in cell growth and disease. Philos Trans R Soc Lond B Biol Sci 353(1368):583–605.
11. Y Miyata, B Chambraud, C Radanyi, J Leclerc, MC Lebeau, JM Renoir, R Shirai, MG Catelli, I Yahara, EE Baulieu. Phosphorylation of the immunosuppressant

FK506-binding protein FKBP52 by casein kinase II: regulation of HSP90-binding activity of FKBP52. Proc Natl Acad Sci USA 94(26):14500–14505, 1997.

12. *http://www.packardinst.com/offices/home_pbbv.htm*
13. G Deng, G Royle, S Wang, K Crain, DJ Loskutoff. Structural and functional analysis of the plasminogen activator inhibitor-1 binding motif in the somatomedin B domain of vitronectin. J Biol Chem 271(22):12716–12723, 1996.
14. *http://www.apbiotech.com/prod/html/apb_prod.html*
15. *http://www.apbiotech.com/prod/html/apb_prod.html*, *http://www.nenlifesci.com/p_srv_13.htm#74*
16. *http://www.corningcostar.com/costar/sub.html?sg=20*, *http://www.polyfiltronics.com/overview.html*, *http://www.greineramerica.com/products/96well.htm*
17. *http://www.wallac.fi/mag2/p6.html*, *http://www.moldev.com/*, *http://www.bmglabtechnologies.com/fluo.htm*
18. *http://www.corningcostar.com/launch384well.html*, *http://www.greineramerica.com/products/384well.htm*
19. *http://www.greineramerica.com/products/1536well.htm*
20. *http://www.cartesiantech.com/prod01.htm*
21. *http://www.fujimed.com/science/index.html*, *http://www.mdyn.com/*
22. *http://www.packardinst.com/cgibin/hazel.exe?action=SERVE&item=prod_serv/mprobe.htm*, *http://www.tecan.com/index_tecan.shtml*, *http://134.217.3.35/beckman/biorsrch/prodinfo/biomek/biowork.asp*
23. *http://134.217.3.35/beckman/biorsrch/prodinfo/biomek/biomek.asp*, *http://www.zymark.com/MARKETIN/drugd/allegr.htm*
24. *www.arqule.com*
25. CM Baldino, DS Casebier, J Caserta, G Slobodkin, C Tu, DL Coffen. Convergent Parallel Synthesis. Synlett 488–490, 1997.
26. *http://www.zymark.com/MARKETIN/drugd/Prsps.htm*, *http://www.gilson.com/gpss.htm*, *http://www.tecan.com/index_tecan.shtml*
27. DA Konings, JR Wyatt, DJ Ecker, SM Freier. Deconvolution of combinatorial libraries for drug discovery: theoretical comparison of pooling strategies. J Med Chem 39(14):2710–2719, 1996.
28. L Wilson-Lingardo, PW Davis, DJ Ecker, N Hebert, O Acevedo, K Sprankle, T Brennan, L Schwarcz, SM Freier, JR Wyatt. Deconvolution of combinatorial libraries for drug discovery: experimental comparison of pooling strategies. J Med Chem 39(14):2720–2726, 1996.
29. *http://www2.perkin-elmer.com/ab/about/dna/dna_seq/bdterm.html*
30. YT Li, JD Henion. J Am Chem Soc 113:6294–6296, 1991.
31. JAE Kraunsoe, RT Aplin, B Green, G Lowe. FEBS Lett 396:108–112, 1996.
32. LYT Li, CH Vestal, JN Kyranos. Proceedings of the 45th ASMS Conference on Mass Spectrometry and Allied Topics, Palm Springs, California, June 1–5, 1997, p. 897.
33. ML Nedved, S Habibi-Goudarzi, B Ganem, JD Henion. Anal Chem 68:4228–4236, 1996.

34. TCJ Pinkerton. J Chromatogr 544:13–23, 1991.
35. Y Hsieh, N Gordon, F Regnier, N Afeyan, SA Martin, GJ Vella. Mol Divers 2: 189–196, 1996.
36. JE Bruce, GA Anderson, R Chen, X Cheng, DC Gale, SA Hofstadler, BL Schwartz, RD Smith. Rapid Commun Mass Spectrom 9:644–650, 1995.
37. J Gao, X Cheng, R Chen, GB Sigal, JE Bruce, BL Schwartz, SA Hofstadler, GA Anderson, RD Smith, GM Whitesides. J Med Chem 39:1949–1956, 1996.
38. YH Chu, YM Dunaevsky, DP Kirby, P Vouros, BL Karger. J Am Chem Soc 118:7827–7835, 1996.
39. YH Chu, DP Kirby, BL Karger. J Am Chem Soc 117:5419–5420, 1995.
40. R Wieboldt, J Zwiegenbaum, JD Henion. Anal Chem 69:1683–1691, 1997.
41. RB Van Breemen, CR Huang, D Nikolic, CP Woodbury, YZ Zhao, DL Venton. Anal Chem 69:2159–2164, 1997.

9
Commercial Resources

Mary Brock and Mark Andrews
Waters Corporation
Milford, Massachusetts

Combinatorial chemistry and high-throughput screening (HTS) have become the fastest growth areas in pharmaceutical development. Combinatorial chemistry is a collection of technologies and disciplines. The ultimate objectives are to prepare compound libraries so that active leads can be identified more quickly, in greater numbers, and at lower costs. Over the past few years there has been an explosion in the number of companies developing, utilizing, and/or offering the various techniques for use in the lab. Press releases announce activities in these areas on a daily basis. It is a challenge to keep up with the current players, whether they are developing or licensing proprietary technologies, establishing collaborative relationships, commercializing products, acquiring or selling business operations, or any of a number of other activities.

Initially the intent of this chapter was to provide the reader with a comprehensive list of commercial suppliers of products (instrumentation and supplies) relating to combinatorial chemistry and HTS. While conducting research into these vendors, it became obvious that a lab would likely consider access to a particular technology, or to an awareness of the potential for collaboration with another company, to be of as much value as a listing of companies that provide specific instrumentation and related products. It also became obvious that a ''comprehensive'' list would be impossible because of the ongoing high level of activity among old and new participants in this area.

Therefore, this chapter is a sampling of companies that could potentially be a resource for your lab. It will provide a snapshot of some of the activities and companies that are currently involved in offering products and services relating to combinatorial chemistry. We offer our apologies to those companies or products that were inadvertently omitted from this listing.

The information provided was obtained from various public sources, including company press releases, company internet web sites, articles appearing in many of the industry trade journals, magazines, product advertising, product brochures, and other company sources. The entries contain information available at the time the chapter was written and are as accurate as the original sources. A listing of a company or its products does not constitute an endorsement.

Abbott Laboratories
100 Abbott Park Road, Abbott Park, IL 60064-3500, USA
Phone: (708) 937-6100; Fax: (708) 937-2013
Toll-free in U.S.: (800) 323-9100
Internet: *http://www.abbott.com*
Stock: ABT (NYSE)

Profile: One of the world's leading health care companies, Abbott is dedicated to improving people's lives through the discovery, development, manufacture, and marketing of health care products and services. Its strategy is to provide innovative technologies that improve the quality of health care while helping customers lower their costs.

Products/technologies: A team of scientists developed what they consider a revolutionary technique for discovering new drugs that could potentially have far-reaching implications for disease management. Called SAR by NMR (Structure–Activity Relationships by Nuclear Magnetic Resonance), Abbott claims the technique can significantly speed the difficult and time-consuming process for identifying molecules that bind to important protein drug targets.

Abbott compares SAR by NMR to combinatorial chemistry as follows: Both methods use a building block approach in the construction of molecules; however, in combinatorial chemistry, thousands to millions of molecules are typically synthesized and tested for activity. In SAR by NMR, less than 10 molecules is required because the chemistry is highly focused on linking molecules demonstrated to bind to the protein target using structural information on how they interact. Patent applications have been filed on the method, but the company plans to make the technology available to other drug developers.

Contact: Ellen Molleston Walvoord, V.P., Investor Relations and Public Affairs

Acacia Bioscience
4136 Lakeside Drive, Richmond, CA 94806, USA
Phone: (510) 669-2330; Fax: (510) 669-2334
E-mail: *info@acaciabio.com*
Internet: *http://www.acaciabio.com*

Profile: Acacia Bioscience is a functional genomics and drug discovery company using advances in genome science and combinatorial chemistry to identify potential lead compounds with higher degrees of therapeutic value.

Products/technologies: Acacia's drug discovery technology, called the Genome Reporter Matrix™, screens molecules against multiple intracellular disease pathways, revealing the in vivo targets of compounds with therapeutic value and increasing the efficiency of the drug discovery process. The Matrix can be used to enhance the screening of combinatorial libraries, detect the potential side effects of a drug candidate, resurrect failed drug candidates, and accelerate the discovery of new drug leads.

Contact: Bruce Cohen, President and CEO

ACADIA Pharmaceuticals
3911 Sorrento Valley Blvd., San Diego, CA 92121, USA
Phone: (619) 558-2871; Fax: (619) 558-2872
E-mail: *receptor@together.net*
Internet: *http://www.acadia-pharm.com*

Profile: ACADIA Pharmaceuticals is a biotechnology company engaged in development and use of high-throughput solutions for drug discovery. Founded in 1993, the company has developed a platform of proprietary breakthrough technologies for the functional characterization of genes encoding potential drug targets. ACADIA pursues drug discovery alliances with major pharmaceutical firms, as well as with biotechnology companies with expertise in genomics and combinatorial chemistry. It continues to develop and expand its technology platform and in-house discovery efforts on novel targets. Research facilities are maintained in both San Diego and Copenhagen, Denmark.

The company changed its name in August 1997 from Receptor Technologies Inc., and moved from Winooski, Vermont, to its California location.

Products/technologies: It has developed a proprietary technology that is referred to as Receptor Selection and Amplification Technology (R-SAT™). R-SAT permits the assaying of drug activity using recombinant receptors. The R-SAT assay is a high throughput assay that can be performed with a wide range of receptors that use a diversity of signal transduction mechanisms. The company claims that precise quantitative data for agonist, partial agonists, as well as neutral antagonists and negative antagonists (inverse agonists) can be obtained. Company offers contract assays, mass screening, reagents, and a novel method for expression cloning based on the R-SAT assay.

Contact: Mark R. Brann, Ph.D., CEO and Chief Scientific Officer

Advanced ChemTech
5609 Fern Valley Road, Louisville, KY 40228-1075, USA
Phone: (502) 969-0000; Fax: (502) 968-1000
Toll-free in U.S.: (800) 456-1403
E-mail: *info@peptide.com*
Internet: *http://www.peptide.com*

Profile: A manufacturer of synthesis instrumentation, biochemicals, and life science products, the company is a pioneer in the field of multiple synthesis technology. It manufactures a diverse line of instrumentation with applications in molecular diversity and rational design for drug discovery and development, including commercial models for high throughput organic small molecule and mixture combinatorial synthesis, as well as numerous proprietary synthesizer models. In addition to its headquarters in Kentucky, subsidiaries and distributors are located in Europe, Japan, Singapore/Southeast Asia, Korea, China/Hong Kong, India, and Taiwan.

Products/technologies: These include the BenchMark Synthesis series of combinatorial and high-throughput organic synthesis systems (384 HTS, 496 MOS, 440 MOS); ReacTech organic synthesizer; plus other instrumentation for combinatorial peptide and organic synthesis; catalog of peptide libraries, custom peptides, and peptide library synthesis services; extensive selection of amino acid derivatives, solid supports reagents, and other starting materials for solid phase organic, peptide, and combinatorial synthesis. The ACT Model 496 Multiple Organic Synthesizer™ is a fully automated dual-arm robotics synthesizer that can make up to 96 different compounds in a single run; reaction blocks with 40, 16, and 8 wells are also available. The ACT Model 357 FBS can be used in automated single or multiple organic and bioorganic molecular synthesis. It can synthesize up to 36 different compounds and can serve as a large-scale, single-product synthesizer with scales of up to 36 g or more of starting resin. Combinatorial libraries can also be prepared by the split-pool method.

Contact: Mark L. Peterson, Ph.D.

Afferent Systems, Inc.
440 Collingwood St., San Francisco, CA 94114, USA
Phone: (415) 647-6659; Fax: (415) 647-6551
E-mail: *info@afferent.com*
Internet: *http://www.afferent.com*

Profile: The company develops software for combinatorial chemistry, including instrument control, and product data generation, storage, and access. It also offers solution phase and solid phase combinatorial synthesis instruments that are low cost and offer a high throughput. Software runs on PCs under Windows 95 and NT 4.0 and on SGI workstations. Free short-term evaluation licenses are offered for those interested in evaluating the software.

Products/technologies: Myriad™ integrated system for combinatorial chemistry informatics combines multiple functional modules under a single graphical user interface. The first three modules are a combinatorial database generator, combinatorial database engine and browser, and a synthetic instrument controller. The Afferent synthesis station is based on Myriad, the Gilson 215 liquid handler, and optionally the Charybdis Calypso reaction block.

Affymax
4001 Miranda Ave., Palo Alto, CA 94304, USA
Phone: (415) 812-8700; Fax: (415) 424-0832
E-mail: *webmaster@affymax.com*
Internet: *http://www.affymax.com*

Profile: Affymax is a wholly owned subsidiary of Glaxo Wellcome.

Products/technologies: ESL (Encoded Synthetic Library) technology enables the identification of individual organic compounds from combinatorial libraries containing hundreds to hundreds of thousands of such compounds. Using ESL, a scientist can synthesize a large library of compounds in combinatorial mixtures, attaching tags at each reaction step. When an active compound is found, the tags allow the scientist to readily identify the specific active structure.

Contact: Gordon Ringold, Ph.D., CEO and Scientific Director, Affymax Research Institute

Affymetrix
3380 Central Expressway, Santa Clara, CA 95051, USA
Phone: (408) 731-5000; Fax: (408) 481-0422
E-mail: *sales@affymetrix.com*
Internet: *http://www.affymetrix.com*
Stock: AFFX (NASDAQ)

Profile: Affymetrix started operations in 1991 as a division of Affymax, N.V. and began operating independently as a wholly owned subsidiary of Affymax in February 1993. In March 1995, Glaxo Wellcome purchased Affymax, including its then 65% interest in Affymetrix. As a result of subsequent financings, Glaxo Wellcome's ownership was approximately 34% of Affymetrix as of August 1997. The company has approximately 150 employees.

Products/technologies: The company's GeneChip technology has potential for genomic applications—including linkage, association studies, and DNA fingerprinting. A collaboration agreement with Whitehead Institute in Cambridge, Massachusetts is focused on the development of a set of single-nucleotide polymorphic (SNP) markers that can be scored on a microchip—this could have uses in studies on gene expression or as a diagnostic screening tool to identify even single-point gene mutations.

Alanex Corporation
3550 General Atomics Ct., San Diego, CA 92121, USA
Phone: (619) 455-3200; Fax: (619) 455-3201

Profile: Alanex began operations as a limited partnership in May 1991. In 1993 the partnership was converted to a California corporation. Alanex was acquired in May 1997 and operates as a privately held subsidiary of Agouron

Pharmaceuticals (traded as AGPH on NASDAQ). The company is involved in the discovery and optimization of small-molecule drugs using its proprietary combinatorial chemistry, computational chemistry, high-throughput screening, and integrated information management system. An important element of the company's strategy is to enter into collaborations with pharmaceutical companies.

Products/technologies: Alanex has developed and integrated its proprietary ChemInformatics system with combinatorial chemistry and high throughput screening. This combined approach is designed to significantly shorten the time required to find lead compounds and optimize them into viable drug candidates. Its proprietary core drug discovery technology, Pharmacophore Directed Parallel Synthesis (PDPS), combines combinatorial chemistry with computational and medicinal chemistries that, when used in conjunction with high-throughput screening and pharmacology, form an integrated drug discovery platform that can be broadly applied to a wide array of biological targets. Its proprietary library design software, LiBrain, is used to maximize the diversity of its exploratory library by selecting for synthesis compounds from Alanex's virtual library of chemical structures.

Contact: Ed Baracchini, Ph.D.

Amersham Pharmacia Biotech
Bjorkgatan 30, S-751 82 Uppsala, Sweden
Phone: 46 (0)18 16 50 00; Fax: 46 (0)18 16 64 58
Internet: *http://www.apbiotech.com*
Stock: Parent companies listed as PH&U on Stockholm Stock Exchange; and PNU (NYSE)

Profile: In 1997 Amersham International plc and Pharmacia & Upjohn, Inc. merged their life science businesses, Amersham Life Science and Pharmacia Biotech, creating the world's largest research-based biotechnology supplier. With over $700 million in combined annual sales and more than 3600 employees, the company will focus on lab and industrial chromatography, industrial DNA synthesis, high throughput drug screening, custom radiochemical synthesis, molecular biology reagents, DNA sequencing and mapping, and electrophoresis. The new company will focus R&D efforts on technologies for matrix/polymer chemistry, high throughput sequencing, drug screening, and microarrays.

Products/technologies: The SMART System is optimized for micropurification and microbore chromatography. It recovers and purifies biologically active material present in subnanogram, nanogram, and picogram amounts. By scaling down column dimensions and optimizing system components, it delivers the low volumes and high concentrations of pure biologically active substances that life science researchers need for further investigation by microanalytical techniques. It is also used in peptide sequencing and protein structure–function applications. The company, along with Molecular Dynamics, developed a microarray system that permits researchers to make and analyze high-density microarrays with increased speed, efficiency, and sensitivity. Data from a single microarray experiment provide researchers with the ability to accurately measure gene expression levels in thousands of samples. In late 1996 the two companies launched the Microarray Technology Access Program to make available preferential, precommercial access to their microarray technology.

Contact: Mike Evans, Vice President

Argonaut Technologies
887 Industrial Road, Suite G, San Carlos, CA 94070, USA
Phone: (415) 598-1350; Fax: (415) 598-1359
E-mail: *info@argotech.com*
Internet: *http://www.argotech.com*

Profile: Argonaut was incorporated in November 1994 and completed a first-round venture financing of $4.5 million in January 1995. A second-round venture financing of $9.5 million was completed in May 1996. In addition to the research and development facility the company occupies in San Carlos, California, it also has facilities in Tucson, Arizona, and Basel, Switzerland, as well as a distributor in Japan.

Products/technologies: The company designs and develops a full range of instruments, reagents, consumables, and software that apply solid phase technologies to the synthesis and purification of small organic molecules. Products are aimed at improving the productivity of the medicinal chemistry laboratory by exploiting the vast productivity gains derived from high-speed parallel synthesis and combinatorial chemistry. Products include ArgoGel resins and ArgoPore resins for solid phase synthesis. The Nautilus 2400 Organic Synthesizer handles complex reactions, has the flexibility to synthesize small molecules using a broad range of chemistries, and handles a wide range of reagents, with capability for temperature control and use of inert atmospheres. The Quest

210 manual synthesizer can run up to 20 parallel reactions and can be used for solid or solution phase chemistries in a wide range of synthesis applications.

Contact: David P. Binkley, Ph.D., President and CEO

ArQule, Inc.
200 Boston Avenue, Suite 3600, Medford, MA 02155, USA
Phone: (781) 395-4100; Fax: (781) 395-1225
E-mail: *jsorvillo@arqule.com*
Internet: *http://www.arqule.com*
Stock: ARQL (NASDAQ)

Profile: ArQule was started in 1993 and completed its IPO in October 1996. The company created a technology platform for the discovery and production of chemical compounds with commercial potential, and provides novel compounds to the pharmaceutical and biotechnology industries.

Products/technologies: Proprietary modular building block technology integrates structure-guided drug design, high-speed parallel chemical synthesis; and information technology to accelerate the identification and optimization of drug development candidates. The company's Mapping Array™ products are proprietary libraries of novel, diverse, small-organic-molecule compounds used for screening against biological targets in lead generation. Directed Array™ products are customized libraries of closely related compounds for lead optimization. Products are based on a technology platform composed of structure-guided drug design, modular building block chemistry, combinatorial chemistry, informatics, and the company's Automated Molecular Assembly Plant (AMAP™). Scalability of products enables the development process to accelerate directly into the lead optimization phase.

Contact: Eric B. Gordon, M.D., President and CEO

Aurora Biosciences Corporation
11149 N. Torrey Pines Road, La Jolla, CA 92037, USA
Phone: (619) 452-5000; Fax: (619) 452-5723
E-mail: *info@aurorabio.com*
Internet: *http://www.auroabio.com*
Stock: ABSC (NASDAQ)

Profile: Founded in 1995, Aurora's primary mission is to advance drug discovery by developing miniaturized, automated systems for ultrahigh-throughput screening based on proprietary, versatile, fluorescence-based assays. The company was incorporated in California in May 1995 and reincorporated in Delaware in January 1996. The company announced its initial public offering of common stock in June 1997.

Products/technologies: Designs and develops proprietary drug discovery systems, services, and technologies to accelerate and enhance the discovery of new medicines. These systems enable very rapid screening of multiple genomic and other molecular targets to quickly and economically identify lead compounds with novel therapeutic potential. Aurora is developing an integrated technology platform comprising a portfolio of proprietary fluorescent assay technologies and an ultrahigh-throughput screening system designed to allow assay miniaturization. According to the company, their ultrahigh-throughput screening system (UHTSS™) can handle hundreds of thousands of compounds every 24 hours.

Contact: Paul Grayson, Vice President, Corporate Development

Axiom Biotechnologies Inc.
3550 General Atomics Court, San Diego, CA 92121-1194, USA
Phone: (619) 455-4500; Fax: (619) 455-4501
E-mail: *jlinton@axiombio.com*
Internet: *http://www.axiombio.com*

Profile: A privately held company established in August 1995, Axiom is a drug discovery company that provides pharmaceutical companies with novel high-throughput pharmacology systems to accelerate the process of drug discovery. Axiom serves the pharmaceutical industry by offering access to its novel systems through research collaborations. In such collaborations Axiom provides research and development expertise, and technology transfer of its proprietary high-throughput pharmacology systems, including instrumentation, cell-based assays, laboratory information management systems, and analytical pharmacoinformatics tools, to accelerate in-house discovery efforts.

Products/technologies: These include the High Throughput Pharmacology System (HT-PS) for drug discovery that uses natural cell lines, which more closely simulate the body's environment, to test compounds for a desired drug activity. The system consists of novel high-throughput pharmacology instrumentation, high-sensitivity cell-based assays, and novel pharmacoinformatics tools for managing and mining high-throughput pharmacology experimental data. Complementing Axiom's HT-PS system is its LIMS system, built on an ORACLE platform that can be customized to meet individual labs' needs.

Contact: James P. Linton, Senior Director, Business Development

AXYS Pharmaceuticals, Inc.
180 Kimball Way, South San Francisco, CA 94080, USA
Phone: (650) 829-1000; Fax: (650) 829-1001
Internet: *http://www.axyspharm.com*
Stock: AXPH (NASDAQ)

Profile: AXYS is involved in the integration of drug discovery technologies from gene identification through clinical development and is focused on the discovery of small-molecule therapeutics.

Products/technologies: Formerly Arris Pharmaceutical Corp., the company announced in January 1998 its plans to commercialize its capabilities in combinatorial chemistry and pharmacogenomics as well as its patented technology. The company licenses its combinatorial chemistry technology to pharmaceutical and biotech companies, and collaborates with companies wishing to screen its libraries.

Contact: Daniel Petree

Beckman Coulter, Inc.
2500 Harbor Blvd. Fullerton, CA 92834-3100, USA
Phone: (714) 871-4848; Fax: (714) 773-8898
Internet: *http://www.beckman.com*
Stock: BEC (NYSE)

Profile: Founded in 1935, the company develops, manufactures, and markets automated systems and supplies for life science research and clinical diagnostic laboratories. Through its acquisition of Sagian, Beckman gained access to high-throughput screening technology and liquid-handling robotics instrumentation. In late 1996 the company formed an alliance with MDL Information Systems to develop an integrated hardware and software system for high-throughput screening used in the drug discovery process. In October1997, Beckman acquired Coulter Corp. Beckman Coulter, Inc. will have combined worldwide sales of more than $1.7 billion.

Products/technologies: These include lab robotics, including ORCA, and other products from the purchase of the lab robotics division of Sagian, and the Biomek 2000 biorobotics workstation; 96-channel pipettor; SAMI software, robotic peripherals, and custom engineering services. New products introduced at the 1997 Pittsburgh Conference (Pittcon) were the Biomek integrated laboratory automation system for high-throughput drug screening using ELISA system, a receptor binding assay system, and a cell-based assay system.

Contact: Jay Steffenhagen, Vice President, Planning/Development

Belmont Research Inc.
84 Sherman Street, Cambridge, MA 02140, USA
Phone: (617) 868-6878; Fax: (617) 868-2654
E-mail: *software-chan@belmont.com*
Internet: *http://www.belmont.com*

Profile: Belmont Research develops, markets, and supports data visualization, data management, and rapid application development software. The company's worldwide customer base includes leading scientific, engineering, and business organizations in information-intensive industries such as pharmaceuticals, microelectronics, biotechnology, and government. Belmont helps pharmaceutical companies solve key database problems such as standardization and reuse of clinical database definitions and components, global tracking and reporting of drug adverse events to national regulatory authorities, tracking of data discrepancies in high-integrity databases, and visualization of chemical activity databases for new drug discovery.

Products/technologies: These include multidimensional data visualization and graphical reporting for applications relating to clinical trials, quality control, and high-throughput screening. The company also provides software consulting services.

Contact: Jim Ong

BioDiversity
Brunel Science Park, Uxbridge, Middlesex, UK
Phone: 44 1895 812020; Fax: 44 0895 811051
E-mail: *biodiv@dial.pipex.com*

Profile: A small UK contract research organization, the company was established to address the problems commonly encountered when natural samples are used for drug screening. The three founding members have expertise in the field isolation of fungi and actinomycetes, and are developing new methods for field and laboratory cultivation of unusual microbial groups.

Products: Prescreening analysis and characterization of microbial samples. Plans to extend and develop its library of organisms and characterized samples so that a database associated with chemical data can be made available for industry.

Contact: Neil Porter, Managing Director

BioFocus
130 Abbott Drive, Sittingbourne Research Centre, Sittingbourne, Kent, ME9 8AZ, UK
Tel: 44 1795 412300; Fax: 44 1795 471123
E-mail: *admin1@biofocus.co.uk*
Internet: *http://www.biofocus.com*
Stock: Traded on the OFEX market

Profile: BioFocus is a European pharmaceutical company formed in early 1997 to specialize in combinatorial chemistry by applying sophisticated technologies to the lead discovery and optimization process. Its scientific team gained its experience in drug discovery and development in one of the world's largest pharmaceutical companies. In August 1997 the company had a successful flotation on the OFEX market.

Products/technologies: The company tailors its combinatorial chemistry methodologies to meet specific client needs. It offers three main levels of products and services, which include a fully integrated lead discovery, expansion, and optimization service; design and synthesis of high-purity, diverse or focused representational arrays; and supply of custom-designed combinatorial synthesis sets.

Contact: Alan Clabon, Marketing and Customer Relations Director

Biotage, Inc.
1500 Avon Street Ext., Charlottesville, VA 22902, USA
Phone: (804) 979-2319; Fax: (804) 979-4743
Internet: *http://www.dyax.com*

Profile: Biotage is a division of the privately held Dyax. In 1995 Biotage, Inc. and Protein Engineering Corporation were merged to form Dyax for the purpose of developing novel proteins and peptides for use in separations, therapeutics, and diagnostics based on proprietary phage display technology. (See also listing for Dyax.)

Products/technologies: An important part of their overall business plan is parallel purification—synthesis and purification; its proprietary technology includes prepacked flash cartridges. Products include the PFP640™ purification system, FLASH 40 compression modules, SGM641™ step gradient module, AFC642™ step gradient module and AFC642 automated fraction collector. The company also offers the Parallex™ HPLC system for high-throughput parallel purification, designed to purify libraries, not just single compounds. It is collaborating with CombiChem in the development of an automated parallel purification system, the CombiSynx PWS-20x, for combinatorial libraries.

Contact: Robert A. Dishman, Ph.D., President and CEO

Bohdan Automation, Inc.
1500 McCormick Blvd., Mundelein, IL 60060, USA
Phone: (847) 680-3939; Fax (847) 680-1199
E-mail: *sales@bohdaninc.com*
Internet: *http://www.bohdaninc.com*

Profile: The company has over 25 years of experience in the field of automation. After many years of working in the industrial automation area, Bohdan began automating laboratory activities for the pharmaceutical, chemical, petrochemical, food, environmental, and other laboratory areas around 1985.

Products/technologies: Pioneered the development of the automated laboratory workstation. Its concept employs a miniaturized version of a gantry style robot that Bohdan developed in 1978. The benchtop chassis, which can be customized to a lab's needs, is used to automate activities in an application-specific or functionally dedicated fashion. Its workstations are designed to

automate all aspects of solution phase and solid phase, small molecule organic synthesis. Bohdan offers a complete family of automated laboratory workstations, such as the Automated RAM™ Synthesizer Workstation, that can support most weighing applications, has a capacity for hundreds of sample containers, and can process samples at the rate of 120 per hour. It also markets a combinatorial chemistry reaction block that accommodates a wide variety of organic solvents and handles both solid phase and solution phase chemistry.

Bruker Instruments Inc.
19 Fortune Drive, Billerica, MA 01821, USA
Phone: (978) 667-9580; Fax: (978) 667-3954
Internet: *http://www.bruker.com*

Profile: Bruker Instruments was incorporated in 1960 and is headquartered in Germany. Since its founding, the company has diversified into a broad spectrum of analytical techniques and methods in both research and QC/QA applications. These include NMR, FTIR, MRI, MS, and EPR.

Products/technologies: Robotic sample introduction for automated MS and LC-MS includes the HP 1100 HPLC unit (automated LC-FTMS) and the Gilson 215 for 96/384 microtiter plates.

Calbiochem-Novabiochem
Boulevard Industrial Park, Padge Road, Beeston, Notts NG9 2JK, UK
Phone: 44 (115) 9430-840; Fax: 44 (115) 943-0951
E-mail: *technical@calbiochem.com*
Internet: *http://www.calbiochem.com*

Profile: Calbiochem, Novabiochem, and Oncogene Research Products are brand names of CN Biosciences (trades as CNBI on NASDAQ). CNBI develops, produces, markets, and distributes a broad array of products used worldwide in disease-related life sciences research at pharmaceutical and biotechnology companies, academic institutions, and government laboratories.

Products: CNBI's product lines include biochemical and biological reagents, antibodies, assays, and research kits that it sells principally through its general and specialty catalogs. Resins and linker products are used in combinatorial chemistry.

Calipher Technologies Corp.
1275 California Avenue, Palo Alto, CA 94304 USA
Phone: (415) 842-1960; Fax: (415) 841-1970
E-mail: *info@calipertech.com*
Internet: *http://www.calipertech.com*

Profile: Founded in June 1995, the company was formed to advance laboratory techniques in step with genomics and combinatorial chemistry. It was among the last companies formed by Avalon Ventures before it closed out its venture capital activities. Caliper's president and CEO was previously a general partner in Avalon.

Products/technologies: LabChips microminiature analytical system technology—their Lab-on-a-Chip technology miniaturizes and automates experiments, such as drug screening and DNA analysis, on microchips the size of a dime. Chips can screen compound libraries against genomic targets, diagnose diseases, perform separation procedures, and conduct bioanalysis of FDA material. The company also has microfluidics technology.

Contact: Lawrence Bock, President and CEO

Cambridge Combinatorial Ltd.
Cambridge, England
Phone: 44 123 462244

Profile: The company was launched in February 1997 to capitalize on the increasing trend to outsource elements of the drug discovery process by large pharmaceutical and emerging biotechnology companies. Its mission is to specialize in the design, production, and supply of chemical structures for the drug discovery industry. Oxford Molecular Group PLC (OMG) was fundamental in the establishment of Cambridge Combinatorial. OMG's CEO is Dr. Tony Marchington, brother of Cambridge Combinatorial CEO Dr. Allan Marchington. Principal shareholders are management, the founding scientists, and the University of Cambridge. In August 1997, Oxford Molecular Group announced that it had taken an option to buy Cambridge Combinatorial. The move is intended to turn OMG from a company that designs software into one that designs new drugs. In February 1998 the company acquired the exclusive worldwide patent rights to laminar combinatorial chemistry technology from Pfizer.

Products/technologies: Initial products will be moderately sized libraries of up to 20,000 compounds in a pure, well-characterized reproducible form. Milligram batches of each component in a library will be produced at the same time to provide the end-user with sufficient material for every stage of testing. Combinatorial libraries include those designed by Cambridge Combinatorial's founding shareholder, the Oxford Molecular Group. Laminar combinatory chemistry technology, acquired from Pfizer, allows the production of thousands of pure chemical compounds in milligram quantities, as single entities of a known structure, synthesized to the customer's individual needs.

Contact: Dr. Allan Marchington, CEO

Cellomics Inc.
635 William Pitt Way, Pittsburgh, PA 15238, USA
Phone: (412) 826-3600; Fax: (412) 826-3850
E-mail: *taylor@cellomics.com*
Internet: *http://www.cellomics.com*

Profile: Cellomics Inc., formerly BioDx Inc., is a privately held corporation founded in October 1996. Cellomics' mission is to improve the efficiency of the drug discovery process by delivering a cell-based screening platform that automates target validation and lead optimization using fluorescence-based assays.

Products/technologies: ArrayScan™ system is the platform for high-content screening (HCS) approaches for screening combinatorial libraries. Integrated platform consists of proprietary, fluorescence reagents and assays; a novel, proprietary, cell-based HCS system; and bioinformatics software. The platform provides deep biological information (time, space, and activity) about a drug candidate's physiological impact on specific cellular targets in living cells. Cellomics comarkets the Carl Zeiss Ultra High Throughput Screening (UHTS) system and can be combined directly with the Cellomics HCS system through an exclusive, worldwide alliance.

Contact: D. Lansing Taylor, Ph,D.

Charybdis Technologies, Inc.
2131 Palomar Airport Rd., Suite 300, Carlsbad, CA 92009, USA
Phone: (619) 431-5160; Fax: (619) 431-5163
E-mail: *srbush@charybtech.com*
Internet: *http://www.carybtech.com*

Profile: According to company statements, Charybdis was started in 1996 to design, develop, and implement the new standard in combinatorial technology and informatics, and to apply this paradigm towards the discovery of novel, potent, and effective chemical entities.

Products/technologies: The Calypso System™, designed for both lead generation and lead optimization, is an integrated system for high-throughput and combinatorial organic synthesis, and is based around the Calypso 96/48/24 reaction block. It is capable of integration with a variety of automation platforms including the Myriad software written by Afferent Systems. The Calypso System gas manifold is a replacement top coverplate for the Calypso reaction block.

Contact: Thomas J. Baiga, President; Steven R. Bush, Director of Information Technologies

ChemBridge Corporation
16981 Via Tazon, Suite G, San Diego, CA 92127, USA
Phone: (619) 451-7400; Fax: (619) 451-7401
Toll-free in U.S.: (800) 964-6143
E-mail: *chem@chembridg.com*
Internet: *http://www.chembridge.com*

Profile: The company is an international provider of chemical tools for drug discovery and development. With main offices in San Diego, California, it also has a research, development, and production infrastructure in Russia. Members of ChemBridge's management team have senior executive and senior research experience with leading U.S., European, and Japanese pharmaceutical and biotech companies.

Products/technologies: Hand-synthesized small molecule screening libraries for lead generation offer DIVERSet™96 Program, SCREEN-Set™ Program, CHERRY-Pick™ Program, and EXPRESS-Pick™ Program; COMBI-

TOOLS™ Program for combinatorial chemistry and parallel synthesis; and custom chemical synthesis and manufacturing.

ChemGenics Pharmaceuticals, Inc.
One Kendall Square, Building 300, Cambridge, MA 02139, USA
Phone: (617) 374-9090; Fax: (617) 225-2997

Profile: ChemGenics is a drug discovery company that applies its two technology platforms, Drug Discovery Genomics™ and Advanced Drug Selection Technologies, to key rate-limiting steps to gene-based drug discovery; it generates novel targets from disease genes and lead structures from molecular diversity. The company was established through an alliance that combined Myco Pharmaceuticals' leadership in gene technologies and its expertise in drug discovery with the advanced separations/analysis tools and drug discovery technologies of PerSeptive Biosystems. In January 1997 it was jointly announced that Millennium Pharmaceuticals would acquire the privately held ChemGenics Pharmaceuticals as part of their mutual strategy to become a world leader in drug discovery. Following the merger, the combined company will employ over 345 people. PerSeptive Biosystems is a principal ChemGenics shareholder. In August 1997, Perkin-Elmer announced its plans to acquire PerSeptive.

Products/technologies: ChemGenics has state-of-the-art, high throughput screening capabilities; it has validated its screening technology in combinatorial, synthetic chemical, and natural products libraries. The company has a proprietary natural products drug source and access to compound libraries through corporate alliances; these complement the combinatorial and pharmaceutical compound libraries that Millennium has accessed through its collaborations with Eli Lilly and Wyeth-Ayerst. ChemGenics also has microbial genetics technologies and has developed biocombinatorial methods involving genetic engineering to enhance chemical diversity.

Contact: Alan L. Crane, Vice President, Business Development

Chemical Computing Group Inc.
1255 University St., Suite 1600, Montreal, Quebec, Canada H3B 3X3
Phone: (514) 393-1055; Fax: (514) 874-9538
E-mail: *hayden@chemcomp.com*

Profile: CCG develops and markets scientific software for protein modeling, 3D bioinformatics, cheminformatics, molecular modeling, combinatorial chemistry, and high-throughput screening.

Products/technologies: QuaSAR-Binary™, a proprietary technology for the analysis of high-throughput screening experimental data. The technology can analyze the results of HTS experiments and the structure of molecules in order to make predictions regarding the biological activity of chemical compounds. It can make timely predictions about the biological activity of chemical compounds from large amounts of binary HTS data; it is based on fundamentals that enable the direct modeling of experimental error and uncertainty. CCG is the supplier of the Molecular Operating Environment (MOE), which the company says is the next generation molecular computing system and cross-platform software system for chemical visualization, simulation, and analysis.

Contact: Bill Hayden, Vice President, Marketing

Chemical Design Ltd.
Roundway House, Cromwell Park, Chipping Norton, Oxfordshire, OX7 5SR, UK
Phone: 44 1608 6444000; Fax: 44 1608 642244
E-mail: *sales@chemdesign.co.uk*
Internet: *http://www.chemdesn.com*

Profile: Chemical Design is one of the oldest molecular modeling and database companies. Founded in 1983, the company reported over 700 installations of its software worldwide in 1997. Hardware sales were a strong component of its business in its early years; however, in 1990, the company moved into database software development and phased out hardware sales. In addition to the UK location, offices are also located in the United States and Germany.

Products/technologies: Produces Chem-X combinatorial chemistry software that is used to facilitate the discovery and design of novel compounds that exhibit a particular therapeutic activity. ChemDiverse software is a module for Chem-X. Chem-X is a complete clientserver system for R-group selection, library design, library registration and enumeration, integrated with biological data and robotics systems. There are also options for reaction, 2D and 3D searching. The HTS chemicals collection on CD-ROM provides a source of diverse molecules for high-throughput screening programs and reagents for combinatorial chemistry; can be used on a wide range of platforms. An entry level Chem-X/BASE system for MS-Windows (including Windows 95) or Apple Macintosh (including Power Macintosh) is contained on the CD-ROM. Additional products include Chem-X/INVENTORY and ChemIdea module.

ChemStar, Ltd.
Geroev Panfilovtsev str. 20, Bldg. 1, Office 300, Moscow 123514, Russia
Phone: (7 095) 948 5468; Fax: (7 095) 977 5919
E-mail: *chemstar@ict.msk.ru*
Internet: *http://www.glasnet.ru/*chemstar/

Profile: ChemStar offers scientific and business cooperation in screening organic substances for the development of new products to be used in pharmaceutical, agricultural, and other fields. Company information indicates that it is associated with leading chemists of the Commonwealth of Independent States.

Products/technologies: Offers an extensive database of organic compounds for screening, as well as samples of organic compounds for high-throughput screening in quantities ranging from 1 to 1000 mg.

Contact: Geroev Panfilovtsev

Chiron Corp.
4560 Horton Street, Emeryville, CA 94608-2916, USA
Phone: (510) 655-8730; Fax: (510) 655-9910
Toll-free in U.S.: (800) 524-4766
E-mail: *Corpcom@cc.chiron.com*
Internet: *http://www.chiron.com*
Stock: CHIR (NASDAQ)

Profile: Founded in May 1981, Chiron is a $1+ billion science-driven health-care company that combines diagnostic, vaccine, and therapeutic strategies for controlling disease. The company has research programs in gene therapy, combinatorial chemistry, cancer, infectious and cardiovascular disease, and critical care through its Chiron Technologies business unit.

Products/technologies: Proprietary small molecule combinatorial chemistry technology, small molecule libraries, and proprietary automated chemical synthesis systems. The Multipin system technology (involving parallel synthesis of individual peptides on a solid phase support of 96 polyethylene pins arrayed in a format similar to microtiter plates) is now also being applied to libraries of small organic molecules. Multipin SPOC kits (solid phase organic chemistry) are sold through Chiron Mimotopes.

Contact: Kimberly Kraemer, Manager, Corporate Communications

Chiroscience Group plc
Cambridge Science Park, Milton Road, Cambridge CB4 4WE, UK
Phone: 44 (0)1223 420430; Fax: 44 (0)1223 420440
E-mail: *info@chiroscience.com*
Internet: *http://www.chiroscience.com*
Stock: CRO (London Stock Exchange)

Profile: Founded in 1992 to capitalize on its founders' expertise in chiral technologies and the anticipated market growth in chiral drugs, the company acquired Darwin Molecular Corp. in December 1996. The move paved the way for Chiroscience's entry into genomics, based on Darwin's integrated gene-based discovery platform that includes advanced bioinformatics, combinatorial chemistry, and cell-based screening capabilities. It continues to expand its technological knowledge in chiral and medicinal chemistry, drug discovery, and disease mechanisms, and has used the knowledge to build a broad drug discovery program. ChiroTech, the company's technology services business, has developed relationships with companies in the pharmaceutical industry by solving complex chiral problems through various partnerships and collaborations.

Products/technologies: Rational drug design capabilities; chiral technology; mass spectrometry Tag technology. ChiroTech offers the ChiroChem Collection for application in combinatorial and medicinal chemistry. The collection of molecules, derived from specifically selected classes of compounds, is being launched using a phased approach. The regular supply of these structures, both novel and technically challenging in their synthesis, will complement the customer drive to establish greater structural diversity. Each new series in the collection will be accompanied by a diagram illustrating the extent of diversity achieved.

Contact: Dr. Brian Gennery, Director of Development

ChiroTech (See Chiroscience Group)

CombiChem
9050 Camino Santa Fe, San Diego, CA 92121, USA
Phone: (619) 530-0484; Fax: (619) 530-9998
E-mail: *cci@combichem.com*
Internet: *http://www.combichem.com*

Profile: Founded in 1994, this privately held company develops and provides integrated discovery chemistry technologies and services in collaboration with pharmaceutical, biotechnology, and fine-chemical companies to accelerate the drug discovery process. In these partnerships, the companies provide the biological capabilities and Combichem provides the chemical expertise of the discovery process. Its mission is to provide a highly efficient and successful approach to the discovery of new drug molecules.

Products/technologies: The company has proprietary technology and products in the areas of instrumentation for automated multiple-parallel synthesis, advanced chemical design software for diversity assessment and optimized chemistries. Products include the Drug Discovery Engine, which accelerates drug discovery through the integration of medicinal chemistry, small library design technology, and automation tools. Lead Generation is a process employed for novel targets with no known leads. Its proprietary Universal Informer Library generates information and launches the drug discovery cycle. With Lead Evolution, hypotheses are generated based on available data; it then matches the hypotheses against its proprietary virtual library and ultimately synthesizes alternate drug templates. Lead Optimization is a proprietary design methodology that constructs libraries around a collaborative partner's lead to improve it before identifying it as a drug development candidate. CombiChem is collaborating with Biotage in the development of an automated parallel purification system, the CombiSynx PWS-20x, for combinatorial libraries.

Contact: Vincent Anido, Jr., Ph.D., President and CEO

ComGenex, Inc.
Hollan Erno u.5, Budapest, H-1136, Hungary
Phone: 36-1-1124-874; Fax: 36-1-214-2310
E-mail: *www@cdk-cgx.hu*
Internet: *http://www.comgenex.com*

Profile: The company was established in Budapest, Hungary, in November 1992 after an incubatory period of 5 years under CompuDrug Chemistry Ltd. ComGenex supplies substances for classical and high-throughput screening. The company is said to have a strong chemical background in solid and solution phase combinatorial chemistry and adapts its products and services to the ever changing and growing needs of the market. ComGenex is a sister company of RoboSynthon, Inc., which markets the MultiReactor™ combinatorial chemistry workstation for parallel organic synthesis manufactured by ComGenex. Offices are also located in Europe, Japan, and the United States.

Products/technologies: The company has more than 15,000 compounds in stock for high-throughput screening; this stock is dynamically changing by about 10–30% of the compounds each month. The supply of compounds is split between external suppliers and in-house synthesis, which is carried out using its Matrix Technology, a proprietary solution phase combinatorial chemistry technology. ComGenex also offers lab services; its labs are located in the Central Chemical Research Institute of the Hungarian Academy of Sciences. Additionally, it provides plant extracts, offering a collection of more than 390 extracts from the geographic region of Hungary.

Contact: Dr. Ferenc Darvas, President and CEO

Daylight Chemical Information Systems, Inc.
27401 Los Altos, Suite 370, Mission Viejo, CA, USA
Phone: (714) 367-9990; Fax: (714) 367-0990
E-mail: *info@daylight.com*
Internet: *http://www.daylight.com*

Profile: Daylight Chemical Information Systems, Inc. was incorporated in 1987 and grew from the MedChem Project at Pomona College. The invention of the SMILES language, by Dave Weininger, first at the U.S. Environmental Protection Agency in the early 1980s and then Pomona, laid the groundwork for the creation of a new chemical information system. Daylight's mission has been to provide high-performance chemical information processing tools to chemists. New software has been developed continually and the list of available and supported software continues to grow. Emphasis has been placed on the Daylight Toolkit, a set of programming libraries constituting a chemical information infrastructure on which custom applications can be built. The Toolkit has been used both by Daylight developers to build supported applications and by customers to build custom in-house applications. In addition to the corporate office in Mission Viejo, the company has a research office in Santa Fe, New Mexico.

Products/technologies: Daylight Software Release 4.5 is an integrated set of programs and libraries providing high performance chemical information processing, including structural entry, display, distributed data storage, retrieval, and searching. Release 4.5 runs under Unix with user access via X-Windows (e.g., Macs, PCs). The company also provides server programs providing network services, and various toolkits (object-oriented programming libraries). Daylight's combinatorial library toolkits are used to register, store, and search designed libraries.

Dyax Corp.
765 Concord Avenue, Cambridge, MA 02138-1044, USA
Phone: (617) 868-0868; Fax: (617) 868-0898
E-mail: *dyax@dyax.com*
Internet: *http://www.dyax.com*

Profile: A privately held company, Dyax was formed in 1996 with the merger of Biotage, Inc., a provider of preparative chromatography products and located in Charlottesville, VA, and Protein Engineering Corp., a Cambridge, MA, biotechnology company. Its ''purification by design'' philosophy incorporates the design and creation of a broad range of systems to effectively meet any purification challenge.

Products/technologies: The company's proprietary Phage Display technology is used for discovery of high-affinity ligands in drug discovery, for development of custom affinity purification ligands, and for development of highly specific diagnostics. Phage display is a method for generating novel molecules with high binding affinity for virtually any target. As a part of its existing separations business, the company is exploiting this technology to develop high-value affinity separation products designed to increase the efficiency and reduce the cost of pharmaceutical purification. Dyax is also applying phage display to the discovery and development of a new generation of diagnostic imaging agents and novel pharmaceutical compounds that bind tightly and specifically to known diagnostic markers and therapeutic targets. Other products include chromatographic systems, columns, and media for the purification process; Proprep sanitary protein gradient LC systems have capacities ranging from 0.05 to 50 liters per minute; Parallex parallel purification systems, include the PFP 640 and the Parallex HPLC system, a high-throughput preparative HPLC system. See also Biotage listing.

Contact: Pamela Hay, Director of Corporate Development

Dynex Technologies
14340 Sullyfield Circle, Chantilly, VA 20151-1683, USA
Phone: (703) 631-7800; Fax: (703) 631-7816
Toll-free in U.S.: (800) 336-4543
E-mail: *webmaster@dynextechnologies.com*
Internet: *http://www.dynextechnologies.com*

Profile: Formerly Dynatech Laboratories, the company has supplied microtiter technology products and services to research and clinical laboratories in the healthcare market for more than 30 years. Its products are marketed to scientists and researchers within universities, hospitals, and industry involved in research assays and clinical diagnostics. Dynex Technologies is a subsidiary of Thermo BioAnalysis Corporation, a Thermo Electron company.

Products/technologies: These include Microtiter® Plastics (plates, strips, accessories); Microtiter plate coating systems; Microtiter detection reader systems, automated Microtiter plate immunoassay processing systems (such as the Dias™ immunoassay system); Microtiter plate washers; and Revelation software.

Contact: Monica Durrant, Director

EVOTEC BioSystems GmbH
Grandweg 64, D-22529 Hamburg, Germany
Phone: 49 40 560810; Fax: 49 40 56081222
E-mail: *evotec@evotec.de*
Internet: *http://www.evotec.de*

Profile: EVOTEC was cofounded in 1993 by Nobel laureate Manfred Eigen, Ph.D., of the Max Planck Society, a pioneer in the research of molecular evolution and one of the first researchers to address molecular diversity through parallelization and miniaturization. The company's technology originates from

Dr. Eigen's work and from other Max Planck Institutes. It is also in collaboration with other researchers at academic institutions throughout Europe. Nearly half of its staff of 70 hold doctorate degrees. The company has been funded through private capital as well as government grants and early revenues from research contracts with international companies. EVOTEC makes its technologies available to third parties under product development or technology and transfer agreements.

Products/technologies: These include instrumentation and optical technology supporting ultrahigh-throughput screening. Its optical system provides accuracy and flexibility in the detection of molecular interaction. The EVOscreen instrument incorporates Evotec's proprietary fluorescence detection technology—a technology sensitive enough to detect single molecules, even within cells, and yield important information about their interactions—with an automated, miniaturized platform designed for rapid screen configuration. Single-cell/single-bead selection techniques combine conical fluorescence scanning and single-bead/living cell retrieval systems, directed molecular evolution methods, and artificial intelligence–guided chemical synthesis of functional compounds.

Contact: Christen Hence, Ph.D., CEO

Genevac Ltd.
The Sovereign Centre, Farthing Road, Ipswich IP1 5AP, UK
Phone: 44 1473 240000; Fax: 44 1473 461176
E-mail: *sales@genevac.co.uk*
Internet: *http://www.genevac.co.uk*

Products/technologies: Atlas high-throughput evaporators are designed to meet the demands of combinatorial chemistry, parallel synthesis, natural products research, and HPLC prep. The new technology and large sample capacity allows rapid evaporation of up to 1 liter or more of even the most difficult solvents in a few hours at low or moderate sample temperatures; suitable for use with most sample formats, including microtiter plates, test tubes, and vials.

Contact: Harry Cole, Sales Manager

Genelabs Technologies Inc.
505 Penobscot Drive, Redwood City, CA 94063, USA
Phone: (650) 369-9500
Internet: *http://www.genelabs.com*
Stock: GNLB (Nasdaq)

Profile: Genelabs is a biopharmaceutical company with research focused on the discovery of small-molecule drugs that act by binding to DNA or RNA to regulate gene expression or inactivate pathogens. In February 1998 the company began acquisition of directed combinatorial chemistry compounds through agreements with Tripos Inc., MDS Panlabs, and SRI International.

Products/technologies: Merlin and Viria assay systems screen combinatorial chemistry libraries for molecules and subunits of molecules that bind to segments of genetic material as small as four base pairs. The database will archive the subunits and the tiny genetic segments they disable. The database created to archive the subunits and the tiny genetic segments they disable will initially enable the rapid design of drugs to counteract pathogens employed in biological warfare; it will also have application as a database for traditional drug discovery.

Contact: Cynthia Edwards, Vice President, Research

Genome Systems Inc.
4633 World Parkway Circle, St. Louis, MO 63114, USA
Phone: (314) 427-3222; Fax: (314) 427-3324
Toll-free in U.S.: (800) 430-0030
E-mail: *info@GenomeSystems.Com*
Internet: *http://www.genomesystems.com*

Profile: For more than 4 years, Genome Systems has supplied researchers with genomic clones and technical support. The company makes the libraries it screens.

Products/technologies: ''Do-it-yourself library screening'' — perform library screening with ''Down-to-the-Well''™ PCRable DNA pools or by using high density colony filters.

Genome Therapeutics Corp.
100 Beaver Street, Waltham, MA 02154, USA
Phone: (718) 893-5007; Fax: (718) 893-8277
E-mail: *webmaster@genomecorp.com*
Internet: *http://www.cric.com*
Stock: GENE (NASDAQ)

Profile: The company was founded in 1961; initial research was conducted under an SBIR grant from the National Science Foundation. The company develops proprietary gene databases and uses genomic research technologies to accelerate the development of novel therapeutics, vaccines, and diagnostics. Through alliances, the company is attempting to bridge the gap between gene discovery and drug discovery.

Products/technologies: These include proprietary high-throughput ''multiplex'' DNA sequencing; positional cloning; bioinformatics; bacterial genomics; PathoGenome sequence database.

Contact: John Richard, Vice President, Business Development

Gilson, Inc.
3000 W. Beltline Hwy., Middleton, WI 53562-0027, USA
Phone: (608) 836-1551; Fax (608) 831-4451
Toll-free in U.S.: (800) 445-7661
E-mail: *sales@gilson.com*
Internet: *http://www.gilson.com*

Profile: Gilson is a manufacturer of specialized analytical instrumentation for scientific research and industrial markets since the early 1950s. The company has developed software and instrumentation for HPLC, LC, and sample preparation technologies. It also produces high-precision pipettes and tips and automated instruments and systems for increasing throughput and productivity in the laboratory.

Products/technologies: Products include automated purification and analysis systems for combinatorial chromatography, including its high-throughput Combinatorial Chromatography System for HPLC purification and analysis. Additional product offerings are the UniPoint System software; automated sample preparation products; autosamplers that accept up to 10 deep well or standard microplates; dual-function autosamplers for small-scale preparative applications; and graphic sample tracking system to eliminate errors in HPLC purification of combinatorial products. Its ASPEC XL4 off-line, high capacity (108 samples/batch) SPE systems are used for purification applications using GC and LC-MS and combinatorial chemistry techniques. Other products are a chemical synthesis workstation and a high-throughput personal synthesizer system for lead generation and lead optimization libraries. Its high-throughput LC-MS Interface System is for sample preparation, injection and fraction collection.

Hewlett-Packard Co.
Centerville Road, Wilmington, DE 19808, USA
Phone: (302) 633-8696; Fax: (302) 633-8916
Toll-free in U.S.: (800) 227-9770
Internet: *http://www.hp.com/go/chem*
Stock: HWP (NYSE)

Profile: A major manufacturer of chemical-analysis instrumentation since 1965, H-P's core competencies are in measurement, computers, and communications. The company combines a full range of data-handling and standard networking products with HP and other vendors' analytical instruments to provide chemists with integrated information to increase productivity. In November 1997, HP and IRORI announced an agreement to develop an integrated drug discovery platform that combines IRORI's microchip synthesis and screening technologies with HP's measurement and information portfolio for chemical analysis.

Products/technologies: Offers an automated combinatorial chemistry analysis system built around the HP 1100 Series HPLC, with the Gilson 233 XL sampling injector. Its GeneArray Scanner focuses a laser onto a 3-µm section of a 20-µm probe array on the Affymetrix chip and scans the chip in as little as 5 min. The company claims the GeneArray can read next-generation GeneChip probe arrays with up to 400,000 DNA probe sequences.

Houghten Pharmaceutials (see Trega Biosciences)

ICAgen, Inc.
4222 Emperor Boulevard, Suite 460, Durham, NC 27703, USA
Phone: (919) 941-5206; Fax: (919) 941-0813
E-mail: *info@icagen.com*
Internet: *http://www.icagen.com*

Profile: Founded in 1992, this privately held company is involved in the research and development of drugs that work by opening and closing ion channels. The company is currently in partnerships for the discovery of novel ion channel modulating drugs, and is the first biopharmaceutical company to integrate combinatorial biology with combinatorial chemistry focused exclusively on ion channel drug discovery and development. ICAgen has discovery programs in cardiovascular, central nervous system, and immunological disorders.

Products/technologies: These include proprietary biological targets and chemical libraries.

Contact: Kay Wagoner, Ph.D., Chief Executive Officer

IGEN International Inc.
16020 Industrial Drive, Gaithersburg, MD 20877, USA
Phone: (301) 984-8000; Fax: (301) 230-0158
E-mail: *gurewitz@igen.com*
Internet: *http://www.igen.com*
Stock: IGEN (NASDAQ)

Profile: Founded in 1982, the company develops, manufactures, and markets diagnostic systems utilizing its patented ORIGEN technology. The technology is used in products developed by IGEN as well as its licensees. ORIGEN is based on the principle of electrochemiluminescence, which uses labels that, when attached to a biological substance and then electrochemically stimulated, emit light at a particular wavelength.

Products/technologies: These include the Origen high-throughput screening system, built around IGEN's ECL modules (ECLMs). ECLM is expected to provide the advantages of rapid assay kinetics and improved assay range, while reducing the volume of critical reagent.

Contact: Herman Spolders, V.P., Business Development and Planning

Incyte Pharmaceuticals, Inc.
3174 Porter Drive, Palo Alto, CA 94304, USA
Phone: (650) 855-0555; Fax: (650) 855-0572
E-mail: *Webmaster@incyte.com*
Internet: *http://www.incyte.com*
Stock: INCY (NASDAQ)

Profile: The company was incorporated in 1991 by a founding team of scientists and venture capitalists; in 1996, Genome Systems Inc. and Combion Inc. were acquired. Incyte designs, develops, and markets genomic databases and services to pharmaceutical companies on a nonexclusive basis. A wholly owned subsidiary, Genome Systems, Inc., is located in St. Louis, Missouri. The combined companies employ more than 500 people.

Products/technologies: Incycte designs, develops, and markets genomic databases and services to pharmaceutical companies. The company has proprietary DNA sequence, gene expression, and microbial genome databases and high-throughput full-length cloning technology. Its LifeSeq® database is claimed by the company to be one of the world's largest sources of genomic data, containing gene sequence and expression information from normal as well as diseased cells and tissues from most of the major tissues of the human body.

Contact: Lisa L. Peterson, Director, Business Development

Institute National de la Propriete Industrielle (INPI)
26 bis, rue de Saint-Petersbourg, 75800 Paris Cedex 08, France
Phone: 01 53 04 53 04; Fax: 01 42 93 59 30
Internet: *http://www.inpi.fr*

Profile: INPI is the French Patent and Trademark Office and the French Register of Commerce & Trade. INPI registers patents, trademarks, industrial designs and companies. One of its missions is to make this information more readily available by providing worldwide access to more than 35 million records of technical and business information. In December 1997, INPI and Derwent Information announced an agreement in principle to create an extensive structure searchable patent information resource for the pharmaceutical and chemical communities. Their new venture is aimed at focusing on the identification of lead compounds by shortening the time scale during the drug and chemical development life cycles. It is particularly relevant to those organizations incorporating combinatorial chemistry methodology into their development process.

Products/technologies: It is intended that a new file will be available to customers by mid-1998 as a fast-alerting, structure-searchable on-line file. INPI and Derwent plan to extend the backfile coverage in parallel with the current coverage. Within a couple of years, they anticipate that retrospective coverage will be a full 20 years, focusing initially on pharmaceutical patents and extending coverage to all technologies later.

Institute for Scientific Information
3501 Market Street, Philadelphia, PA 19104, USA
Phone: (215) 336-4474; Fax: (215) 386-2911
Toll-free in U.S.: (800) 336-4474
E-mail: *custserv@isinet.com*
Internet: *http://www.isinet.com*

Profile: ISI was founded in 1958 and is a database publisher that indexes bibliographic data, cited references, and author abstracts from scientific, technical, and medical sources. The company publishes *Current Contents*® and the *Science Citation Index*®. Its objective is to provide immediate desktop access to the most significant scientific literature, in its entirety, to anyone, anytime, anywhere.

Products/technologies: Index Chemicus® database allows access to important reaction diagrams, full bibliographic information, full-length author abstracts, and a variety of searchable indices; published in one substructure- and text-searchable database.

Irori Quantum Microchemistry
11025 North Torrey Pines Road, La Jolla, CA 92037, USA
Phone: (619) 546-1300; Fax: (619) 546-3083
E-mail: *info@irori.com*
Internet: *http://www.irori.com*

Profile: Established in 1995, the company is venture-funded. It is an early-stage development microelectronic and microchemistry company focusing on applying memory technology to solid-state matrices. Irori applies its SMART technology to the development of small-molecule combinatorial libraries. In November 1997, IRORI and Hewlett-Packard announced an agreement to develop an integrated drug discovery platform that combines IRORI's microchip synthesis and screening technologies with HP's measurement and information portfolio for chemical analysis.

Products/technologies: SMART system (single or multiple, addressable, radiofrequency tags) and MEMs (memory enhanced matrices); SMART Microspheres, which are used to record information about individual chemical reactions as members of a combinatorial library are synthesized. Current microchip-based products from IRORI include a combinatorial chemistry module, a robotic sorter (AutoSortTM-10K), and a chemistry cleavage station (AccuCleaveTM-96). Initial bioassay and high-throughput screening systems are scheduled for launch during late 1998 with diagnostic/sensory industry to follow.

Contact: Michael P. Nova, M.D., President and CEO

Isco, Inc.
4700 Superior Street, Lincoln, NE 68505, USA
Phone: (402) 464-0231; Fax: (402) 464-4543
E-mail: *info.sid@isco.com*
Internet: *http://www.isco.com*
Stock: ISKO (NASDAQ/NMS)

Profile: Founded in 1958, the company designs, manufactures, and markets instruments used by engineers, technicians, and scientists in the laboratory and in the field to address concerns such as research and development in chemistry and biotechnology, water pollution, process monitoring, quality control, and environmental testing.

Products/technologies: Automated flash systems for combinatorial chemistry are designed for either high-throughput or high productivity. Both high-throughput systems, the 10-channel Parallel System and the 5-channel Parallel System can be used with either a single solvent or multiple solvents. The initial 5- and 10-channel systems offer the ability to collect fractions based on time. The high productivity systems can run up to 16 or 32 columns with Isco's sequential automated flash systems. One version is engineered for single-use columns, the other for reusable columns or cartridges. Additional products include LC, HPLC, optical detectors, fraction collectors, pumps, electrophoresis apparatus, SFE systems and syringe pumps.

Contact: Douglas M. Grant, President and CEO

LC Packings
80 Carolina Street, San Francisco, CA 94103, USA
Phone: (415) 552-1855; Fax: (415) 552-1859
Toll-free in U.S.: (800) 621-2625
E-mail: *info@lcpackings.com*
Internet: *http://www.lcpackings.com*

Profile: LC Packings is a privately held company based in San Francisco, Amsterdam and Zurich. It was founded in Zurich in 1987 with the goal to develop, manufacture, and commercialize packed microcolumns for use in HPLC. The company offers a complete range of products for use in microseparation techniques such as micro, capillary and nano LC, including electrochromatography, capillary electrophoresis, and LC-MS.

Products/technologies: The mixing routine of the FAMOS™ microsampling HPLC workstation enables automated method development for sample handling techniques of single beads to produce combinatorial libraries; high-throughput is achieved by multiple analysis per vial.

LEAP Technologies
P. O. Box 969, Carrboro, NC 27410, USA
Phone: (919) 929-8814; Fax: (919) 929-8956
Toll-free in U.S.: (800) 229-8814
E-mail: *info@leaptec.com*
Internet: *http://www.leaptec.com*

Profile: LEAP Technologies has been in the liquid sample loading business since 1989. The company provides front-end automation for chromatography (LC and GC), mass spectroscopy, elemental analysis, dissolution testing, and other analytical techniques. It specializes in applications that demand reliability, flexibility, precision, and high-throughput. LEAP works with major chromatography and mass spectroscopy companies to provide total integrated solutions for critical automation applications.

Products/technologies: LC products include the CTC HTS PALS high throughput screening LC-MS sample loader; CTC A200LC autosampler for column-switching LC-MS; FLUX Rheos 4000 (true low-flow low-dead-volume-LC pump); FLUX compact degassers. For GC, the company offers the CTC COMBI PAL GC sampler for headspace and liquid injections, CTC A200SE GC autosampler. Products for dissolution testing include the HIDRA high-yield fiberoptic UV-Vis dissolution system; the flow-through dissolution systems (USP Apparatus #4), and the PAL automated sampling system.

Eli Lilly and Company
Lilly Corporate Center, Indianapolis, IN 46285, USA
Phone: (317) 276-2000; Fax: (317) 276-2095
Internet: *http://www.lilly.com*
Stock: LLY (NYSE)

Profile: Founded in 1876 by Colonel Eli Lilly, the company was incorporated in 1901. It is a global research–based pharmaceutical corporation dedicated to creating and delivering innovative pharmaceutical-based healthcare solutions that enable people to live longer, healthier, and more active lives.

Products/technologies: In 1994, Eli Lilly felt they had secured a solid position in the field of combinatorial chemistry when they acquired Sphinx Pharmaceuticals, which Lilly viewed as having prominent achievements in this technology. In March 1997 the company signed a contract with Taisho Pharmaceutical Co. to introduce technology in the field of combinatorial chemistry, including the construction of a large-scale, diversified library of chemical compounds, and to efficiently synthesize derivatives.

Contact: Stephen Stitle, Vice President, Corporate Affairs

LJL BioSystems, Inc.
404 Tasman Drive, Sunnyvale, CA 94043, USA
Phone: (408) 541-8755; Fax: (408) 541-8786
Toll-free in U.S.: (888) 611-4555
E-mail: *Htinfo@ljlbio.com*
Internet: *http://www.ljlbio.com*

Profile: A privately held company founded in 1988, LJL develops and manufactures automated assay systems for high-demand, high-throughput applications in clinical, research, and drug discovery applications. Its plan is to develop and market systems to accelerate drug discovery and to supply instrumentation, high-value, application-specific reagents and related consumables, applications software for database management, and informatics for the acceleration of high-throughput screening (HTS). The company is a registered medical device manufacturer, operates under GMP, and is ISO 9001 certified. In 1993, LJL was ranked forty-first in the Inc. 500 list of fastest growing private companies.

Products/technologies: These include high-throughput screening systems designed to alleviate the screening bottlenecks resulting from recent advances in genomics and combinatorial chemistry; proprietary, high performance reagents, and instrumentation for a large majority of the biosassays used in the HTS market. Through a licensing agreement LJL will incorporate FluorRx's fluorescence lifetime (FLT) sensing technology into LJL's HTS product systems. FLT technology measures the excited state lifetime, rather than the intensity, of proprietary fluorophores by means of a robust and sensitive method of fluorescence lifetime measurement, termed *phase modulation*. In September 1997 the company announced its first HTS product system, Criterion™. The system consists of Analyst™, an automated, multimode assay detection instrument designed to seamlessly connect to existing robotics systems; a spectrum of applications-specific reagents and assay kits; assay development and application support services. Analyst is capable of performing all four major types of optical detection assays currently in use (i.e., fluorescence, fluorescence polarization, time resolved fluorescence, and luminescence assays). It reads assays in 96- or smaller assay volume, 384-well microtiter plates with no loss of sensitivity or dynamic range.

Contact: William G. Burton, Ph.D., V.P. of Technology and Business Development

Lynx Therapeutics
3832 Bay Center Pl., Hayward, CA 94545, USA
Phone: (510) 670-9300; Fax: (510) 670-9302
E-mail: *postmaster@lynxcalif.com*

Profile: The company was founded in 1992 by the founder and former CEO and Chairman of Applied Biosystems, now a division of Perkin-Elmer. Its objective was to acquire, develop, integrate, and deploy novel molecular, chemical, and information technologies that provide a strong competitive advantage; and to leverage existing and future technology platforms through corporate collaborations and contracts in order to fund limited internal product development programs. Recently formed a Germany biotech startup company in Heidelberg with pharmaceutical giant BASF. According to the company, the joint venture, BASF-Lynx Bioscience AG, is a first for Germany and has the goal of serving as the launching pad for a biotech industry in Germany. It is also meant to serve as a genomics drug discovery arm for BASF and to finance the further development of Lynx's technologies.

Products/technologies: Massively parallel signature sequencing technology has the power to examine the genetic activity of a single cell.

Contact: Sam Eletr, PhD., Chairman and CEO

MDL Information Systems, Inc.
14600 Catalina St., San Leandro, CA 94577, USA
Phone: (510) 895-1313; Fax: (510) 614-3608
E-mail: *webmaster@mdli.com*
Internet: *http://www.mdli.com*
Stock: MDLI (NASDAQ)

Profile: Founded in 1978, MDL supplies integrated scientific information management systems, databases, and services that are used worldwide in pharmaceutical and chemical companies and in industries that use chemical products. The company's name was changed from Molecular Design Limited in 1993, the year it had its IPO. It is an information management software firm that has a combinatorial chemistry line of products. In late 1996 the company formed an alliance with Beckman Instruments to develop an integrated hardware and software system for high-throughput screening used in the drug discovery process. MDL is a wholly owned subsidiary of Elsevier Science.

Products/technologies: These include software programs, such as the Available Chemicals Director, which is a collection of chemical products offered by more than 200 chemical suppliers and totaling over 400,000 compounds. Other products include a line of systems for high-throughput discovery (incorporates Molecular Informatics' BioMerge software, a genomic data management product); ISIS, a client/server system that helps researchers select diverse chemicals for discovery; MDL SCREEN, a client/server product that manages the process of high-throughput screening; and MDL Central Library, a client/server system for managing combinatorial chemistry libraries.

Contact: Thomas D. Jones, President and COO

MDS Panlabs
11804 North Creek Parkway South, Bothell, WA 98011-8805, USA
Phone: (206) 487-8200; Fax: (206) 487-3878
E-mail: *info@mdsintl.com*
Internet: *http://www.mdsintl.com*

Profile: MDS is a health and life sciences company whose products and services are used in the prevention, diagnosis, and management of illness and disease. The company has six operating divisions: MDS Laboratory Services, MDS Ingram & Bell, MDS Nordion, MDS Communicare, MDS Sciex, and MDS Pharmaceutical Services. MDS Pharmaceutical Services provides drug discovery and development services to help the pharmaceutical and biotechnology industries to speed new drugs to market. MDS Panlabs is part of this division and provides discovery of diversified R&D support services. The Pharmaceutical Services division also offers clinical trial services, contract research, and specialized research in quality control testing and analytical R&D.

Products/technologies: For product discovery assistance, the company offers high-throughput bioactivity screening for rapid identification of new drug leads; chemical and natural product libraries for drug discovery; and human DNA products from cell culture for use as disease models. In product support, MDS has *PharmaScreen*™ pharmacology profiles for early evaluation of new drug candidates; and rapid analoging for new lead molecules. Bioprocess development services include recombinant DNA microbes for fermentation manufacturing; strain, media, and process optimization studies to reduce production costs; high-yielding cultures for production of penicillin, cephalosporin, and erythromycin antibiotics; and process safety assessment and validation.

Micromass Limited
Floats Road, Wythenshawe, Manchester M23 9LZ, UK
Phone: 44 161 945 4170; Fax: 44 161 998 8915
E-mail: *mark.mcdowall@micromass.co.uk*
Internet: *http://www.micromass.co.uk*

Profile: Micromass is a worldwide developer, manufacturer, and distributor of mass spectrometry (MS) instruments. In business since 1969, the company is a technology leader in the areas of magnetic sector, time-of-flight and quadrupole geometry mass spectrometers. When MS is combined with HPLC, LC-MS gives scientists a powerful tool for analyzing newly synthesized compounds and measuring the metabolites of drugs under investigation. Waters Corporation, the leader in HPLC technology, acquired Micromass in September 1997. Waters trades as WAT on the NYSE.

Products/technologies: Some product offerings include systems for rapid analysis of combinatorial libraries and purification of new drug candidates. These systems are based on Waters HPLC components, Micromass Platform LC or Platform LCZ benchtop API mass spectrometer and MassLynx™ software with special options for combinatorial chemistry, such as OpenLynx™ Diversity and FractionLynx™. The Waters LC-MS Solution for Drug Discovery provides a single-quadrupole solution for combinatorial analysis. Additionally, Micromass offers triple quadrupole, time-of-flight, magnetic sector, and hybrid geometry analyzers.

Contact: Mark McDowall, International Marketing Manager

Millennium Pharmaceuticals, Inc.
640 Memorial Drive, Cambridge, MA 02139-4815, USA
Phone: (617) 679-7000; Fax: (617) 374-9379
E-mail: *heller@mpi.com*
Internet: *http://www.bio.com/companies/Millennium.html*
Stock: LMNM (NASDAQ)

Profile: The company was founded in 1993. PerSeptive Biosystems is one of its principal investors. In January 1997 it was jointly announced that Millennium Pharmaceuticals would acquire privately held ChemGenics Pharmaceuticals as part of their mutual strategy to become a world leader in drug discovery. Following the merger, the combined company will employ more than 345 people. PerSeptive Biosystems is also a principal ChemGenics shareholder. In August 1997, Perkin-Elmer announced its plans to acquire PerSeptive.

Products/technologies: In addition to the technologies that ChemGenics will bring (as noted earlier in this chapter), Millennium employs large-scale genetics, genomics, and bioinformatics in an integrated, broad-based drug discovery program applicable to all major human diseases both independently, and in strategic alliances with leading pharmaceutical companies.

Contact: Beverly Holley, Director of Investor and Public Relations

Millipore Corporation
80 Ashby Road, Bedford, MA 01730-2271, USA
Phone: (781) 275-9200; Fax: (781) 533-3301
E-mail: Philip *Onigman@millipore.com*
Internet: *http://www.millipore.com*
Stock: MIL (NYSE)

Profile: Millipore makes products utilizing separation technology for the analysis, identification, and purification of fluids. Products include disc and cartridge filters and housings, filter-based test kits, precision pumps, and other ancillary equipment and supplies. The company's products are used in the pharmaceutical and biotechnology industries in sterilization, including virus reduction and sterility testing of products such as antibiotics and protein solutions; cell harvesting; and isolation of compounds from complex mixtures. Millipore, which acquired W. R. Grace's Amicon Separation Sciences, also makes protein purification tools and semiconductor manufacturing equipment through its subsidiary, Tylan General.

Products/technologies: MultiScreen® Assay Plate System is a 96-well plate configuration with membrane filters individually sealed to the bottom of each well. This allows for incubation and flow-through washes. These plates are now available with PTFE membranes and a plastic base that is solvent-resistant.

Contact: Philip Onigman

Molecular Devices Corporation
1311 Orleans Drive, Sunnyvale, CA 94089, USA
Phone: (408) 747-1700; Fax: (408) 747-3601
E-mail: *info@moldev.com*
Internet: *http://www.moldev.com*
Stock: NDCC (NASDAQ)

Profile: Founded in 1983, the company designs, develops, manufactures and markets proprietary, high-performance, bioanalytical measurement systems, including consumables, that are designed to accelerate and improve the cost effectiveness of the drug discovery and development process. Its systems have applications in many aspects of the therapeutic development process, from drug discovery and clinical research through manufacturing and quality control. In 1996 the company acquired NovelTech Systems, Inc., which designs, develops, manufactures, and markets high-throughput drug screening products.

Products/technologies: Sells bioanalytical systems for research in the life sciences, supporting combinatorial chemistry, genomics, and drug discovery. Its line of microplate reader systems is aimed at the life science market. The Fluorescence Imaging Plate Reader (FLIPR) is for high-throughput screening using live cell targets. SPECTRAmax PLUS is a high-throughput spectrophotometer that combines the power and performance of a spectrophotometer with the flexibility and throughput of a microplate reader in one system. It uses new technology called PathCheck™.

Contact: Andrew Galligan

Molecular Dynamics, Inc.
928 East Arques Avenue, Sunnyvale, CA 94086-4520, USA
Phone: (408) 773-1222; Fax: (408) 773-1493
Toll-free in U.S.: (800) 333-5703
E-mail: *info@mdyn.com*
Internet: *http://www.mdyn.com*
Stock: MDYN (NASDAQ)

Profile: The company is a developer, manufacturer, and international marketer of systems that accelerate genetic discovery and analysis. The company's products increase scientists' ability to visualize, quantify, and analyze genetic information. Molecular Dynamics is a founding member of the Genetic Analysis Technology Consortium (GATC), formed to develop and publish global standards and to provide a unified technology platform to design, process, read, and analyze DNA arrays. Molecular Dynamics and Amersham Pharmacia Biotech developed a microarray system that permits researchers to make and analyze high-density microarrays with increased speed, efficiency, and sensitivity. Data from a single microarray experiment provides researchers with the ability to accurately measure gene expression levels in thousands of samples. In late 1996, the two companies launched the Microarray Technology Access Program to make available preferential, precommercial access to their microarray technology.

Products/technologies: Combines advanced laser scanning, electro-optical, and software technologies to produce optical scanners and confocal microscopes. These instrument systems are designed to improve the productivity of molecular and cell biologists by automating routine procedures, increasing speed and accuracy of quantitation, and enabling new bioanalytical techniques. MegaBACE 1000 Genetic Analysis System, a high-throughput system, uses novel capillary technology for rapid and highly parallel analysis of genetic samples. The company also has laser fluorescence confocal scanning technology for DNA analysis, as well as innovative microfluidic technologies.

Contact: Jay Flatley, President and CEO

Molecular Informatics (See PE Molecular Informatics)

Molecular Simulations Inc.
9685 Scranton Road, San Diego, CA 92121-3752, USA
Phone: (619) 458-9990; Fax: (619) 458-0136
E-mail: *solutions@msi.com*
Internet: *http://www.msi.com*

Profile: Founded in 1984, this privately held company provides molecular modeling and simulation software for both life and materials science research. The company employs more than 280 people (approximately half of whom are Ph.D. scientists); it operates sales offices around the world and a research and development facility in Cambridge, England. MSI's Combinatorial Chemistry Consortium addresses the full scope of the combinatorial chemistry process and is focused on maximizing the productivity of library design and analysis. In February 1998, Molecular Simulations Inc. and Pharmacopeia Inc. announced a definitive agreement whereby Pharmacopeia will acquire all of the outstanding stock of MSI. The transaction is expected to be completed in the second quarter of 1998; upon completion MSI will become a wholly owned subsidiary.

Products/technologies: MSI's software combines the core atomistic simulation technologies of molecular mechanics and quantum mechanics with molecular visualization, modeling and instrument simulation. Products include C2 Diversity combinatorial chemistry software that provides a guideline for design and analysis of combinatorial libraries; and SAR information in library design of large (100,000+) enumerated libraries. In October 1997, the company announced the first commercial release of WebLab ViewerPro, which brings molecular visualization and analysis to desktop personal computers. The product runs on Windows 95 or NT, Power Macintosh, and Silicon Graphics platforms, and is used to build and view molecules, proteins, and crystalline materials, helping scientists to understand molecular structures and properties. Cerius 3.0 drug design software contains features for drug design, including the integration of organic and aromatic chemistry for the improved rendering of pharmaceutical compounds, faster rotation speeds, and a combinatorial chemistry module. MSI implemented a web-based architecture that will expand the availability of MSI software products and libraries to any scientist using a web browser on a desktop computer.

Contact: Brenda Pfeiffer, Corporate Communications

Neugenesis
Manoa Innovation Center, 2800 Woodlawn Dr., Suite 251, Honolulu, HI 96822-1865, USA
Phone: (808) 539-3801; Fax: (808) 539-3804 or 539-3625
E-mail: *info@neugenesis.com*
Internet: *http://www.neugenesis.com*

Profile: A private company, Neugenesis was founded in 1992 by several key individuals who recognized the potential commercial value of *Neurospora* for biotechnology applications. It has nine employees and occupies a 1600 square foot facility for its headquarters and R&D facilities. The company uses proprietary combinatorial biology technology to discover and develop new biotherapeutics, both alone and in partnership with other biopharmaceutical companies.

Products/technologies: These include a combinatorial biology drug discovery platform for the development, improvement, and production of therapeutic proteins; recombinant protein production systems for monomeric, dimeric, and heteromeric proteins. Neugenesis' combinatorial biology technology (the

CombiKARYON™) is a way to create diversity in complex proteins. Two important features are directed vs. random libraries to reduce the level of screening required and provide more efficient libraries overall; and shape/space enrichment whereby new proteins created through the Neugenesis system will exhibit improvements yet maintain the core structure of the protein.

Contact: Lynn M. Uehara, VP, Business Development

NeuralMed, Inc.
2525 Meridian Parkway, Suite 240, Durham, NC 27713, USA
Phone: (919) 549-0270; Fax: (919) 549-0271
E-mail: *info@neuralmed.com*
Internet: *http://www.neuralmed.com*

Profile: The company was founded in 1994 in Pittsburgh, Pennsylvania, by a group of engineers, statisticians, and business executives from several companies specializing in data mining applications. In early 1996, NeuralMed relocated to Research Triangle Park, North Carolina, to better access the scientific and engineering talent of the area. Its staff has expertise in data mining technologies, software systems development, system support, and consulting. Also, it has a broad range of experience in both the private and public sectors, including international expertise and university teaching.

Products/technologies: Has delivered various applications, including combinatorial chemistry predictions, by use of five neural network algorithms for building predictive models.

Contact: Tom Stallings, President and CEO

Neurogen Corp.
35 N.E. Industrial Road, Branford, CT 06405, USA
Phone: (203) 488-8201; Fax: (203) 481-8683
E-mail: *info@nrgn.com*
Internet: *http://www.bio.com/co/Neurogen.html*
Stock: NRGN (NASDAQ)

Profile: Organized around scientists from Yale University and industry, the company was founded in 1988. It is engaged in the design and development of a new generation of drugs that the company expects will provide improved treatment for a broad variety of neuropsychiatric disorders, including anxiety, obesity, schizophrenia, sleep disorders, dementia, depression, stress-related disorders and epilepsy.

Products/technologies: These include a combinatorial chemistry library. Its AIDD (Accelerated Intelligent Drug Design) program is a proprietary blend of combinatorial chemistry with high throughput screening, robotics, and informatics. New drug candidates are developed through the integration of cutting edge neurobiology, medicinal chemistry and molecular biology, and AIDD.

Contact: Stephen Davis

NeXstar Pharmaceuticals, Inc.
2860 Wilderness Pl., Suite 200, Boulder, CO 80301, USA
Phone: (303) 444-5893; Fax: (303) 444-0672
E-mail: *Kdoherty@NeXstar.com*
Internet: *http://www.nexstar.com*
Stock: NXTR (NASDAQ)

Profile: Founded in 1995 with the merger of NeXagen and Vestar, the company utilizes proprietary compounds to develop therapeutics and diagnostics to serve unmet medical needs. From NeXagen, NeXstar acquired the SELEX (Systematic Evolution of Ligands by Exponential Enrichment) combinatorial chemistry technology.

Products/technologies: Proprietary SELEX combinatorial chemistry technology that can produce aptamers, which are modified oligonucleotides that bind to cancer-specific markers. SELEX is a combinatorial chemistry technology that can create an extremely large library of aptamers of oligonucleotides (synthetic oligonucleotides) to identify those oligonucleotides with the highest specificity and affinity for the target. NeXstar's Blended SELEX technology converts molecules with low specificity but potentially useful biological activity into high-specificity drug candidates. A planned extension of SELEX, Parallel SELEX will be used to develop orally available small-molecule drug candidates.

Contact: Michael Burke, Vice President, Business Development

Oncogene Science, Inc.
See OSI Pharmaceuticals, Inc.

Oak Samples Trading, Ltd.
5 Sapiernoye pole St., 252042 Kiev, Ukraine
Phone: 380 44 269 3467; Fax: 380 44 269-3467 (local night hours)
E-mail: *sdl@public.ua.net*
Internet: *http://www.osc.edu/ccl/commercial/OST.html*

Profile: Oak Samples Trading(OST) company was founded in 1994 to establish reliable links between Ukrainian chemists and their foreign colleagues. The main objectives of OST were to facilitate research information exchange and to encourage investments in Ukranian science. Today the company's focus is on providing samples for high-throughput screening. During the course of their research, Ukrainian chemists synthesize many new compounds that are potentially active in agricultural and pharmaceutical screens.

Products/technologies: OST's collection of small molecules was gathered from about 20 institutions throughout the Ukraine. These molecules are the results of the creative efforts of nearly 100 Ukrainian chemists. The OST collection of compounds is available as ISIS/Base-formatted diskettes (Windows) for preview prior to purchase and will be available through FTP and WWW services.

Ontogen Corp.
2325 Camino Vida Roble, Carlsbad, CA 92009, USA
Phone: (619) 930-0100; Fax: (619) 930-0200
Internet: *http://www.ontogen.com*

Profile: Ontogen is a pharmaceutical company focusing on advancing techniques for the development of small molecule therapeutics to treat cancer and diseases of the immune system. The company has approximately 45 employees.

Products/technologies: For automation of combinatorial chemistry on solid phase in a modular approach, the OntoBlock central piece of hardware consists of 96 2-ml polymeric reaction vessels containing beads on which the synthesis takes place. From there the test components are stored, are characterized, or undergo high-throughput screening. Characterization is done by high speed electrospray mass spectrometry. The synthesis station uses reagent dispensers configured with coaxial needles to displace reagents by pressurization with inert gas.

Contact: Barry E. Toyonaga, Ph.D., President and CEO

OSI Pharmaceuticals, Inc.
106 Charles Lindbergh Blvd., Uniondale, NY 11553-3649, USA
Phone: (516) 222-0023; Fax: (516) 222-0114
E-mail: *mhaines@itg.com*
Internet: *http://www.osip.com*
Stock: OSIP (NASDAQ)

Profile: OSI Pharmaceuticals is a drug discovery company utilizing a platform of proprietary, broadly enabling technologies to facilitate the rapid and cost-effective discovery and development of novel, small molecule compounds for the treatment of major human diseases. The company conducts the full range of drug discovery activities, from target identification to drug candidate. OSI has a number of fully funded collaborations with various major pharmaceutical companies. The company was founded in 1983 by nine scientists from NCI. During 1996 it acquired MYCOsearch, Inc., and Aston Molecules, a UK-based provider of discovery and pharmaceutical development services to the world-wide pharmaceutical industry. Effective October 1, 1997, the company's name was changed to OSI Pharmaceuticals, Inc. from Oncogene Science, Inc. In late September 1997, OSI opened its new MYCOsearch Natural Products Drug Discovery Center in Durham, North Carolina.

Products/technologies: The proprietary gene transcriptional technologies are used to develop biopharmaceutical products for the diagnosis and treatment of cancer and other cell control–linked diseases. MYCOsearch, a division of OSI, added its Natural Products compound library, other chemical libraries, combinatorial and medicinal chemistry, informatics, molecular modeling, and pharmaceutical development operations. Additionally, the combinatorial chemistry technologies developed by Aston Molecules adds to the company's ability to optimize lead compounds identified in the company's live cell–based high throughput screening systems.

Contact: Matthew D. Haines, Director, Corporate Communications

Oxford Asymmetry Limited
151 Milton Park, Abingdon, Oxon OX14 4SD, England
Phone: 44(0)1235 861561; Fax: 44(0)1235 863139
E-mail: *sales@oa-od.com*
Internet: *http://www.oa-od.com*

Profile: A privately held company, it comprises Oxford Diversity, Oxford Asymmetry, and Oxford USA. Oxford Diversity is the company's combinatorial chemistry business; it has drug discovery experience in molecular modeling, organic chemistry, and combinatorial techniques leading to the identification of active leads. The company specializes in custom manufacturing and drug development for the pharmaceutical industry. In September 1997, Oxford Asymmetry expanded its pilot plant capacity and doubled its laboratory space. In February 1998 the company announced plans to raise funds through a listing on the London Stock Exchange.

Products/technologies: Offers a range of advanced chemical products and services required for rapid drug discovery and development, including everything from the production of diverse chemical libraries to the supply of multitonnes of bulk drug substance; and the Prospector series of combinatorial libraries.

Oxford Molecular Group, PLC (OMG)
The Medawar Centre, Oxford Science Park, Oxford OX4 4GA, UK
Phone: 44 1865 784600; Fax: 44 1865 784601
E-mail: *products@oxmol.com*
Internet: *http://www.oxmol.com*
Stock: OMG (London Stock Exchange)

Profile: The company, whose name was changed in 1996 from IntelliGenetics, is a worldwide provider of integrated solutions for drug discovery for companies and universities. OMG was fundamental in the establishment of a new company, Cambridge Combinatorial, in February 1997 (see separate listing). In August 1997, OMG announced that it had taken an option to buy Cambridge Combinatorial. The move is intended to turn OMG from a company that designs software into one that designs new drugs. Its Collaborative Discovery division, along with its associate companies, Cambridge Combinatorial Limited and Cambridge Drug Discovery, provide research services and consulting in drug design, combinatorial chemistry and high-throughput screening.

Products/technologies: OMG offers a wide range of computer-aided molecular design (CAMD) software, advanced bioinformatics tools, cheminformatics tools, and collaborative research services. Some products now on the market include PC/GENE® sequence analysis software, InteleGenetics® Suite, GeneWorks®, GENESEQ™ patent sequence DNA and protein database, MPSRCH™, BIONET®, MacVector™ sequence analysis software, RS3 Discovery software for corporate chemical information management. DIVA™, a visualization and analysis tool for chemical and biological data stored in large databases, reduces the time needed for analyzing large sets of data. OMIGA, the first in a family of integrated, enterprise-wide software tools and a comprehensive sequence analysis tool, is available for Windows 95 and Windows NT 4 platforms. Through its Collaborative Discovery Division, the company provides software and service solutions to the drug discovery industry.

Contact: Dr. Tony Marchington, CEO

Packard Instrument Company, Inc.
800 Research Parkway, Meriden, CT 06450, USA
Phone: (203) 238-2351; Fax: (203) 639-2172
Toll-free in U.S.: (800) 323-1891
E-mail: *webmaster@packardinst.com*
Internet: *http://www.packardinst.com*

Profile: Packard is a subsidiary of Canberra Industries, Meriden, Connecticut. Its instruments, reagents, and applications support are sold to organizations that conduct research and develop products in life science, diagnostics, and environmental analysis. The company employs more than 550 people worldwide with 50 employees at its corporate headquarters and 275 at its manufacturing facility in Downers Grove, Illinois.

Products/technologies: The company offers a complete line of microplate reader instrumentation. Discovery, a nonisotopic high-throughput screening analysis system, is based on a Nobel Prize–winning chemistry designed to make drug discovery faster, safer and more cost effective. Technology can be applied to a screening process called Homogeneous Time Resolved Fluorescence (HTRF), which can replace older, conventional radioisotopic processes, reducing exposure risks and waste disposal costs. The patented HTRF chemistry is from CIS Biointernational of France. Using Packard's high capacity microplate analyzer Discovery, designed specifically for automated drug screening, HTRF can screen 50,000 samples per day, compared to only 10,000 using conventional technology. HTRF incorporates photon-counting electronics, dual photomultiplier tubes, and time-delayed measurements to eliminate background fluorescence. MultiPROBE® VersaTip™ technology will sense small sample volumes in microplates.

Contact: Richard McKernan, President

Pangea Systems, Inc.
1999 Harrison Street, Suite 1100; Oakland, CA 94612, USA
Phone: (510) 628-0100
E-mail: *info-pr@PangeaSystems.com*
Internet: *http://www.pangeasystems.com*

Profile: A bioinformatics company, Pangea applies information technology to biological research. Software applications integrate data with analysis and visualization tools for biological and chemical information. Incorporated in 1993, it is a privately held company.

Products/technologies: EcoCyc 4.0 is a bioinformatics metabolic pathways dataset for E. coli and is available through the world wide web; links approximately 2000 more genes than the previously available version. PathoLogic™ creates metabolic pathway datasets of other pathogenic organisms. GeneWorld® is an automated, high throughput application for analysis of DNA and protein sequences; GeneMill™ is a workflow management application designed for DNA sequencing laboratories. GeneThesaurus™ is a sequence and annotation data warehouse containing information from multiple sources and integrates common public gene and protein sequence and protein structure databases.

Contact: Marie Martin

Panlabs, International
11804 North Creek Parkway, S., Bothell, WA 98011-8805, USA
Phone: (206) 487-8200; Fax: (206) 487-3787
E-mail: *panlabs@panlabs.com*
Internet: *http://www.panlabs.com*
Stock (parent company): MHG.A, MHG.B (TSE)

Profile: Founded in 1970, this contract research organization (CRO) became a wholly owned subsidiary of MDS Health Group Ltd., Canada, in 1995. In mid-1995 Panlabs joined with Tripos in a collaboration to offer pioneering services in drug discovery to pharmaceutical, biotechnology, and chemical research organizations.

Product/technologies: Provides discovery services in the form of screening assays, screening materials from chemical libraries, plants, and microbes with robotic screening capacity of 100,000 compounds per week and pharmacological profiling of lead compounds on a contractual basis. Products and services include PharmaScreen®, DiscoveryScreen®, ProfilingScreen™, ImmunoScreen™, pharmacology discovery and profiling, high throughput screening (HTS). The company has proprietary technologies to synthesize, extract and deliver the chemical libraries in 96-well plates for screening. Along with Tripos, the company markets Optiverse™ screening libraries for HTS programs.

Contact: Dr. Christopher Ball, Chairman and CEO

Perkin-Elmer Corp. / PE SCIEX / PE Applied Biosystems
761 Main Ave., Norwalk, CT 06859, USA
Phone: (203) 762-1000; Fax: (203) 762-6000
Internet: *http://www.perkin-elmer.com*
http://www.pesciex.com
Stock: PKN (NYSE)

Profile: P-E is a world leader in the development, manufacture, and marketing of life science systems and analytical instruments used in markets such as biotechnology, pharmaceuticals, environmental testing, chemicals, food, agriculture, and chemical manufacturing. The company had sales of more than $1.3 billion in fiscal year 1997 and employs 5700 people worldwide. Perkin-Elmer Sciex Instruments (PE SCIEX) is a joint venture between MDS SCIEX and P-E. PE SCIEX is a leader in the development of mass spectrometers for use in the analytical and life sciences. In 1996, PE signed an exclusive licensing agreement for the development of LXR Biotechnology's patented scanning laser digital imaging (SLDI) technology; the Scanning Laser Digital Imager images cells and tissues. In July 1997, PE Applied Biosystems, which makes automated DNA analysis systems, signed an agreement with Tecan U.S. to develop and sell systems for automating combinatorial chemistry. PE Applied Biosystems will focus on marketing and support of high speed parallel synthesis systems developed by both companies. Tropix, an operating unit of PE Applied Biosystems Division, established an additional laboratory in 1997 to develop customer screening assays for the pharmaceutical industry. In November 1997, P-E acquired Molecular Information, Inc., a developer of bioinformatics software. P-E acquired an additional 14.5% stake in Tecan AG in December 1997, increasing its shares to 52%. In January 1998, P-E acquired PerSeptive Biosystems. (See separate listings for acquired companies).

Products/technologies: P-E offers integrated, high-throughput, and high-performance systems for DNA amplification and analysis, and LC-MS systems used by pharmaceutical companies in research and combinatorial chemistry. The company has developed a comprehensive set of enabling tools for pharmaceutical research—tools that compress the time and reduce the cost of pharmaceutical drug discovery and development. Perkin-Elmer Cetus has a worldwide marketing agreement through SCIEX with MDS Health Group Limited, Canada, for its API III MS analyzer. Under development are various research projects relating to software, software systems, and automated data management and analysis systems to speed the drug discovery process. Other products include the PE-SCIEX API 150 MCA (Mass Chromatographic Analyzer) LC-MS, Open MS™ control; and purpose-designed workstations, such as drug discovery or combinatorial chemistry built around the MCA. Tropix, which has developed proprietary drug discovery assay systems on a limited basis, now makes this service available to the worldwide pharmaceutical industry.

Contact: Tony L. White, Chairman, President and CEO

PE Molecular Informatics
1800 Old Pecos Trail, Suite M, Santa Fe, NM 87505, USA
Phone: (505) 995-4475; Fax: (505) 982-7690
Internet: *http://www.molinfo.com*

Profile: Acquired by Perkin-Elmer in November 1997, Molecular Informatics provides comprehensive bioinformatics software systems to manage, integrate, and analyze the vast amounts of information flowing into the scientific discovery process. The company is a pioneer in the development of infrastructure software for pharmaceutical, biotechnology, and agrochemical industries that perform genome data collection and management, genome mapping, drug discovery, and clinical and diagnostic genetic research. The company is a spin-off of the National Center for Genome Resources, Santa Fe, New Mexico.

Products/technologies: Currently provides two products for the management of genomic data. The BioMerge™ system consists of an advanced, object-oriented relational database and a group of programs that organize public, proprietary, and third-party data in a single database. The system provides for editing and annotating sequence data. An optional Auditor database captures a history of changes made to your database. Customers can purchase automatic daily updates of public data with the ''Genomes Today'' option. BioLIMS™, developed for the Applied Biosystems (ABI) division of Perkin-Elmer Corporation, manages automated DNA sequencing for ABI's Prism™ DNA sequencing instruments. With BioLIMS, sequence analysis results are entered into a central database, rather than as thousands of separate files.

Contact: Edward Cantrell

PerSeptive Biosystems, Inc.
500 Old Connecticut Path, Framingham, MA 01701, USA
Phone: (508) 383-7700; Fax: (508) 383-7851
Toll-free in U.S.: (800) 899-5858
E-mail: *tsupport@pbio.com*
Internet: *http://www.pbio.com*

Profile: The company was founded in 1988 to commercialize Perfusion Chromatography® products based on discoveries in particle coatings and design. Numerous acquisitions followed; the most recent being that of PerSeptive Technologies II Corporation in March 1996, which strengthened PerSeptive's position in drug discovery. Millennium Pharmaceuticals, which is partially owned by PerSeptive, announced in January 1997 that it would acquire privately held ChemGenics Pharmaceuticals as part of their strategy to become a world leader in drug discovery. PerSeptive Biosystems is also a principal ChemGenics shareholder. In August 1997, Perkin-Elmer (see separate listing) announced its plans to acquire the company; the transaction was completed in January 1998.

Products/technologies: These include Perfusion Chromatography® technology; advanced workstations, chemistry and other tools for the biotechnology industry; robotics interface device for preparing biological samples for analysis by mass spectrometry. The MARINER™ LC/MS Biospectrometry™ Workstation for combinatorial chemistry is targeted for high-throughput applications in characterizing proteins and peptides, in metabolic studies, in combinatorial chemistry strategies, and in high-throughput screening. Selectronics, a proprietary drug discovery technology, combines multistep chromatography, affinity selection, and mass spectrometry for high-throughput selection of new drug lead compounds from a variety of chemical sources including chemical combinatorial libraries, natural product libraries, and synthetic chemical files.

Contact: John F. Smith, President

Pharmacopeia, Inc.
101 College Road E., Princeton, NJ 08540, USA
Phone: (609) 452-3600; Fax: (609) 452-2434
Internet: *http://www.pcop.com*
Stock: PCOP (NASDAQ)

Profile: Founded in 1993, the company is active in the field of drug discovery using small-molecule combinatorial chemistry. It went public in December 1995. In February 1998, Pharmacopeia and Molecular Simulations, Inc. announced a definitive agreement whereby Pharmacopeia will acquire all of the outstanding stock of MSI.

Products/technologies: These include discovery libraries of combinatorial chemistry collections of novel small molecules for screening. Using Encoded Combinatorial Libraries on Polymeric Support (ECLIPS™), its proprietary tagging technology, the company generates large libraries consisting of millions of diverse individually labeled small molecules. Using these libraries, Pharmacopeia is developing three potential profit centers. These include the licensing of libraries to pharmaceutical companies for evaluation in multiple drug discovery programs; the identification and optimization of lead compounds for specific targets provided by customers; and the licensing to pharmaceutical companies of drug development candidates developed in the company's internal drug discovery programs.

Contact: Lewis J. Shuster, Executive Vice President, Corporate Development and Chief Financial Officer

Pharm-Eco Laboratories, Inc.
128 Spring Street, Lexington, MA 02173-7800, USA
Phone: (781) 861-9303; Fax: (781) 861-9386
E-mail: *main@pharmeco.com*
Internet: *http://www.pharmeco.com*

Profile: This privately held company was founded in 1972. It is a full-service drug synthesis and chemical services company that performs a variety of laboratory, process scale-up, and manufacturing tasks including development of processes and synthesis routes for new medicinal products, validating bulk pharmaceutical processes, and authoring Drug Master Files. The company also has a pilot plant/small volume manufacturing site in North Andover, Massachusetts.

Products/technologies: This contract research and development group offers GMP production; Drug Master Files; validation of analytical methods; multistep organic synthesis development; and consultancy services, such as during clinical trials.

Contact: Susie Boast

Polymer Laboratories Ltd.
Essex Rd., Church Stretton, Shrophire SY6 6AX, UK
Phone: 44 01694 723581
E-mail: *PL@polymerlabs.com*
Internet: *http://www.polymerlabs.com*

Profile: The company was founded in 1976 to develop techniques and instrumentation for the characterization of polymer systems, and to develop high-technology polymer products for chromatography, and for diagnostic and pharmaceutical applications. The company operates from offices in the USA, UK, and Europe. Primary products are in the specialized areas for gel permeation/size exclusion chromatography (GPC/SEC) including columns, standards, advanced high temperature systems and detectors, and GPC/SEC software.

Products/technologies: PL sells a range of combinatorial chemistry resins consisting of polystyrene resins with chloromethyl functionality (PL-CMS) for use in solid phase synthesis of peptides, small-molecule libraries, and other solid-supported organic reactions. Larger particles with higher loading can also be manufactured using the same approach. PL-CMS resins are available in sizes up to 300 μm in diameter with up to 4 meq/g of loading.

Contact: Dr. Frank Warner, President

Porvair Sciences Ltd.
Unit 6, Shepperton Business Park, Govett Avenue, Shepperton, Middlesex TW17 8BA, UK
Phone: 44 1932 240255; Fax: 44 1932 254393
E-mail: *porvair-paul@classic.msn.com*
Internet: *http://www.porvair-sciences.com*

Profile: The company has used its expertise in plastics technologies and molding processes to engineer the Microlute plate to meet the needs of researchers and laboratory analysts who require SPE sample preparation.

Products/technologies: Porvair sells the Microlute™ Solid Phase Extraction in a Microplate system that provides 96 solid phase extractions in one compact unit (using any brand of sorbent). It can be automated using most standard liquid handling and robotic systems. Porvair's 384-well plate is compatible with most automated liquid handling instruments, readers for EIA, fluorescence, luminescence, and scintillation assays as well as robotic-handling devices.

Contact: Tony Castleman

Protogene Laboratories, Inc.
4030 Fabianway, Palo Alto, CA 94303, USA
Phone: (415) 842-1888; Fax: (415) 857-9229
Internet: *http://www.protogene.com*

Profile: Was founded as a spin off of an earlier company by the same name. The company's parallel array synthesis technology was originally developed for DNA synthesis by one of the company founders and others at Stanford University's Human Genome Center under a Department of Energy Human Genome grant. After commercializing the technology and becoming the world's largest manufacturer of DNA, the DNA business was sold to its marketing partner, Life Technologies, Inc. The company announced a collaboration with Ribozyme Pharmaceuticals, Inc. in December 1996; one goal is to create an instrument that will synthesize thousands of RPI's proprietary nuclease resistant ribozymes, combinatorial libraries of other compounds, and methods that allow for low-level quantitative detection of RNA in cells.

Products/technologies: Protogene Laboratories focuses on a world scale approach to synthesis facilities on systems for synthesizing DNA on surfaces, and on combinatorial and high-throughput synthesis of drug libraries.

Contact: Dr. Robert Molinari, President and CEO

Receptor Technologies, Inc.
(See ACADIA Pharmaceuticals)

Regis Technologies, Inc.
8210 Austin Avenue, Morton Grove, IL 60053, USA
Phone: (847) 967-6000; Fax: (847) 967-5876
Toll-free in U.S.: (800) 323-8144
E-mail: *whelko@aol.com*
Internet: *http://www.registech.com*

Profile: Founded in 1956, this privately held company has served as a partner in the development, scale-up, and manufacture of complex, multistep, custom synthetic products for clients in the pharmaceutical, diagnostic, and biotech industries. Over the years the company has merged its synthetic proficiency and chromatographic capabilities to offer a preparative separation service designed for the resolution of compounds on a gram to kilogram scale. It develops proprietary separations technology for the life sciences industry and provides contract manufacturing for pharmaceuticals.

Products/technologies: IAM.PC Drug Discovery chromatography column (immobilized artificial membrane, or IAM) is a screening method for the high-throughput estimation of drug permeability. IAM chromatography phases prepared from PC analogs closely mimic the surface of a biological cell membrane, which makes IAM useful for the study of drug–membrane interactions.

Contact: Louis Glunz, Ph.D., President

Ribozyme Pharmaceuticals, Inc.
2950 Wilderness Pl., Boulder, CO 80301, USA
Phone: (303) 449-6500; Fax: (303) 449-6995
E-mail: *webmaster@rpi.com*
Internet: *http://www.rpi.com*
Stock: RYZM (NASDAQ)

Profile: Founded in 1992, the company commercializes patented ribozyme technology in the human therapeutic, agricultural, animal health, and diagnostic fields, through both internal efforts and collaborations with strategic partners. It announced a collaboration with Protogene Laboratories, Inc. in December 1996; one goal is to create an instrument that will synthesize thousands of RPI's proprietary nuclease resistant ribozymes, combinatorial libraries of other compounds, and methods that allow for low-level quantitative detection of RNA in cells.

Products/technologies: Proprietary technology for the synthesis of ribozymes; involved in the application of ribozymes to human therapeutics—automated parallel array technologies.

Contact: Ralph E. Christoffersen, Ph.D., President and CEO

Robosynthon, Inc.
1105 Grandview, S. San Francisco, CA 94080, USA
Phone: (650) 244-0799; Fax: (650) 244-0795
E-mail: *info@robosynthon.com*
Internet: *http://www.robosynthon.com*

Profile: The company is a sister company of ComGenex, Inc., which manufacturers the MultiReactor™ workstation for parallel organic synthesis marketed by Robosynthon.

Products/technologies: The MultiReactor is the first in a line of combinatorial chemistry workstations for parallel organic synthesis. It allows chemists to synthesize 24 reactions simultaneously with precise temperature and magnetic stirring in all tubes. The MaxiReactor™ scales syntheses discovered with the MultiReactor. It can run six synthetic reactions simultaneously, handle up to 500 ml and gram quantities, claims better accuracy and uniformity than oil baths, has independently controlled mixing and heating zones, and offers data logging for regulatory compliance. Robosynthon also sells a wide range of classical synthetic glassware.

Rosetta Inpharmatics
12040 115th Avenue, NE, Kirkland, WA 98034, USA
Phone: (425) 823-7300; Fax: (425) 821-5354

Profile: A biotechnology startup, the company raised initial funding in 1997 from U.S. and international investors. It plans to build ''biochips'' holding as many as 100,000 strands of synthesized genetic material to be used along with proprietary analytical techniques to help major pharmaceutical companies avoid costly false starts in developing new drugs. The two initial target areas are cancer and immunological diseases.

Products/technologies: Its key pieces of technology are an ink-jet oligonucleotide synthesizer and analytical tools developed by the company's cofounders to assess a drug candidate's desirable and undesirable effects. The synthesizer will be used to build chip-like arrays with thousands of tiny wells. Each well holds a string of genetic material, built up by directing the device to sequentially spray the four nucleotides, or building blocks of DNA, much as a color ink-jet printer sprays four shades of ink. With such arrays, the company will test drug candidates to assess their effects on various segments of the genome.

Contact: Stephen Friend, President and Chief Scientific Officer

Shimadzu Scientific Instruments
7102 Riverwood Drive, Columbia, MD 21046, USA
Phone: (410) 381-1227; Fax: (410) 381-1222
Toll-free in U.S.: (800) 477-1227
E-mail: *webmaster@shimadzu.com*
Internet: *http://www.shimadzu.com*

Profile: Shimadzu Corporation, headquartered in Kyoto, Japan, was established in 1875. It is one of the world's largest manufacturers of analytical instrumentation. It entered the North American market in 1975 with the formation of Shimadzu Scientific Instruments.

Products/technologies: Offerings include the MTP-40 autosampler for direct sampling from wells to vials in combinatorial projects; and the MTP-10A autosampler for direct sampling from 96-well microplates eliminating the need to transfer samples from wells to vials in combinatorial chemistry projects.

SIBIA Neurosciences, Inc.
505 Coast Blvd. South, Suite 300, LaJolla, CA 92037-4641, USA
Phone: (619) 452-5892; Fax: (619) 452-1609
E-mail: *mdunn@sbia.com*
Internet: *http://www.sibia.com*
Stock: SIBI (NASDAQ)

Profile: The company is engaged in the discovery and development of novel small-molecule therapeutics for the treatment of neurodegenerative, neuropsychiatric, and neurological disorders. It is involved in the development of proprietary drug discovery platforms that combine key tools necessary for modern drug discovery, including genomics, high-throughput screening, advanced combinatorial chemistry, and pharmacology. The company's proprietary molecular targets and drug candidates, together with its drug discovery technologies and research expertise, have enabled the company to establish several corporate collaborations.

Products/technologies: Proprietary high-throughput screening technology consists of automated optical detection systems for primary screening assays and the profiling of compounds to identify new drug leads.

Contact: Michael J. Dunn, Vice President, Business Development

SIDDCO (See Systems Integration Drug Discovery Co., Inc.)

Soane Biosciences, Inc.
3906 Trust Way, Hayward, CA 94545-3716, USA
Phone: (510) 293-1855
E-mail: *mfranklin@soane.com*
Internet: *http://www.soane.com*

Profile: In 1997 the company purchased substantially all the assets of privately held Seurat Analytical Systems, located in Sunnyvale, California.

Products/technologies: The high-throughput sample-handling devices for drug discovery applications provide a means of transferring vast numbers of samples in parallel from the macro world of microtiter plates to the micro world of chips.

Contact: Roger Noesky, CEO

Spark Holland B.V.
P. de Keyserstraat 8, NL-7825 VE Emmen, The Netherlands
Phone: 31 591-631700; Fax: 31 591-630035
E-mail: *sales@spark.nl*

Profile: The company is a maker of HPLC equipment and specializes in selling autosamplers and automated SPE instruments.

Products/technologies: Midas autosampler with 96-tray capacity.

SRI International
333 Ravenswood Avenue, Menlo Park, CA 94025, USA
Phone: (415) 859-3000; Fax: (415) 859-2889
Toll-free in U.S.: (800) 982-8655
E-mail: *bdd@sri.com*
Internet: *http://www.sri.com/pharmaceutical*

Profile: Founded as Stanford Research Institute in 1946, SRI is a private, nonprofit contract research organization and employs over 2400 individuals worldwide. In addition to its California location, the company also has offices in Europe and Asia.

Products/technologies: SRI has combinatorial chemistry methods for finding and optimizing advanced lubricants and coatings formulations that claim to allow scientists to formulate and characterize hundreds of thousands of different fluids, identifying those with desired rheological properties. SRI said this represents a modification of combinatorial technology, including miniature biosensors, developed over the last several years for drug screening and clinical diagnostics. The technology is capable of preparing and analyzing different variations of mixtures with up to 20 compounds; the picoliter samples are then simultaneously measured for viscosity and yield stress. SRI also developed a parallel screening technique based on the emission and absorption of visible light. The technique produces spectral information on reaction products and can be used for screening catalytic activity. The company is attempting to extend the method to use ultraviolet and near-infrared light, which would broaden the types of compounds and information that can be screened.

Contact: Pieter Bax, Ph.D., V.P., Biopharmaceutical Development Division

Structural Bioinformatics, Inc.
10929 Technology Place, San Diego, CA 92127, USA
Phone: (619) 675-2400; Fax: (619) 451-3828
E-mail: *webmaster@strubix.com*
Internet: *http://www.strubix.com*

Profile: Founded in 1996, this privately held company is a developer of leading edge supercomputer-based technology for rapid conversion of novel gene sequence information into protein structural information and drug lead compounds.

Products/technologies: A supercomputational operating system makes possible the immediate and practical use of genomic (gene sequence) data in a broad range of structure-based drug discovery and design processes leading to the rapid design and identification of small-molecule lead compounds. Its proprietary SBd-Base allows comparisons of pharmacologically important protein shapes that cannot be made through simple gene sequence comparisons but are essential for finding highly selective drugs. It plans to expand its 4-D protein structural database and rapid nonpeptide drug lead molecule generation technologies. The company provides access to the technology through partnerships and subscriptions.

Contact: Susan K. Brugess

Supelco
Supelco Park, Bellefonte, PA 16823, USA
Phone: (814) 359-3441; Fax: (800) 447-3044

Toll-free in U.S.: (800) 247-6628
E-mail: *supelco@sial.com*
Internet: *http://www.supelco.sial.com/supelco.html*

Profile: Supelco is a member of the Sigma-Aldrich family and has 30 years experience in providing chromatography products for analysis and purification. Aldrich combines its own products with those of Sigma, Fluka, and Supelco to provide a broad range of products for the combinatorial market.

Products/technologies: These include Visiprep™ solid phase extraction vacuum manifolds to manipulate combinatorial chemistry reaction mixtures; a rare chemicals library of over 60,000 products; CombiKits™ to create a customized kit of preweighed building blocks; SUPELCOSIL ABZ+ HPLC columns; and a chemical directory on CD-ROM.

Synopsys Scientific Systems, Ltd.
175 Woodhouse Lane, Leeds LS2 3AR, UK
Phone: 44 (0) 113245 3339; Fax: 44 (0) 113243 8733
E-mail: *sales@synopsys.co.uk*
Internet: *http://www.snopsys.co.uk/*

Profile: The company provides scientific information products and services to the fine-chemical, pharmaceutical, agrochemical, biotechnology, and academic research communities. Synopsys operates worldwide, with offices and distributors in Europe, the United States, and Japan. It specializes in chemical reaction databases that deliver added-value information and refined data.

Products/technologies: These include the Accord range of chemical software. Architecture is designed to be compatible with existing chemical and non-chemical software, yet be scalable and focused to meet future needs. Database products are available for use with chemical database systems such as REACCS, ISIS, and ORAC. Database products are also available for solid phase synthesis, methods in organic synthesis (MOS), BioCatalysis, Protecting Groups, and BIOSTER. The Protecting Groups database comprises over 24,000 carefully selected protecting group reactions. It provides chemists with ready access to the full range of protecting group chemistry and assists users in identifying the most appropriate conditions to perform a protection or deprotection step. BIOSTER database is a compilation of over 2000 bioisosteric transformations—including drugs, agrochemicals, and enzyme inhibitors.

Systems Integration Drug Discovery Co., Inc. (SIDDCO)
9000 S. Rita Road, Bldg. 40, Tucson, AZ 85747, USA
Phone: (520) 663-4001; Fax: (520) 663-0795
E-mail: *reising@siddco.com*

Profile: Founded in late 1996, SIDDCO is a privately held company that is based on a mission to integrate and apply state-of-the-art approaches to drug discovery. SIDDCO is a combinatorial chemistry consortium that pools the resources of the consortium partners to fund a critical mass of chemists to develop the combinatorial chemistry technology database and methodologies. SIDDCO provides a dedicated team of medicinal chemists to each partner to exploit the shared technology and satisfy the drug discovery and drug optimization needs of specific projects.

Products/technologies: Combinatorial chemistry and high-density miniaturized high-throughput assay technologies are integrated and applied in a comprehensive drug discovery program to serve the needs of SIDDCO and its industrial partners. Multiarray plate screening (MAPS) converts genomic sequence information directly into screening assays for discovering drugs and increases the number of biochemical assays that can be efficiently conducted. Chemical libraries, validated library technology, and automated methods may be used by consortium partners on a consortium-exclusive basis. SIDDCO also provides a dedicated team of medicinal chemists (SWOT = SWift Optimization Team) to each consortium partner pursuing the confidential design and synthesis of libraries for specific targets and the optimization of leads, royalty-free.

Contact: Pete Reisinger, Vice President, Operations

TECAN Group
Feldbachstrasse 80, CH-8634 Hombrechtikon, Switzerland
Phone: 41 (0)55 254 81 11; Fax: 41 (0)55 244 38 83
E-mail: *tecan@tecan.ch*
Internet: *http://www.tecan.com*
Stock: Traded on Zurich stock exchange TECI (EBS)

Profile: Founded in 1980, the company is involved in the development and manufacture of robotic liquid-handling systems, automated sample processors, microplate photometry, and major components. Tecan U.S., the North American headquarters in Research Triangle Park, North Carolina, has been developing new product concepts for pharmaceutical drug discovery and human genomic markets since 1994. This has led to a broad line of laboratory automation and to developments in the field of automated parallel synthesis of medicinal compounds using both solution and solid phase chemistry. Cavro Scientific Instruments, Inc. is a Tecan Group company. In December 1997, Perkin Elmer Corporation acquired a 14.5% stake in Tecan AG increasing its share to 52%.

Products/technologies: The company sells the CombiTec parallel synthesis system, an organic chemical synthesizer that includes a robotic sample processor, and reaction blocks of 8–56 chambers. Other products include the TRAC system for high-throughput screening with more than 100 microplates; the GENESIS Series Robotic Sample Processor (RSP); fully automated microplate-based systems; and the Cavro RSP 9000 Robotic Sample Processor (XYZ module).

Contact: Hansruedi Merz, Sales Manager

Terrapin Technologies, Inc.
750 Gateway Blvd., South San Francisco, CA 94080, USA
Phone: (650) 244-9303; Fax: (650) 244-9388
E-mail: *inquiry@trpntech.com*
Internet: *http://www.terrapintech.com*

Profile: Founded in 1986, Terrapin is a privately held pharmaceutical discovery company that targets unmet medical needs in diabetes, allergy/asthma, and oncology/hematology. Research programs focus on the discovery of new intracellular targets as well as novel therapeutic compounds. The company is engaged in collaborations relating to combinatorial chemistry.

Products/technologies: These include enabling technology and an integrated approach to drug discovery that allows for the rapid identification and optimization of therapeutic leads from large chemical libraries. It has a compound library and a proprietary molecular fingerprinting technology (TRAP™). Applying computational, combinatorial, and medicinal chemistry techniques, the company refines selected compounds, enhancing properties such as potency and specificity.

Contact: Sharon Tetlow, V.P., Finance and Administration

The Automation Partnership
Melbourn Science Park, Melbourn, Royston, Hertfordshire SG6 6HB UK
Phone: 44 1763 262026; Fax: 44 1763 262613
E-mail: *info@autoprt.co.uk*
Internet: *http://www.autoprt.co.uk*

Profile: The Automation Partnership had previously been the Automation Division of The Technology Partnership until June 1995 when it was spun out as a wholly owned subsidiary. In 1997, TTP Group was set up as a holding company with The Automation Partnership, The Technology Partnership plc, and Signal Computing Ltd. as group members. The company's Haystack™ system automates the storage, tracking, retrieval, and dispensation of proprietary compounds in support of combinatorial chemistry and for high-throughput drug discovery.

Products/technologies: Haystack Automated Compound Storage stores dry compounds in vials; stores microtubes of liquid compounds; dispenses powders in Haywain™ for dissolving; retrieves samples on demand for biochemical assays; tracks and schedules all compounds, samples and operations; works very fast around the clock, is modular and fail-safe. Large systems have held 0.5 million dry compounds and 2.5 million liquid samples. Modules can be configured to the needs of the customer. The database software, intrinsic to any configuration, interfaces to customers' existing systems. Modules that are available include dry compound store (variable size), liquid sample store (variable size), microtiter plate store (variable size), bar coded glass bottle and caps, Haywain dispensing unit, manual weigh station, solubilization and aliquotting unit. The system can be combined with OPTIMA™ substance management and customer ordering software from EMAX Solution Partners. The software provides the user interface for registration, dispatch, and build order functions, as well as removal orders, inventory changes, inquiries, and alterations.

Contact: Judy Kramer

The Technology Partnership plc
Melbourn Science Park, Melbourn, Royston, Herts SG8 6EE, UK
Phone: 44 1763 262626; Fax: 44 1763 261582
E-mail: *cgb@techprt.co.uk*
Internet: *http://www.ttpgroup.co.uk*

Profile: The Technology Partnership (TTP) is a product development and engineering company based near Cambridge, UK. TTP develops new products, improves existing products, and supplies automated manufacturing systems. TTP operates in a wide range of market sectors including pharmaceuticals, novel printing technology, office products, and digital mobile phones. TTP is known for its Cellmate™ automated vaccine manufacturing system and Haystack™ automated compound handling system used in drug discovery. As well as providing new products for the world's leading companies, TTP has a policy of investing a significant percentage of its revenue in technologically advanced in-house developments. These then create business opportunities that may be exploited as joint ventures, licensing agreements, or spinoff companies. TTP employs more than 300 staff, who own 75% of the company. A member of the TTP Group plc, its first U.S. office was set up as TAP, Inc. in Killingworth, CT, in early 1997.

Products/technologies: Myriad™ automated synthesis systems are based on a series of robotic processing modules. These modules split the synthesis process into efficient, high-throughput units. The chemist can also develop new reactions on a Personal Synthesizer™, fully compatible with the Myriad system. Synthesizer products include the Chemistry Developer (low-cost system for reaction development); the Personal Synthesizer (24-vessel system used for lead optimization and process development); and the Core System (high-throughput synthesizer designed for production of large libraries.

Contact: Christopher Graeme-Barber

ThermoQuest Corp.
355 River Oaks Parkway, San Jose, CA 95134, USA
Phone: (408) 577-1053
E-mail: *webmaster@thermoquest.com*
Internet: *http://www.thermo.com/subsid/tmq.html*
Stock: TMQ (ASE)

Profile: The company develops, manufactures, and sells mass spectrometers, liquid chromatographs, and gas chromatographs for the environmental, pharmaceutical, and industrial marketplaces. These analytical instruments are used in the quantitative and qualitative chemical analysis of organic and inorganic compounds at ultratrace levels of detection. ThermoQuest is a public subsidiary of Thermo Instrument Systems, Inc. (AMEX: THI), a Thermo Electron (NYSE: TMO) company.

Products/technologies: ThermoQuest sells the Navigator LC-MS (Finnigan) with additional modules for combinatorial and biochemical applications; and the LCQ or TSQ for LC-MS-MS.

3-Dimensional Pharmaceuticals, Inc.
Eagleview Corporate Center, 655 Stockton Drive, Suite 104, Exton, PA 19341, USA
Phone: (610) 458-8959; Fax: (610) 458-8258
E-mail: *webmaster@3dp.com*
Internet: *http://www.3dp.com*

Profile: This privately held company was founded in 1993 to integrate advanced technologies in structure-based design, combinatorial chemistry, and chemi-informatics for the cost-effective discovery of orally active pharmaceuticals. The company has developed a system capable of generating new drugs through computer-controlled robotic synthesis and analysis of chemical libraries. Current drug discovery efforts are focused on cardiovascular disease and cancer.

Products/technologies: 3DP's integrated drug discovery technology combines the selection power of DirectedDiversity®, a proprietary method of directing combinatorial chemistry toward specific molecular targets, with the precision of structure-based design, to discover and optimize new drugs rapidly. According to the company, this technology platform allows it to discover and refine drugs active against a wide range of therapeutic targets more quickly than conventional approaches. The process allows parallel and iterative chemistry; it is an operating system for gathering information from many parts of the drug discovery pipeline. The company also has a chemi-informatics database.

Contact: Thomas P. Stagnaro, President and CEO

Torrey Pines Institute for Molecular Studies (TPIMS)
3550 General Atomics Court, San Diego, CA 92121, USA
Phone: (619) 455-3803; Fax: (619) 455-3804
E-mail: *Houghten@tpims.org*
Internet: *http://www.tpims.org*

Profile: TPIMS is a nonprofit biomedical research institute focused on the development of combinatorial chemistry techniques that can be applied to all compound types. It was founded in 1988 to continue earlier research begun at The Scripps Research Institute in La Jolla, California, and in 1989 it began its research activities. Less than one year after beginning its operations, TPIMS scientists had developed a method for synthesizing and screening combinatorial libraries of tens of millions of peptides and other nonpeptide compounds. As a result of this early research, TPIMS became an internationally recognized research center in the field of molecular diversity and combinatorial chemistry. Research is funded by the National Institutes of Health, the National Science Foundation, the U.S. Army, and by a variety of pharmaceutical companies.

Products/technologies: These include PositionalScan, a patented new technology that permits the rapid identification of highly active individual compounds from compound libraries. It is a screening method that allows highly active compounds to be identified in a very short time. The technology is licensed to Houghten Pharmaceuticals, Inc. (HPI) for use in its internal and collaborative combinatorial chemistry programs. In 1994 the concept of "libraries from libraries" was introduced. Originally the permethylation of a positional scanning hexapeptide library was described. This technique permitted preexisting combinatorial libraries to be transformed in a single synthetic step to other compound classes having very different physical/chemical properties. Peralkylation of other combinatorial libraries using other alkylation reagents has also been performed. In 1995 a description of the exhaustive reduction of a positional scanning hexapeptide library was published. These techniques have been combined to produce a large number of combinatorial libraries from preexisting libraries; additionally, these techniques have been applied to other reaction schemes and pharmacophores. Combinatorial libraries include cyclic peptide libraries and combinatorial libraries spaced around cyclic peptide templates; peptidomimetics; and heterocyclic combinatorial libraries.

Contact: Dr. Richard A. Houghten, President and Chief Technical Officer

TransCell Technologies, Inc.
8 Cedar Brook Drive, Cranbury, NJ 08512, USA
Phone: (609) 655-6932; Fax: (609) 655-6930
E-mail: *george@transcell.com*
Internet: *http://www.transcell.com*

Profile: Founded in 1991, Transcell Technologies is a privately held, majority-owned subsidiary of Interneuron Pharmaceuticals, Inc. (IPIC on NASDAQ). Its business strategy is to form strategic alliances with major corporate partners in the area of new drug discovery, and to license and apply the technology. The company is researching the production of combinatorial libraries of synthetic carbohydrates and glycoconjugates for drug discovery. TransCell has broadly applicable, proprietary technologies that can be applied to corporate collaborations in a number of areas, including infectious diseases, cancer, and metabolic diseases. It is developing new pharmaceuticals based on carbohydrate combinatorial chemistry.

Products/technologies: It has two core technologies: a cell permeation enhancer to improve drug delivery, and the rapid synthesis of oligosaccharides for the creation of libraries of oligosaccharides and glycoconjugates for use in identifying drug candidates.

Contact: George W. Brodhead, VP, Business Development

Trega Biosciences, Inc.
3550 General Atomics Ct., San Diego, CA 92121, USA
Phone: (619) 455-3814; Fax: (619) 455-2544
Internet: *http://www.trega.com*
Stock: TRGA (NASDAQ)

Profile: Founded in 1992, the company develops combinatorial chemistry libraries for internal R & D programs and for corporate collaborative development programs for the discovery of novel small-molecule drug therapies. The company's name was changed on May 1, 1997, from Houghten Pharmaceuticals. The company has leveraged its technology platform by entering into pharmaceutical alliances, enabling partners to access Trega's technologies in exchange for licensing fees, potential milestone payments, and royalties, and by establishing joint discovery alliances with biotechnology companies.

Products/technologies: Utilizes its combinatorial chemistry libraries for the discovery of novel, small-molecule drug therapies in its internal discovery programs, in collaboration with pharmaceutical companies and in joint discovery alliances with biotechnology firms. Results of their combinatorial research agreement with Torrey Pines Institute for Molecular Studies (TPIMS) have included a range of inventions including PositionalScan, a method for identifying active compounds from mixture-based libraries.

Contact: Susan Adler, Senior Director, Business Development

Tripos, Inc.
1699 S. Hanley Road, St. Louis, MO 63144, USA
Phone: (314) 647-1099; Fax: (314) 647-9241
Toll-free in U.S.: (800) 323-2960
E-mail: *info@tripos.com*
Internet: *http://www.tripos.com*

Profile: Founded in 1979, the company is a developer of molecular modeling and computational chemistry software that focuses on pharmaceuticals and biotechnology. Its Accelerated Discovery Services division provides a range of research services, including library design, to comprehensive discovery collaborations. In November 1997, the company announced the acquisition of Receptor Research Ltd., a UK-based company specializing in custom synthesis using solid phase chemistry for biological applications. The acquisition will allow Tripos to directly apply its molecular informatics, design, and analysis expertise in new compound discovery collaborations in the pharmaceutical, biotechnology, agrochemical, and related industries.

Products/technologies: Chemical diversity technology is used to design libraries with high structural variation. The company supplies drug discovery services, chemical compounds, and scientific software to facilitate the discovery of new therapeutic and bioactive compounds in pharmaceutical, biotechnology, chemical, and agrochemical industries. ChemSpace proprietary technology is used for the design and analysis of combinatorial libraries and related data; the ChemSpace virtual database of over one trillion synthetically accessible, small organic chemical structures provides a platform from which screening and focused chemical libraries can be designed and managed. Combinatorial chemistry software applications add on to its traditional Sybyl software for molecular modeling and analysis, along with Unison Windows software for desktop data access and management. Legion is a combinatorial chemistry application, and Selector provides software tools for managing and analyzing chemical diversity. Tripos/Panlabs offers Optiverse™, a standard compound screening library, that includes integrated design and synthesis for optimal lead discovery and lead refinement.

Contact: Dr. John P. McAlister, President and CEO

Varian Associates, Inc.
3050 Hansen Way, Palo Alto, CA 94304-1000, USA
Phone: (650) 493-4000; Fax: (281) 240-6752
Toll-free in U.S.: (800) 926-3000
Internet: *http://www.varian.com*
Stock: VAR (NYSE)

Profile: Varian is a billion dollar high-technology company. As a diversified, international electronics company, it designs, manufactures, and markets high-tech systems and components for applications in worldwide markets. It is organized around three core businesses that include semiconductor equipment, healthcare systems, and instruments. Its instruments group supplies analytical and research instrumentation and related equipment for chemical composition studies of myriad substances. Additionally, this operation manufactures vacuum products, helium leak detectors, and solid phase extraction sample preparation products that are used for chemical isolation and purification. The company acquired the assets of Dynatech Precision Sampling Corporation, a manufacturer of several autosampling devices for LC and GC. In early 1997 it also acquired the LC line of Rainin Instrument Company, Inc.

Products/technologies: These include solid phase extraction products in 96-well format; and Mercury™, which is a small, low-cost NMR spectrometer. The acquisition of Rainin allows Varian to compete in the preparative HPLC market and gives it a lower cost product line to complement its research grade LC systems.

Versicor Inc.
270 E. Grand Avenue, South San Francisco, CA 94080, USA
Phone: (415) 829-7000; Fax: (415) 635-0973

Profile: Versicor Inc., a majority-owned subsidiary of Sepracor Inc. (Nasdaq: SEPR), is a second-generation company that integrates biology/genomics, combinatorial chemistry, high-throughput screening, and informatics in a balanced format to discover new anti-infective drugs against resistant pathogens.

Products/technologies: Screening and synthesis of chemical libraries, high-throughput assay development, and lead validation.

Contact: Jonae R. Barnes, Manager, Corporate Communications

Waters Corporation
34 Maple Street, Milford, MA 01757, USA
Phone: (508) 478-2000; Fax: (508) 872-1990
Toll-free in U.S.: (800) 252-HPLC
E-mail: *brian_j_murphy@waters.com*
Internet: *http://www.waters.com*
Stock: WAT (NYSE)

Profile: Waters is the largest company in the analytical instrument industry devoted to HPLC (instrumentation, consumables, and associated postsale support services) and related technologies. The company has held the leadership position in HPLC for more than 35 years, with the pharmaceutical industry being Waters' largest single market. Its HPLC and LC-MS systems are ideally suited for combinatorial analysis. In addition to leading in the $2.1 billion HPLC market, the company has also expanded its product and technology offerings through several recent acquisitions. Acquisitions in 1997 included Micromass Limited (Manchester, England), a technology leader in the areas of magnetic sector, time-of-flight, and quadrupole geometry mass spectrometers, and the largest supplier of benchtop MS instruments, including systems for combinatorial screening and rapid purification of drug candidates. Also acquired in 1997 was YMC Inc. (Wilmington, NC), the North American distributor of YMC brand columns and bulk packing materials. Sales to international customers account for more than 60% of total revenues.

Products/technologies: The Waters LC-MS solutions for drug discovery consist of a Waters Alliance™ HPLC system (with optional 996-photodiode array or 2487 UV/Vis detectors), Waters 2700 Sample Manager (for 96-well microtiter plates), Micromass Platform LC or Platform LCZ benchtop API mass spectrometer, and MassLynx™ software with special options for combinatorial chemistry, such as OpenLynx™ Diversity and FractionLynx™. Waters also provides columns and sample preparation products, such as Symmetry®, SymmetryShield®, and Oasis HLB cartridges and extraction plates, that are designed for high-throughput drug assays and preparative applications.

Contact: Brian J. Murphy, Corporate Communications

Xenometrix, Inc.
2425 North 55th Street, Suite 111, Boulder, CO 80301-5700, USA
Phone: (303) 447-1773; Fax: (303) 447-1758
Toll-free in U.S.: (800) 4DNATOX
E-mail: *Jwillows@xeno.com*
Internet: *http://www.xeno.com*.

Profile: A biotechnology company with proprietary technology, Xenometrix helps the pharmaceutical industry improve the effectiveness of drug discovery and development.

Products/technologies: Unique gene response profiles provide mechanistic information that can be used to optimize drug leads, as well as automated assays for screening drug leads for safety. Tools are offered as kits for customers to perform in their own laboratories or as a service through Xenometrix's Customer Research Laboratory.

Contact: Stephen J. Sullivan, President and CEO

Xenova Group plc
240 Bath Road, Slough, Berkshire SL1 4EF, UK
Phone: 44 1753 692229; Fax: 44 1753 692613
Internet: *http://www.xenova.co.uk*
Stock: XNVAY (NASDAQ)

Profile: Founded in 1986, Xenova went public in 1994. The company provides rapid, efficient, and cost-effective routes to discovering novel drugs where there is no obvious chemical starting point.

Products/technologies: The company has diverse libraries of natural compounds prepared from plant, fungal, and microbial sources; and proprietary processes for the isolation and characterization of individual compounds, which are compatible with modern high-throughput screening technologies.

Contact: Dr. Louis Nisbet, CEO

YMC, Inc.
3233 Burnt Mill Drive, Wilmington, NC 28403, USA
Phone: (910) 762-7154; Fax: (910) 343-0907
Toll-free in U.S.: (800) YMC-6311
E-mail: *ymcinc@ymc-hplc.com*
Internet: *http://www.ymc-hplc.com*

Profile: YMC, Inc. manufactures and supplies YMC brand HPLC columns and bulk packing materials. It was founded as a joint venture company in 1985 with Yamamura Chemical Laboratory Co. Ltd. (YMC Co. Ltd.) as the North American distributor. Waters Corporation has held an equity position in YMC Co. Ltd. since 1987, and in 1997 it acquired YMC, Inc. Waters trades as WAT on the NYSE.

Products/technologies: The CombiChrom™ columns provide HPLC analysis and purification of diverse mixtures. Also, CombiScreen™ columns are used for one-step scale-up for fast analysis and high throughput; and CombiPrep™ columns are designed for one-step scale-up for purification of complex mixtures.

Contact: Jim Carroll

Zymark Corporation
Zymark Center, Hopkinton, MA 01748, USA
Phone: (508) 435-9500; Fax (508) 435-3439
E-mail: *solutions@zymark.com*
Internet: *http://www.zymark.com*

Profile: Founded in 1981, Zymark is a leader in laboratory automation for the pharmaceutical, biotechnology, and chemical industries. Having pioneered the technology, the company serves drug discovery research, bioanalytical development, formulations development, and quality control functions. In 1994 Zymark transitioned from a technology-based product business to a laboratory automation solutions business. The focus is on providing solutions to companies to improve lab productivity by integrating technology, support, consulting services, and project management. In September 1996, Zymark became a member of the Berwind Group of companies. The Berwind Group is a highly diversified, family-owned enterprise, consisting of a number of independently managed businesses.

Products/technologies: In the area of discovery research, the company offers microassay technologies including high-throughput screening sstems for a range of biological assays (the Zymate immunoassay system, cell-based assay systems, and receptor binding systems); compound preparation and distribution systems, and automated organic synthesis systems. Zymark's RapidPlate-96 workstation simultaneously transfers to and from 96 wells of a microplate. The company also has designed a software interface, the PCS productivity, communications, and scheduling software that can be customized to customer needs. The Allegro ultrahigh-throughput screening technology applies automation and assay technologies in a production line architecture that allows screening of 100,000 assay points per day.

Contact: Kenneth N. Rapp, Vice President, Sales

Index